CRYPTOGRAPHY
Theory and Practice

CRYPTOGRAPHY
Theory and Practice

Douglas R. Stinson
Computer Science and Engineering Department
and Center for Communication and Information Science
University of Nebraska, Lincoln

CRC Press
Boca Raton Boston London New York Washington, D.C.

Library of Congress Cataloging-in-Publication Data

Stinson, D. R. (Douglas Robert), 1956–
    Cryptography : theory and practice / D.R. Stinson
       p. cm.—(Discrete mathematics and its applications)
    Includes bibliographical references and index.
    ISBN 0-8493-8521-0
    1. Coding theory. 2. Cryptography. I. Title. II. Series.
QA268.S75 1995
005.8′2—dc20
                                        95-5237
                                         CIP

© 1995 by CRC Press LLC

No claim to original U.S. Government works
International Standard Book Number 0-8493-8521-0
Library of Congress Card Number 95-5237
Printed in the United States of America   8 9 0
Printed on acid-free paper

The CRC Press Series on Discrete Mathematics and Its Applications

Discrete mathematics is becoming increasingly applied to computer science, engineering, the physical sciences, the natural sciences, and the social sciences. Moreover, there has also been an explosion of research in discrete mathematics in the past two decades. Both trends have produced a need for many types of information for people who use or study this part of the mathematical sciences. The CRC Press Series on Discrete Mathematics and Its Applications is designed to meet the needs of practitioners, students, and researchers for information in discrete mathematics. The series includes handbooks and other reference books, advanced textbooks, and selected monographs. Among the areas of discrete mathematics addressed by the series are logic, set theory, number theory, combinatorics, discrete probability theory, graph theory, algebra, linear algebra, coding theory, cryptology, discrete optimization, theoretical computer science, algorithmics, and computational geometry.

Kenneth H. Rosen, Series Editor
Distinguished Member of Technical Staff
AT&T Bell Laboratories
Holmdel, New Jersey
e-mail:krosen@arch4.ho.att.com

Advisory Board

Charles Colbourn
Department of Combinatorics and Optimization, University of Waterloo

Jonathan Gross
Department of Computer Science, Columbia University

Andrew Odlyzko
AT&T Bell Laboratories

Preface

My objective in writing this book was to produce a general, comprehensive text-book that treats all the essential core areas of cryptography. Although many books and monographs on cryptography have been written in recent years, the majority of them tend to address specialized areas of cryptography. On the other hand, many of the existing general textbooks have become out-of-date due to the rapid expansion of research in cryptography in the past 15 years.

I have taught a graduate level cryptography course at the University of Nebraska-Lincoln to computer science students, but I am aware that cryptography courses are offered at both the undergraduate and graduate levels in mathematics, computer science and electrical engineering departments. Thus, I tried to design the book to be flexible enough to be useful in a wide variety of approaches to the subject.

Of course there are difficulties in trying to appeal to such a wide audience. But basically, I tried to do things in moderation. I have provided a reasonable amount of mathematical background where it is needed. I have attempted to give informal descriptions of the various cryptosystems, along with more precise pseudo-code descriptions, since I feel that the two approaches reinforce each other. As well, there are many examples to illustrate the workings of the algorithms. And in every case I try to explain the mathematical underpinnings; I believe that it is impossible to really understand how a cryptosystem works without understanding the underlying mathematical theory.

The book is organized into three parts. The first part, Chapters 1–3, covers private-key cryptography. Chapters 4–9 concern the main topics in public-key cryptography. The remaining four chapters provide introductions to four active research areas in cryptography.

The first part consists of the following material: Chapter 1 is a fairly elementary introduction to simple "classical" cryptosystems. Chapter 2 covers the main elements of Shannon's approach to cryptography, including the concept of perfect secrecy and the use of information theory in cryptography. Chapter 3 is a lengthy discussion of the **Data Encryption Standard**; it includes a treatment of differential cryptanalysis.

The second part contains the following material: Chapter 4 concerns the **RSA Public-key Cryptosystem**, together with a considerable amount of background on number-theoretic topics such as primality testing and factoring. Chapter 5 discusses some other public-key systems, the most important being the **ElGamal System** based on discrete logarithms. Chapter 6 deals with signature schemes, such as the **Digital Signature Standard**, and includes treatment of special types of signature schemes such as undeniable and fail-stop signature schemes. The subject of Chapter 7 is hash functions. Chapter 8 provides an overview of the numerous approaches to key distribution and key agreement protocols. Finally, Chapter 9 describes identification schemes.

The third part contains chapters on selected research-oriented topics, namely, authentication codes, secret sharing schemes, pseudo-random number generation, and zero-knowledge proofs.

Thus, I have attempted to be quite comprehensive in the "core" areas of cryptography, as well as to provide some more advanced chapters on specific research areas. Within any given area, however, I try to pick a few representative systems and discuss them in a reasonable amount of depth. Thus my coverage of cryptography is in no way encyclopedic.

Certainly there is much more material in this book than can be covered in one (or even two) semesters. But I hope that it should be possible to base several different types of courses on this book. An introductory course could cover Chapter 1, together with selected sections of Chapters 2–5. A second or graduate course could cover these chapters in a more complete fashion, as well as material from Chapters 6–9. Further, I think that any of the chapters would be a suitable basis for a "topics" course that might delve into specific areas more deeply.

But aside from its primary purpose as a textbook, I hope that researchers and practitioners in cryptography will find it useful in providing an introduction to specific areas with which they might not be familiar. With this in mind, I have tried to provide references to the literature for further reading on many of the topics discussed.

One of the most difficult things about writing this book was deciding how much mathematical background to include. Cryptography is a broad subject, and it requires knowledge of several areas of mathematics, including number theory, groups, rings and fields, linear algebra, probability and information theory. As well, some familiarity with computational complexity, algorithms and NP-completeness theory is useful. I have tried not to assume too much mathematical background, and thus I develop mathematical tools as they are needed, for the most part. But it would certainly be helpful for the reader to have some familiarity with basic linear algebra and modular arithmetic. On the other hand, a more specialized topic, such as the concept of entropy from information theory, is introduced from scratch.

I should also apologize to anyone who does not agree with the phrase "Theory and Practice" in the title. I admit that the book is more theory than practice. What I mean by this phrase is that I have tried to select the material to be included in the book both on the basis of theoretical interest and practical importance. So, I may include systems that are not of practical use if they are mathematically elegant or illustrate an important concept or technique. But, on the other hand, I do describe the most important systems that are used in practice, e.g., **DES** and other U. S. cryptographic standards.

I would like to thank the many people who provided encouragement while I wrote this book, pointed out typos and errors, and gave me useful suggestions on material to include and how various topics should be treated. In particular, I would like to convey my thanks to Mustafa Atici, Mihir Bellare, Bob Blakley, Carlo Blundo, Gilles Brassard, Daniel Ducharme, Mike Dvorsky, Luiz Frota-Mattos, David Klarner, Don Kreher, Keith Martin, Vaclav Matyas, Alfred Menezes, Luke O'Connor, William Read, Phil Rogaway, Paul Van Oorschot, Scott Vanstone, Johan van Tilburg, Marc Vauclair and Mike Wiener. Thanks also to Mike Dvorsky for helping me prepare the index.

Douglas R. Stinson

To my children,
Michela and Aiden

Contents

1

Classical Cryptography

1.1 Introduction: Some Simple Cryptosystems

The fundamental objective of cryptography is to enable two people, usually referred to as Alice and Bob, to communicate over an insecure channel in such a way that an opponent, Oscar, cannot understand what is being said. This channel could be a telephone line or computer network, for example. The information that Alice wants to send to Bob, which we call "plaintext," can be English text, numerical data, or anything at all — its structure is completely arbitrary. Alice encrypts the plaintext, using a predetermined key, and sends the resulting ciphertext over the channel. Oscar, upon seeing the ciphertext in the channel by eavesdropping, cannot determine what the plaintext was; but Bob, who knows the encryption key, can decrypt the ciphertext and reconstruct the plaintext.

This concept is described more formally using the following mathematical notation.

DEFINITION 1.1 *A cryptosystem is a five-tuple $(\mathcal{P}, \mathcal{C}, \mathcal{K}, \mathcal{E}, \mathcal{D})$, where the following conditions are satisfied:*

1. *$\mathcal{P}$ is a finite set of possible plaintexts*
2. *$\mathcal{C}$ is a finite set of possible ciphertexts*
3. *$\mathcal{K}$, the keyspace, is a finite set of possible keys*
4. *For each $K \in \mathcal{K}$, there is an encryption rule $e_K \in \mathcal{E}$ and a corresponding decryption rule $d_K \in \mathcal{D}$. Each $e_K : \mathcal{P} \to \mathcal{C}$ and $d_K : \mathcal{C} \to \mathcal{P}$ are functions such that $d_K(e_K(x)) = x$ for every plaintext $x \in \mathcal{P}$.*

The main property is property 4. It says that if a plaintext x is encrypted using e_K, and the resulting ciphertext is subsequently decrypted using d_K, then the original plaintext x results.

Alice and Bob will employ the following protocol to use a specific cryptosys-

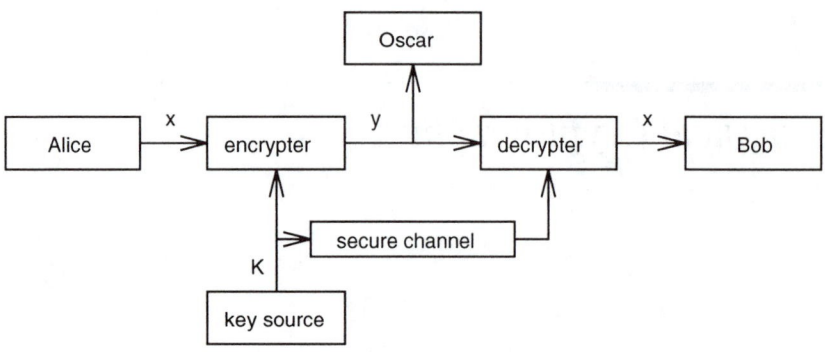

FIGURE 1.1
The Communication Channel

tem. First, they choose a random key $K \in \mathcal{K}$. This is done when they are in the same place and are not being observed by Oscar, or, alternatively, when they do have access to a secure channel, in which case they can be in different places. At a later time, suppose Alice wants to communicate a message to Bob over an insecure channel. We suppose that this message is a string

$$\mathbf{x} = x_1 x_2 \ldots x_n$$

for some integer $n \geq 1$, where each plaintext symbol $x_i \in \mathcal{P}$, $1 \leq i \leq n$. Each x_i is encrypted using the encryption rule e_K specified by the predetermined key K. Hence, Alice computes $y_i = e_K(x_i)$, $1 \leq i \leq n$, and the resulting ciphertext string

$$\mathbf{y} = y_1 y_2 \ldots y_n$$

is sent over the channel. When Bob receives $y_1 y_2 \ldots y_n$, he decrypts it using the decryption function d_K, obtaining the original plaintext string, $x_1 x_2 \ldots x_n$. See Figure 1.1 for an illustration of the communication channel.

Clearly, it must be the case that each encryption function e_K is an injective function (i.e., one-to-one), otherwise, decryption could not be accomplished in an unambiguous manner. For example, if

$$y = e_K(x_1) = e_K(x_2)$$

where $x_1 \neq x_2$, then Bob has no way of knowing whether y should decrypt to x_1 or x_2. Note that if $\mathcal{P} = \mathcal{C}$, it follows that each encryption function is a permutation. That is, if the set of plaintexts and ciphertexts are identical, then each encryption function just rearranges (or permutes) the elements of this set.

1.1.1 The Shift Cipher

In this section, we will describe the **Shift Cipher**, which is based on modular arithmetic. But first we review some basic definitions of modular arithmetic.

DEFINITION 1.2 *Suppose a and b are integers, and m is a positive integer. Then we write $a \equiv b \pmod{m}$ if m divides $b - a$. The phrase $a \equiv b \pmod{m}$ is read as "a is congruent to b modulo m." The integer m is called the modulus.*

Suppose we divide a and b by m, obtaining integer quotients and remainders, where the remainders are between 0 and $m - 1$. That is, $a = q_1 m + r_1$ and $b = q_2 m + r_2$, where $0 \leq r_1 \leq m - 1$ and $0 \leq r_2 \leq m - 1$. Then it is not difficult to see that $a \equiv b \pmod{m}$ if and only if $r_1 = r_2$. We will use the notation $a \bmod m$ (without parentheses) to denote the remainder when a is divided by m, i.e., the value r_1 above. Thus $a \equiv b \pmod{m}$ if and only if $a \bmod m = b \bmod m$. If we replace a by $a \bmod m$, we say that a is *reduced* modulo m.

REMARK Many computer programming languages define $a \bmod m$ to be the remainder in the range $-m + 1, \ldots, m - 1$ having the same sign as a. For example, $-18 \bmod 7$ would be -4, rather than 3 as we defined it above. But for our purposes, it is much more convenient to define $a \bmod m$ always to be non-negative. ∎

We can now define arithmetic modulo m: $\mathbb{Z}_m$ is defined to be the set $\{0, \ldots, m-1\}$, equipped with two operations, $+$ and $\times$. Addition and multiplication in $\mathbb{Z}_m$ work exactly like real addition and multiplication, except that the results are reduced modulo m.

For example, suppose we want to compute 11×13 in $\mathbb{Z}_{16}$. As integers, we have $11 \times 13 = 143$. To reduce 143 modulo 16, we just perform ordinary long division: $143 = 8 \times 16 + 15$, so $143 \bmod 16 = 15$, and hence $11 \times 13 = 15$ in $\mathbb{Z}_{16}$.

These definitions of addition and multiplication in $\mathbb{Z}_m$ satisfy most of the familiar rules of arithmetic. We will list these properties now, without proof:

1. addition is *closed*, i.e., for any $a, b \in \mathbb{Z}_m$, $a + b \in \mathbb{Z}_m$

2. addition is *commutative*, i.e., for any $a, b \in \mathbb{Z}_m$, $a + b = b + a$

3. addition is *associative*, i.e., for any $a, b, c \in \mathbb{Z}_m$, $(a + b) + c = a + (b + c)$

4. 0 is an *additive identity*, i.e., for any $a \in \mathbb{Z}_m$, $a + 0 = 0 + a = a$

5. the *additive inverse* of any $a \in \mathbb{Z}_m$ is $m - a$, i.e., $a + (m - a) = (m - a) + a = 0$ for any $a \in \mathbb{Z}_m$

6. multiplication is *closed*, i.e., for any $a, b \in \mathbb{Z}_m$, $ab \in \mathbb{Z}_m$

7. multiplication is *commutative*, i.e., for any $a, b \in \mathbb{Z}_m$, $ab = ba$

FIGURE 1.2
Shift Cipher

Let $\mathcal{P} = \mathcal{C} = \mathcal{K} = \mathbb{Z}_{26}$. For $0 \leq K \leq 25$, define

$$e_K(x) = x + K \bmod 26$$

and

$$d_K(y) = y - K \bmod 26$$

$(x, y \in \mathbb{Z}_{26})$.

8. multiplication is *associative*, i.e., for any $a, b, c \in \mathbb{Z}_m$, $(ab)c = a(bc)$

9. 1 is a *multiplicative identity*, i.e., for any $a \in \mathbb{Z}_m$, $a \times 1 = 1 \times a = a$

10. multiplication *distributes* over addition, i.e., for any $a, b, c \in \mathbb{Z}_m$, $(a+b)c = (ac) + (bc)$ and $a(b + c) = (ab) + (ac)$.

Properties 1, 3–5 say that $\mathbb{Z}_m$ forms an algebraic structure called a *group* with respect to the addition operation. Since property 2 also holds, the group is said to be *abelian*.

Properties 1–10 establish that $\mathbb{Z}_m$ is, in fact, a *ring*. We will see many other examples of groups and rings in this book. Some familiar examples of rings include the integers, $\mathbb{Z}$; the real numbers, $\mathbb{R}$; and the complex numbers, $\mathbb{C}$. However, these are all infinite rings, and our attention will be confined almost exclusively to finite rings.

Since additive inverses exist in $\mathbb{Z}_m$, we can also subtract elements in $\mathbb{Z}_m$. We define $a - b$ in $\mathbb{Z}_m$ to be $a + m - b \bmod m$. Equivalently, we can compute the integer $a - b$ and then reduce it modulo m.

For example, to compute $11 - 18$ in $\mathbb{Z}_{31}$, we can evaluate $11 + 13 \bmod 31 = 24$. Alternatively, we can first subtract 18 from 11, obtaining -7 and then compute $-7 \bmod 31 = 24$.

We present the **Shift Cipher** in Figure 1.2. It is defined over $\mathbb{Z}_{26}$ since there are 26 letters in the English alphabet, though it could be defined over $\mathbb{Z}_m$ for any modulus m. It is easy to see that the **Shift Cipher** forms a cryptosystem as defined above, i.e., $d_K(e_K(x)) = x$ for every $x \in \mathbb{Z}_{26}$.

REMARK   For the particular key $K = 3$, the cryptosystem is often called the **Caesar Cipher**, which was purportedly used by Julius Caesar.   ∎

We would use the **Shift Cipher** (with a modulus of 26) to encrypt ordinary English text by setting up a correspondence between alphabetic characters and residues modulo 26 as follows: $A \leftrightarrow 0$, $B \leftrightarrow 1, \ldots, Z \leftrightarrow 25$. Since we will be using this correspondence in several examples, let's record it for future use:

A	B	C	D	E	F	G	H	I	J	K	L	M
0	1	2	3	4	5	6	7	8	9	10	11	12

N	O	P	Q	R	S	T	U	V	W	X	Y	Z
13	14	15	16	17	18	19	20	21	22	23	24	25

A small example will illustrate.

Example 1.1
Suppose the key for a **Shift Cipher** is $K = 11$, and the plaintext is

```
wewillmeetatmidnight.
```

We first convert the plaintext to a sequence of integers using the specified correspondence, obtaining the following:

$$\begin{array}{cccccccccc} 22 & 4 & 22 & 8 & 11 & 11 & 12 & 4 & 4 & 19 \\ 0 & 19 & 12 & 8 & 3 & 13 & 8 & 6 & 7 & 19 \end{array}$$

Next, we add 11 to each value, reducing each sum modulo 26:

$$\begin{array}{cccccccccc} 7 & 15 & 7 & 19 & 22 & 22 & 23 & 15 & 15 & 4 \\ 11 & 4 & 23 & 19 & 14 & 24 & 19 & 17 & 18 & 4 \end{array}$$

Finally, we convert the sequence of integers to alphabetic characters, obtaining the ciphertext:

```
HPHTWWXPPELEXTOYTRSE.
```

To decrypt the ciphertext, Bob will first convert the ciphertext to a sequence of integers, then subtract 11 from each value (reducing modulo 26), and finally convert the sequence of integers to alphabetic characters. ⬜

REMARK   In the above example we are using upper case letters for ciphertext and lower case letters for plaintext, in order to improve readability. We will do this elsewhere as well. ∎

If a cryptosystem is to be of practical use, it should satisfy certain properties. We informally enumerate two of these properties now.

1. Each encryption function e_K and each decryption function d_K should be efficiently computable.

2. An opponent, upon seeing a ciphertext string **y**, should be unable to determine the key K that was used, or the plaintext string **x**.

The second property is defining, in a very vague way, the idea of "security." The process of attempting to compute the key K, given a string of ciphertext **y**, is called *cryptanalysis*. (We will make these concepts more precise as we proceed.) Note that, if Oscar can determine K, then he can decrypt **y** just as Bob would, using d_K. Hence, determining K is at least as difficult as determining the plaintext string **x**.

We observe that the **Shift Cipher** (modulo 26) is not secure, since it can be cryptanalyzed by the obvious method of *exhaustive key search*. Since there are only 26 possible keys, it is easy to try every possible decryption rule d_K until a "meaningful" plaintext string is obtained. This is illustrated in the following example.

Example 1.2
Given the ciphertext string

$$\text{JBCRCLQRWCRVNBJENBWRWN},$$

we successively try the decryption keys d_0, d_1, etc. The following is obtained:

```
jbcrclqrwcrvnbjenbwrwn
iabqbkpqvbqumaidmavqvm
hzapajopuaptlzhclzupul
gyzozinotzoskygbkytotk
fxynyhmnsynrjxfajxsnsj
ewxmxglmrxmqiweziwrmri
dvwlwfklqwlphvdyhvqlqh
cuvkvejkpvkogucxgupkpg
btujudijoujnftbwftojof
astitchintimesavesnine
```

At this point, we have determined the plaintext and we can stop. The key is $K = 9$.   ∎

On average, a plaintext will be computed after trying $26/2 = 13$ decryption rules.

FIGURE 1.3
Substitution Cipher

Let $\mathcal{P} = \mathcal{C} = \mathbb{Z}_{26}$. $\mathcal{K}$ consists of all possible permutations of the 26 symbols $0, 1, \ldots, 25$. For each permutation $\pi \in \mathcal{K}$, define

$$e_\pi(x) = \pi(x),$$

and define

$$d_\pi(y) = \pi^{-1}(y),$$

where π^{-1} is the inverse permutation to π.

As the above example indicates, a necessary condition for a cryptosystem to be secure is that an exhaustive key search should be infeasible; i.e., the keyspace should be very large. As might be expected, a large keyspace is not sufficient to guarantee security.

1.1.2 The Substitution Cipher

Another well-known cryptosystem is the **Substitution Cipher**. This cryptosystem has been used for hundreds of years. Puzzle "cryptograms" in newspapers are examples of **Substitution Ciphers**. This cipher is defined in Figure 1.3.

Actually, in the case of the **Substitution Cipher**, we might as well take $\mathcal{P}$ and $\mathcal{C}$ both to be the 26-letter English alphabet. We used $\mathbb{Z}_{26}$ in the **Shift Cipher** because encryption and decryption were algebraic operations. But in the **Substitution Cipher**, it is more convenient to think of encryption and decryption as permutations of alphabetic characters.

Here is an example of a "random" permutation, π, which could comprise an encryption function. (As before, plaintext characters are written in lower case and ciphertext characters are written in upper case.)

a	b	c	d	e	f	g	h	i	j	k	l	m
X	N	Y	A	H	P	O	G	Z	Q	W	B	T

n	o	p	q	r	s	t	u	v	w	x	y	z
S	F	L	R	C	V	M	U	E	K	J	D	I

Thus, $e_\pi(a) = X$, $e_\pi(b) = N$, etc. The decryption function is the inverse permutation. This is formed by writing the second lines first, and then sorting in alphabetical order. The following is obtained:

A	B	C	D	E	F	G	H	I	J	K	L	M
d	l	r	y	v	o	h	e	z	x	w	p	t

N	O	P	Q	R	S	T	U	V	W	X	Y	Z
b	g	f	j	q	n	m	u	s	k	a	c	i

Hence, $d_\pi(A) = d$, $d_\pi(B) = l$, etc.

As an exercise, the reader might decrypt the following ciphertext using this decryption function:

$$\text{MGZVYZLGHCMHJMYXSSFMNHAHYCDLMHA.}$$

A key for the **Substitution Cipher** just consists of a permutation of the 26 alphabetic characters. The number of these permutations is 26!, which is more than 4.0×10^{26}, a very large number. Thus, an exhaustive key search is infeasible, even for a computer. However, we shall see later that a **Substitution Cipher** can easily be cryptanalyzed by other methods.

1.1.3 The Affine Cipher

The **Shift Cipher** is a special case of the **Substitution Cipher** which includes only 26 of the 26! possible permutations of 26 elements. Another special case of the **Substitution Cipher** is the **Affine Cipher**, which we describe now. In the **Affine Cipher**, we restrict the encryption functions to functions of the form

$$e(x) = ax + b \bmod 26,$$

$a, b \in \mathbb{Z}_{26}$. These functions are called *affine functions*, hence the name **Affine Cipher**. (Observe that when $a = 1$, we have a **Shift Cipher**.)

In order that decryption is possible, it is necessary to ask when an affine function is injective. In other words, for any $y \in \mathbb{Z}_{26}$, we want the congruence

$$ax + b \equiv y \pmod{26}$$

to have a unique solution for x. This congruence is equivalent to

$$ax \equiv y - b \pmod{26}.$$

Now, as y varies over $\mathbb{Z}_{26}$, so, too, does $y - b$ vary over $\mathbb{Z}_{26}$. Hence, it suffices to study the congruence $ax \equiv y \pmod{26}$ ($y \in \mathbb{Z}_{26}$).

We claim that this congruence has a unique solution for every y if and only if $\gcd(a, 26) = 1$ (where the gcd function denotes the greatest common divisor of its arguments). First, suppose that $\gcd(a, 26) = d > 1$. Then the congruence $ax \equiv 0 \pmod{26}$ has (at least) two distinct solutions in $\mathbb{Z}_{26}$, namely $x = 0$ and $x = 26/d$. In this case $e(x) = ax + b \bmod 26$ is not an injective function and hence not a valid encryption function.

For example, since $\gcd(4, 26) = 2$, it follows that $4x + 7$ is not a valid encryption function: x and $x + 13$ will encrypt to the same value, for any $x \in \mathbb{Z}_{26}$.

Let's next suppose that $\gcd(a, 26) = 1$. Suppose for some x_1 and x_2 that

$$ax_1 \equiv ax_2 \ (\text{mod } 26).$$

Then

$$a(x_1 - x_2) \equiv 0 \ (\text{mod } 26),$$

and thus

$$26 \mid a(x_1 - x_2).$$

We now make use of a property of division: if $\gcd(a, b) = 1$ and $a \mid bc$, then $a \mid c$. Since $26 \mid a(x_1 - x_2)$ and $\gcd(a, 26) = 1$, we must therefore have that

$$26 \mid (x_1 - x_2),$$

i.e., $x_1 \equiv x_2 \ (\text{mod } 26)$.

At this point we have shown that, if $\gcd(a, 26) = 1$, then a congruence of the form $ax \equiv y \ (\text{mod } 26)$ has, at most, one solution in $\mathbb{Z}_{26}$. Hence, if we let x vary over $\mathbb{Z}_{26}$, then $ax \bmod 26$ takes on 26 distinct values modulo 26. That is, it takes on every value exactly once. It follows that, for any $y \in \mathbb{Z}_{26}$, the congruence $ax \equiv y \ (\text{mod } 26)$ has a unique solution for y.

There is nothing special about the number 26 in this argument. The following result can be proved in an analogous fashion.

THEOREM 1.1
The congruence $ax \equiv b \ (\text{mod } m)$ has a unique solution $x \in \mathbb{Z}_m$ for every $b \in \mathbb{Z}_m$ if and only if $\gcd(a, m) = 1$.

Since $26 = 2 \times 13$, the values of $a \in \mathbb{Z}_{26}$ such that $\gcd(a, 26) = 1$ are $a = 1$, 3, 5, 7, 9, 11, 15, 17, 19, 21, 23, and 25. The parameter b can be any element in $\mathbb{Z}_{26}$. Hence the **Affine Cipher** has $12 \times 26 = 312$ possible keys. (Of course, this is much too small to be secure.)

Let's now consider the general setting where the modulus is m. We need another definition from number theory.

DEFINITION 1.3 *Suppose $a \geq 1$ and $m \geq 2$ are integers. If $\gcd(a, m) = 1$, then we say that a and m are relatively prime. The number of integers in $\mathbb{Z}_m$ that are relatively prime to m is often denoted by $\phi(m)$ (this function is called the Euler phi-function).*

A well-known result from number theory gives the value of $\phi(m)$ in terms of the prime power factorization of m. (An integer $p > 1$ is *prime* if it has no positive divisors other than 1 and p. Every integer $m > 1$ can be *factored* as a product of powers of primes in a unique way. For example, $60 = 2^2 \times 3 \times 5$ and $98 = 2 \times 7^2$.)

We record the formula for $\phi(m)$ in the following theorem.

THEOREM 1.2
Suppose

$$m = \prod_{i=1}^{n} p_i^{e_i},$$

where the p_i's are distinct primes and $e_i > 0$, $1 \le i \le n$. Then

$$\phi(m) = \prod_{i=1}^{n} (p_i^{e_i} - p_i^{e_i - 1}).$$

It follows that the number of keys in the **Affine Cipher** over $\mathbb{Z}_m$ is $m\phi(m)$, where $\phi(m)$ is given by the formula above. (The number of choices for b is m, and the number of choices for a is $\phi(m)$, where the encryption function is $e(x) = ax + b$.) For example, when $m = 60$, $\phi(60) = 2 \times 2 \times 4 = 16$ and the number of keys in the **Affine Cipher** is 960.

Let's now consider the decryption operation in the **Affine Cipher** with modulus $m = 26$. Suppose that $\gcd(a, 26) = 1$. To decrypt, we need to solve the congruence $y \equiv ax + b \pmod{26}$ for x. The discussion above establishes that the congruence will have a unique solution in $\mathbb{Z}_{26}$, but it does not give us an efficient method of finding the solution. What we require is an efficient algorithm to do this. Fortunately, some further results on modular arithmetic will provide us with the efficient decryption algorithm we seek.

We require the idea of a multiplicative inverse.

DEFINITION 1.4   *Suppose $a \in \mathbb{Z}_m$. The multiplicative inverse of a is an element $a^{-1} \in \mathbb{Z}_m$ such that $aa^{-1} \equiv a^{-1}a \equiv 1 \pmod{m}$.*

By similar arguments to those used above, it can be shown that a has a multiplicative inverse modulo m if and only if $\gcd(a, m) = 1$; and if a multiplicative inverse exists, it is unique. Also, observe that if $b = a^{-1}$, then $a = b^{-1}$. If p is prime, then every non-zero element of $\mathbb{Z}_p$ has a multiplicative inverse. A ring in which this is true is called a *field*.

In a later section, we will describe an efficient algorithm for computing multiplicative inverses in $\mathbb{Z}_m$ for any m. However, in $\mathbb{Z}_{26}$, trial and error suffices to find the multiplicative inverses of the elements relatively prime to 26: $1^{-1} = 1$, $3^{-1} = 9$, $5^{-1} = 21$, $7^{-1} = 15$, $11^{-1} = 19$, $17^{-1} = 23$, and $25^{-1} = 25$. (All of these can be verified easily. For example, $7 \times 15 = 105 \equiv 1 \bmod 26$, so $7^{-1} = 15$.)

Consider our congruence $y \equiv ax + b \pmod{26}$. This is equivalent to

$$ax \equiv y - b \pmod{26}.$$

Since $\gcd(a, 26) = 1$, a has a multiplicative inverse modulo 26. Multiplying both sides of the congruence by a^{-1}, we obtain

$$a^{-1}(ax) \equiv a^{-1}(y - b) \pmod{26}.$$

FIGURE 1.4
Affine Cipher

Let $\mathcal{P} = \mathcal{C} = \mathbb{Z}_{26}$ and let

$$K = \{(a, b) \in \mathbb{Z}_{26} \times \mathbb{Z}_{26} : \gcd(a, 26) = 1\}.$$

For $K = (a, b) \in \mathcal{K}$, define

$$e_K(x) = ax + b \bmod 26$$

and

$$d_K(y) = a^{-1}(y - b) \bmod 26$$

$(x, y \in \mathbb{Z}_{26})$.

By associativity of multiplication modulo 26,

$$a^{-1}(ax) \equiv (a^{-1}a)x \equiv 1x \equiv x.$$

Consequently, $x \equiv a^{-1}(y - b) \pmod{26}$. This is an explicit formula for x, that is, the decryption function is

$$d(y) = a^{-1}(y - b) \bmod 26.$$

So, finally, the complete description of the **Affine Cipher** is given in Figure 1.4. Let's do a small example.

Example 1.3
Suppose that $K = (7, 3)$. As noted above, $7^{-1} \bmod 26 = 15$. The encryption function is

$$e_K(x) = 7x + 3,$$

and the corresponding decryption function is

$$d_K(y) = 15(y - 3) = 15y - 19,$$

where all operations are performed in $\mathbb{Z}_{26}$. It is a good check to verify that $d_K(e_K(x)) = x$ for all $x \in \mathbb{Z}_{26}$. Computing in $\mathbb{Z}_{26}$, we get

$$
\begin{aligned}
d_K(e_K(x)) &= d_K(7x + 3) \\
&= 15(7x + 3) - 19 \\
&= x + 45 - 19 \\
&= x.
\end{aligned}
$$

FIGURE 1.5
Vigenere Cipher

Let m be some fixed positive integer. Define $\mathcal{P} = \mathcal{C} = \mathcal{K} = (\mathbb{Z}_{26})^m$. For a key $K = (k_1, k_2, \ldots, k_m)$, we define

$$e_K(x_1, x_2, \ldots, x_m) = (x_1 + k_1, x_2 + k_2, \ldots, x_m + k_m)$$

and

$$d_K(y_1, y_2, \ldots, y_m) = (y_1 - k_1, y_2 - k_2, \ldots, y_m - k_m),$$

where all operations are performed in $\mathbb{Z}_{26}$.

To illustrate, let's encrypt the plaintext *hot*. We first convert the letters h, o, t to residues modulo 26. These are respectively 7, 14, and 19. Now, we encrypt:

$$
\begin{aligned}
7 \times 7 + 3 \bmod 26 &= 52 \bmod 26 &= 0 \\
7 \times 14 + 3 \bmod 26 &= 101 \bmod 26 &= 23 \\
7 \times 19 + 3 \bmod 26 &= 136 \bmod 26 &= 6.
\end{aligned}
$$

So the three ciphertext characters are $0, 23$, and 6, which corresponds to the alphabetic string AXG. We leave the decryption as an exercise for the reader. □

1.1.4   The Vigenere Cipher

In both the **Shift Cipher** and the **Substitution Cipher**, once a key is chosen, each alphabetic character is mapped to a unique alphabetic character. For this reason, these cryptosystems are called *monoalphabetic*. We now present in Figure 1.5 a cryptosystem which is not monoalphabetic, the well-known **Vigenere Cipher**. This cipher is named after Blaise de Vigenere, who lived in the sixteenth century.

Using the correspondence $A \leftrightarrow 0$, $B \leftrightarrow 1$, ..., $Z \leftrightarrow 25$ described earlier, we can associate each key K with an alphabetic string of length m, called a *keyword*. The **Vigenere Cipher** encrypts m alphabetic characters at a time: each plaintext element is equivalent to m alphabetic characters.

Let's do a small example.

Example 1.4
Suppose $m = 6$ and the keyword is $CIPHER$. This corresponds to the numerical equivalent $K = (2, 8, 15, 7, 4, 17)$. Suppose the plaintext is the string

```
thiscryptosystemisnotsecure.
```

We convert the plaintext elements to residues modulo 26, write them in groups of six, and then "add" the keyword modulo 26, as follows:

19	7	8	18	2	17	24	15	19	14	18	24
2	8	15	7	4	17	2	8	15	7	4	17
21	15	23	25	6	8	0	23	8	21	22	15

18	19	4	12	8	18	13	14	19	18	4	2
2	8	15	7	4	17	2	8	15	7	4	17
20	1	19	19	12	9	15	22	8	25	8	19

20	17	4
2	8	15
22	25	19

The alphabetic equivalent of the ciphertext string would thus be:

```
VPXZGIAXIVWPUBTTMJPWIZITWZT.
```

To decrypt, we can use the same keyword, but we would subtract it modulo 26 instead of adding. ⏹

Observe that the number of possible keywords of length m in a **Vigenere Cipher** is 26^m, so even for relatively small values of m, an exhaustive key search would require a long time. For example, if we take $m = 5$, then the keyspace has size exceeding 1.1×10^7. This is already large enough to preclude exhaustive key search by hand (but not by computer).

In a **Vigenere Cipher** having keyword length m, an alphabetic character can be mapped to one of m possible alphabetic characters (assuming that the keyword contains m distinct characters). Such a cryptosystem is called *polyalphabetic*. In general, cryptanalysis is more difficult for polyalphabetic than for monoalphabetic cryptosystems.

1.1.5 The Hill Cipher

In this section, we describe another polyalphabetic cryptosystem called the **Hill Cipher**. This cipher was invented in 1929 by Lester S. Hill. Let m be a positive integer, and define $\mathcal{P} = \mathcal{C} = (\mathbb{Z}_{26})^m$. The idea is to take m linear combinations of the m alphabetic characters in one plaintext element, thus producing the m alphabetic characters in one ciphertext element.

For example, if $m = 2$, we could write a plaintext element as $x = (x_1, x_2)$ and a ciphertext element as $y = (y_1, y_2)$. Here, y_1 would be a linear combination of x_1 and x_2, as would y_2. We might take

$$y_1 = 11x_1 + 3x_2$$

$$y_2 = 8x_1 + 7x_2.$$

Of course, this can be written more succinctly in matrix notation as follows:

$$(y_1, y_2) = (x_1, x_2) \begin{pmatrix} 11 & 8 \\ 3 & 7 \end{pmatrix}.$$

In general, we will take an $m \times m$ matrix K as our key. If the entry in row i and column j of K is $k_{i,j}$, then we write $K = (k_{i,j})$. For $x = (x_1, \ldots, x_m) \in \mathcal{P}$ and $K \in \mathcal{K}$, we compute $y = e_K(x) = (y_1, \ldots, y_m)$ as follows:

$$(y_1, y_2, \ldots, y_m) = (x_1, x_2, \ldots, x_m) \begin{pmatrix} k_{1,1} & k_{1,2} & \cdots & k_{1,m} \\ k_{2,1} & k_{2,2} & \cdots & k_{2,m} \\ \vdots & \vdots & & \vdots \\ k_{m,1} & k_{m,2} & \cdots & k_{m,m} \end{pmatrix}.$$

In other words, $y = xK$.

We say that the ciphertext is obtained from the plaintext by means of a *linear transformation*. We have to consider how decryption will work, that is, how x can be computed from y. Readers familiar with linear algebra will realize that we use the inverse matrix K^{-1} to decrypt. The ciphertext is decrypted using the formula $x = yK^{-1}$.

Here are the definitions of necessary concepts from linear algabra. If $A = (a_{i,j})$ is an $\ell \times m$ matrix and $B = (b_{j,k})$ is an $m \times n$ matrix, then we define the *matrix product* $AB = (c_{i,k})$ by the formula

$$c_{i,k} = \sum_{j=1}^{m} a_{i,j} b_{j,k}$$

for $1 \leq i \leq \ell$ and $1 \leq k \leq n$. That is, the entry in row i and column k of AB is formed by taking the ith row of A and the kth column of B, multiplying corresponding entries together, and summing. Note that AB is an $\ell \times n$ matrix.

This definition of matrix multiplication is associative (that is, $(AB)C = A(BC)$ but not, in general, commutative (it is not always the case that $AB = BA$, even for square matrices A and B).

The $m \times m$ *identity matrix*, denoted by I_m, is the $m \times m$ matrix with 1's on the main diagonal and 0's elsewhere. Thus, the 2×2 identity matrix is

$$I_2 = \begin{pmatrix} 1 & 0 \\ 0 & 1 \end{pmatrix}.$$

I_m is termed an identity matrix since $AI_m = A$ for any $\ell \times m$ matrix A and $I_m B = B$ for any $m \times n$ matrix B. Now, the *inverse matrix* to an $m \times m$ matrix A (if it exists) is the matrix A^{-1} such that $AA^{-1} = A^{-1}A = I_m$. Not all matrices have inverses, but if an inverse exists, it is unique.

With these facts at hand, it is easy to derive the decryption formula given above: since $y = xK$, we can multiply both sides of the formula by K^{-1}, obtaining

$$yK^{-1} = (xK)K^{-1} = x(KK^{-1}) = xI_m = x.$$

(Note the use of the associativity property.)

We can verify that the encryption matrix above has an inverse in $\mathbb{Z}_{26}$:

$$\begin{pmatrix} 11 & 8 \\ 3 & 7 \end{pmatrix}^{-1} = \begin{pmatrix} 7 & 18 \\ 23 & 11 \end{pmatrix}$$

since

$$\begin{pmatrix} 11 & 8 \\ 3 & 7 \end{pmatrix} \begin{pmatrix} 7 & 18 \\ 23 & 11 \end{pmatrix} = \begin{pmatrix} 11 \times 7 + 8 \times 23 & 11 \times 18 + 8 \times 11 \\ 3 \times 7 + 7 \times 23 & 3 \times 18 + 7 \times 11 \end{pmatrix}$$

$$= \begin{pmatrix} 261 & 286 \\ 182 & 131 \end{pmatrix}$$

$$= \begin{pmatrix} 1 & 0 \\ 0 & 1 \end{pmatrix}.$$

(Remember that all arithmetic operations are done modulo 26.)

Let's now do an example to illustrate encryption and decryption in the **Hill Cipher**.

Example 1.5
Suppose the key is

$$K = \begin{pmatrix} 11 & 8 \\ 3 & 7 \end{pmatrix}.$$

From the computations above, we have that

$$K^{-1} = \begin{pmatrix} 7 & 18 \\ 23 & 11 \end{pmatrix}.$$

Suppose we want to encrypt the plaintext *july*. We have two elements of plaintext to encrypt: $(9, 20)$ (corresponding to ju) and $(11, 24)$ (corresponding to ly). We compute as follows:

$$(9, 20) \begin{pmatrix} 11 & 8 \\ 3 & 7 \end{pmatrix} = (99 + 60, 72 + 140) = (3, 4)$$

and

$$(11, 24) \begin{pmatrix} 11 & 8 \\ 3 & 7 \end{pmatrix} = (121 + 72, 88 + 168) = (11, 22).$$

Hence, the encryption of *july* is $DELW$. To decrypt, Bob would compute:

$$(3,4)\begin{pmatrix} 7 & 18 \\ 23 & 11 \end{pmatrix} = (9,20)$$

and

$$(11,22)\begin{pmatrix} 7 & 18 \\ 23 & 11 \end{pmatrix} = (11,24).$$

Hence, the correct plaintext is obtained.   ▯

At this point, we have shown that decryption is possible if K has an inverse. In fact, for decryption to be possible, it is necessary that K has an inverse. (This follows fairly easily from elementary linear algebra, but we will not give a proof here.) So we are interested precisely in those matrices K that are invertible.

The invertibility of a (square) matrix depends on the value of its determinant. To avoid unnecessary generality, we will confine our attention to the 2×2 case.

DEFINITION 1.5   *The determinant of the 2×2 matrix $A = (a_{i,j})$ is the value*

$$\det A = a_{1,1}a_{2,2} - a_{1,2}a_{2,1}.$$

REMARK   The determinant of an $m \times m$ square matrix can be computed by elementary row operations: see any text on linear algebra.   ∎

Two important properties of determinants are that $\det I_m = 1$; and the multiplication rule $\det(AB) = \det A \times \det B$.

A real matrix K has an inverse if and only if its determinant is non-zero. However, it is important to remember that we are working over $\mathbb{Z}_{26}$. The relevant result for our purposes is that a matrix K has an inverse modulo 26 if and only if $\gcd(\det K, 26) = 1$.

We briefly sketch the proof of this fact. First suppose that $\gcd(\det K, 26) = 1$. Then $\det K$ has an inverse in $\mathbb{Z}_{26}$. Now, for $1 \leq i \leq m$, $1 \leq j \leq m$, define K_{ij} to be the matrix obtained from K by deleting the ith row and the jth column. Define a matrix K^* to have as its (i,j)-entry the value $(-1)^{i+j}\det K_{ji}$. (K^* is called the *adjoint matrix* of K.) Then it can be shown that

$$K^{-1} = (\det K)^{-1}K^*.$$

Hence, K is invertible.

Conversely, suppose K has an inverse, K^{-1}. By the multiplication rule for determinants, we have

$$1 = \det I = \det(KK^{-1}) = \det K \det K^{-1}.$$

Hence, $\det K$ is invertible in $\mathbb{Z}_{26}$.

REMARK The above formula for K^{-1} is not very efficient computationally, except for small values of m (say $m = 2, 3$). For larger m, the preferred method of computing inverse matrices would involve elementary row operations. ∎

In the 2×2 case, we have the following formula:

THEOREM 1.3
Suppose $A = (a_{i,j})$ is a 2×2 matrix over $\mathbb{Z}_{26}$ such that $\det A = a_{1,1}a_{2,2} - a_{1,2}a_{2,1}$ *is invertible. Then*

$$A^{-1} = (\det A)^{-1} \begin{pmatrix} a_{2,2} & -a_{1,2} \\ -a_{2,1} & a_{1,1} \end{pmatrix}.$$

Let's look again at the example considered earlier. First, we have

$$\det \begin{pmatrix} 11 & 8 \\ 3 & 7 \end{pmatrix} = 11 \times 7 - 8 \times 3 \bmod 26$$

$$= 77 - 24 \bmod 26$$

$$= 53 \bmod 26$$

$$= 1.$$

Now, $1^{-1} \bmod 26 = 1$, so the inverse matrix is

$$\begin{pmatrix} 11 & 8 \\ 3 & 7 \end{pmatrix}^{-1} = \begin{pmatrix} 7 & 18 \\ 23 & 11 \end{pmatrix},$$

as we verified earlier.

We now give a precise description of the **Hill Cipher** over $\mathbb{Z}_{26}$ in Figure 1.6.

1.1.6 The Permutation Cipher

All of the cryptosystems we have discussed so far involve substitution: plaintext characters are replaced by different ciphertext characters. The idea of a permutation cipher is to keep the plaintext characters unchanged, but to alter their positions by rearranging them. The **Permutation Cipher** (also known as the **Transposition Cipher**) has been in use for hundreds of years. In fact, the distinction between the **Permutation Cipher** and the **Substitution Cipher** was pointed out as early as 1563 by Giovanni Porta. A formal definition is given in Figure 1.7.

As with the **Substitution Cipher**, it is more convenient to use alphabetic characters as opposed to residues modulo 26, since there are no algebraic operations being performed in encryption or decryption.

Here is an example to illustrate:

FIGURE 1.6
Hill Cipher

Let m be some fixed positive integer. Let $\mathcal{P} = \mathcal{C} = (\mathbb{Z}_{26})^m$ and let

$$\mathcal{K} = \{m \times m \text{ invertible matrices over } \mathbb{Z}_{26}\}.$$

For a key K, we define
$$e_K(x) = xK$$
and
$$d_K(y) = yK^{-1},$$
where all operations are performed in $\mathbb{Z}_{26}$.

FIGURE 1.7
Permutation Cipher

Let m be some fixed positive integer. Let $\mathcal{P} = \mathcal{C} = (\mathbb{Z}_{26})^m$ and let $\mathcal{K}$ consist of all permutations of $\{1, \ldots, m\}$. For a key (i.e., a permutation) π, we define

$$e_\pi(x_1, \ldots, x_m) = (x_{\pi(1)}, \ldots, x_{\pi(m)})$$

and

$$d_\pi(y_1, \ldots, y_m) = (y_{\pi^{-1}(1)}, \ldots, y_{\pi^{-1}(m)}),$$

where π^{-1} is the inverse permutation to π.

Example 1.6
Suppose $m = 6$ and the key is the following permutation π:

1	2	3	4	5	6
3	5	1	6	4	2

Then the inverse permutation π^{-1} is the following:

1	2	3	4	5	6
3	6	1	5	2	4

Now, suppose we are given the plaintext

shesellsseashellsbytheseashore.

We first group the plaintext into groups of six letters:

$$\texttt{shesel} \mid \texttt{lsseas} \mid \texttt{hellsb} \mid \texttt{ythese} \mid \texttt{ashore}$$

Now each group of six letters is rearranged according to the permutation π, yielding the following:

$$\texttt{EESLSH} \mid \texttt{SALSES} \mid \texttt{LSHBLE} \mid \texttt{HSYEET} \mid \texttt{HRAEOS}$$

So, the ciphertext is:

$$\texttt{EESLSHSALSESLSHBLEHSYEETHRAEOS.}$$

The ciphertext can be decrypted in a similar fashion, using the inverse permutation π^{-1}. ⬜

In fact, the **Permutation Cipher** is a special case of the **Hill Cipher**. Given a permutation of π of the set $\{1, \ldots, m\}$, we can define an associated $m \times m$ permutation matrix $K_\pi = (k_{i,j})$ according to the formula

$$k_{i,j} = \begin{cases} 1 & \text{if } i = \pi(j) \\ 0 & \text{otherwise.} \end{cases}$$

(A *permutation matrix* is a matrix in which every row and column contains exactly one "1," and all other values are "0." A permutation matrix can be obtained from an identity matrix by permuting rows or columns.)

It is not difficult to see that Hill encryption using the matrix K_π is, in fact, equivalent to permutation encryption using the permutation π. Moreover, $K_\pi^{-1} = K_{\pi^{-1}}$, i.e., the inverse matrix to K_π is the permutation matrix defined by the permutation π^{-1}. Thus, Hill decryption is equivalent to permutation decryption.

For the permutation π used in the example above, the associated permutation matrices are

$$K_\pi = \begin{pmatrix} 0 & 0 & 1 & 0 & 0 & 0 \\ 0 & 0 & 0 & 0 & 0 & 1 \\ 1 & 0 & 0 & 0 & 0 & 0 \\ 0 & 0 & 0 & 0 & 1 & 0 \\ 0 & 1 & 0 & 0 & 0 & 0 \\ 0 & 0 & 0 & 1 & 0 & 0 \end{pmatrix}$$

and

$$K_\pi^{-1} = \begin{pmatrix} 0 & 0 & 1 & 0 & 0 & 0 \\ 0 & 0 & 0 & 0 & 1 & 0 \\ 1 & 0 & 0 & 0 & 0 & 0 \\ 0 & 0 & 0 & 0 & 0 & 1 \\ 0 & 0 & 0 & 1 & 0 & 0 \\ 0 & 1 & 0 & 0 & 0 & 0 \end{pmatrix}.$$

The reader can verify that the product of these two matrices is the identity.

1.1.7   Stream Ciphers

In the cryptosystems we have studied to this point, successive plaintext elements are encrypted using the same key, K. That is, the ciphertext string $\mathbf{y}$ is obtained as follows:

$$\mathbf{y} = y_1 y_2 \ldots = e_K(x_1) e_K(x_2) \ldots.$$

Cryptosystems of this type are often called *block ciphers*.

An alternative approach is to use what are called stream ciphers. The basic idea is to generate a keystream $\mathbf{z} = z_1 z_2 \ldots$, and use it to encrypt a plaintext string $\mathbf{x} = x_1 x_2 \ldots$ according to the rule

$$\mathbf{y} = y_1 y_2 \ldots = e_{z_1}(x_1) e_{z_2}(x_2) \ldots.$$

A stream cipher operates as follows. Suppose $K \in \mathcal{K}$ is the key and $x_1 x_2 \ldots$ is the plaintext string. The function f_i is used to generate z_i (the ith element of the keystream), where f_i is a function of the key, K, and the first $i - 1$ plaintext characters:

$$z_i = f_i(K, x_1, \ldots, x_{i-1}).$$

The keystream element z_i is used to encrypt x_i, yielding $y_i = e_{z_i}(x_i)$. So, to encrypt the plaintext string $x_1 x_2 \ldots$, we would successively compute

$$z_1, y_1, z_2, y_2, \ldots.$$

Decrypting the ciphertext string $y_1 y_2 \ldots$ can be accomplished by successively computing

$$z_1, x_1, z_2, x_2 \ldots.$$

Here is a formal mathematical definition:

DEFINITION 1.6   *A Stream Cipher is a tuple* $(\mathcal{P}, \mathcal{C}, \mathcal{K}, \mathcal{L}, \mathcal{F}, \mathcal{E}, \mathcal{D})$, *where the following conditions are satisfied:*

1. $\mathcal{P}$ *is a finite set of possible plaintexts*

2. $\mathcal{C}$ *is a finite set of possible ciphertexts*

3. $\mathcal{K}$, *the keyspace, is a finite set of possible keys*

4. $\mathcal{L}$ *is a finite set called the keystream alphabet*

5. $\mathcal{F} = (f_1, f_2, \ldots)$ *is the keystream generator. For* $i \geq 1$,

$$f_i : \mathcal{K} \times \mathcal{P}^{i-1} \to \mathcal{L}.$$

6. *For each* $z \in \mathcal{L}$, *there is an encryption rule* $e_z \in \mathcal{E}$ *and a corresponding decryption rule* $d_z \in \mathcal{D}$. $e_z : \mathcal{P} \to \mathcal{C}$ *and* $d_z : \mathcal{C} \to \mathcal{P}$ *are functions such that* $d_z(e_z(x)) = x$ *for every plaintext* $x \in \mathcal{P}$.

We can think of a block cipher as a special case of a stream cipher where the keystream is constant: $z_i = K$ for all $i \geq 1$.

Here are some special types of stream ciphers together with illustrative examples. A stream cipher is *synchronous* if the keystream is independent of the plaintext string, that is, if the keystream is generated as a function only of the key K. In this situation, we think of K as a "seed" that is expanded into a keystream $z_1 z_2 \ldots$.

A stream cipher is *periodic* with period d if $z_{i+d} = z_i$ for all integers $i \geq 1$. The **Vigenere Cipher** with keyword length m can be thought of as a periodic stream cipher with period m. In this case, the key is $K = (k_1, \ldots, k_m)$. K itself provides the first m elements of the keystream: $z_i = k_i$, $1 \leq i \leq m$. Then the keystream just repeats itself from that point on. Observe that in this stream cipher setting for the **Vigenere Cipher**, the encryption and decryption functions are identical to those used in the **Shift Cipher**: $e_z(x) = x + z$ and $d_z(y) = y - z$.

Stream ciphers are often described in terms of binary alphabets, i.e., $\mathcal{P} = \mathcal{C} = \mathcal{L} = \mathbb{Z}_2$. In this situation, the encryption and decryption operations are just addition modulo 2:

$$e_z(x) = x + z \bmod 2$$

and

$$d_z(y) = y + z \bmod 2.$$

If we think of "0" as representing the boolean value "false" and "1" as representing "true," then addition modulo 2 corresponds to the exclusive-or operation. Hence, encryption (and decryption) can be implemented very efficiently in hardware.

Let's look at another method of generating a (synchronous) keystream. Suppose we start with $(k_1, \ldots, k_m)$ and let $z_i = k_i$, $1 \leq i \leq m$ (as before), but we now generate the keystream using a linear recurrence relation of degree m:

$$z_{i+m} = \sum_{j=0}^{m-1} c_j z_{i+j} \bmod 2,$$

where $c_0, \ldots, c_{m-1} \in \mathbb{Z}_2$ are predetermined constants.

REMARK This recurrence is said to have *degree* m since each term depends on the previous m terms. It is *linear* because z_{i+m} is a linear function of previous terms. Note that we can take $c_0 = 1$ without loss of generality, for otherwise the recurrence will be of degree $m - 1$. ∎

Here, the key K consists of the $2m$ values $k_1, \ldots, k_m, c_0, \ldots, c_{m-1}$. If

$$(k_1, \ldots, k_m) = (0, \ldots, 0),$$

then the keystream consists entirely of 0's. Of course, this should be avoided, as the ciphertext will then be identical to the plaintext. However, if the constants $c_0, \ldots, c_{m-1}$ are chosen in a suitable way, then any other initialization vector $(k_1, \ldots, k_m)$ will give rise to a periodic keystream having period $2^m - 1$. So a "short" key can give rise to a keystream having a very long period. This is certainly a desirable property: we will see in a later section how the **Vigenere Cipher** can be cryptanalyzed by exploiting the fact that the keystream has short period.

Here is an example to illustrate.

Example 1.7

Suppose $m = 4$ and the keystream is generated using the rule

$$z_{i+4} = z_i + z_{i+1} \bmod 2$$

($i \geq 1$). If the keystream is initialized with any vector other than $(0, 0, 0, 0)$, then we obtain a keystream of period 15. For example, starting with $(1, 0, 0, 0)$, the keystream is

$$1, 0, 0, 0, 1, 0, 0, 1, 1, 0, 1, 0, 1, 1, 1, \ldots.$$

Any other non-zero initialization vector will give rise to a cyclic permutation of the same keystream. ☐

Another appealing aspect of this method of keystream generation is that the keystream can be produced efficiently in hardware using a *linear feedback shift register*, or LFSR. We would use a shift register with m stages. The vector $(k_1, \ldots, k_m)$ would be used to initialize the shift register. At each time unit, the following operations would be performed concurrently:

1. k_1 would be tapped as the next keystream bit

2. $k_2, \ldots, k_m$ would each be shifted one stage to the left

3. the "new" value of k_m would be computed to be

$$\sum_{j=0}^{m-1} c_j k_{j+1}$$

(this is the "linear feedback").

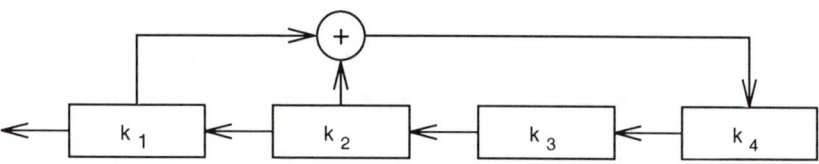

FIGURE 1.8
A Linear Feedback Shift Register

FIGURE 1.9
Autokey Cipher

Let $\mathcal{P} = \mathcal{C} = \mathcal{K} = \mathcal{L} = \mathbb{Z}_{26}$. Let $z_1 = K$, and $z_i = x_{i-1}$ $(i \geq 2)$. For $0 \leq z \leq 25$, define

$$e_z(x) = x + z \bmod 26$$

and

$$d_z(y) = y - z \bmod 26$$

$(x, y \in \mathbb{Z}_{26})$.

Observe that the linear feedback is carried out by tapping certain stages of the register (as specified by the constants c_j having the value "1") and computing a sum modulo 2 (which is an exclusive-or). This is illustrated in Figure 1.8, where we depict the LFSR that will generate the keystream of Example 1.7.

An example of a non-synchronous stream cipher that is known as the **Autokey Cipher** is given in Figure 1.9. It is apparently due to Vigenere.

The reason for the terminology "autokey" is that the plaintext is used as the key (aside from the initial "priming key" K). Here is an example to illustrate:

Example 1.8
Suppose the key is $K = 8$, and the plaintext is

> rendezvous.

We first convert the plaintext to a sequence of integers:

$$17 \quad 4 \quad 13 \quad 3 \quad 4 \quad 25 \quad 21 \quad 14 \quad 20 \quad 18$$

The keystream is as follows:

$$8 \quad 17 \quad 4 \quad 13 \quad 3 \quad 4 \quad 25 \quad 21 \quad 14 \quad 20$$

Now we add corresponding elements, reducing modulo 26:

$$25 \quad 21 \quad 17 \quad 16 \quad 7 \quad 3 \quad 20 \quad 9 \quad 8 \quad 12$$

In alphabetic form, the ciphertext is:

ZVRQHDUJIM.

Now let's look at how Alice decrypts the ciphertext. She will first convert the alphabetic string to the numeric string

$$25 \quad 21 \quad 17 \quad 16 \quad 7 \quad 3 \quad 20 \quad 9 \quad 8 \quad 12$$

Then she can compute

$$x_1 = d_8(25) = 25 - 8 \bmod 26 = 17.$$

Next,

$$x_2 = d_{17}(21) = 21 - 17 \bmod 26 = 4,$$

and so on. Each time she obtains another plaintext character, she also uses it as the next keystream element. □

Of course, the **Autokey Cipher** is insecure since there are only 26 possible keys.

In the next section, we discuss methods that can be used to cryptanalyze the various cryptosystems we have presented.

1.2   Cryptanalysis

In this section, we discuss some techniques of cryptanalysis. The general assumption that is usually made is that the opponent, Oscar, knows the cryptosystem being used. This is usually referred to as *Kerckhoff's principle*. Of course, if Oscar does not know the cryptosystem being used, that will make his task more difficult. But we do not want to base the security of a cryptosystem on the (possibly shaky) premise that Oscar does not know what system is being employed. Hence, our goal in designing a cryptosystem will be to obtain security under Kerckhoff's principle.

First, we want to differentiate between different levels of attacks on cryptosystems. The most common types are enumerated as follows.

Ciphertext-only

The opponent possesses a string of ciphertext, $\mathbf{y}$.

Known plaintext

The opponent possesses a string of plaintext, $\mathbf{x}$, and the corresponding ciphertext, $\mathbf{y}$.

Chosen plaintext

The opponent has obtained temporary access to the encryption machinery. Hence he can choose a plaintext string, $\mathbf{x}$, and construct the corresponding ciphertext string, $\mathbf{y}$.

Chosen ciphertext

The opponent has obtained temporary access to the decryption machinery. Hence he can choose a ciphertext string, $\mathbf{y}$, and construct the corresponding plaintext string, $\mathbf{x}$.

In each case, the object is to determine the key that was used. We note that a chosen ciphertext attack is relevant to public-key cryptosystems, which we discuss in the later chapters.

We first consider the weakest type of attack, namely a ciphertext-only attack. We also assume that the plaintext string is ordinary English text, without punctuation or "spaces." (This makes cryptanalysis more difficult than if punctuation and spaces were encrypted.)

Many techniques of cryptanalysis use statistical properties of the English language. Various people have estimated the relative frequencies of the 26 letters by compiling statistics from numerous novels, magazines, and newspapers. The estimates in Table 1.1 were obtained by Beker and Piper.

On the basis of the above probabilities, Beker and Piper partition the 26 letters into five groups as follows:

1. E, having probability about 0.120

2. T, A, O, I, N, S, H, R, each having probabilities between 0.06 and 0.09

3. D, L, each having probabilities around 0.04

4. $C, U, M, W, F, G, Y, P, B$, each having probabilities between 0.015 and 0.028

5. V, K, J, X, Q, Z, each having probabilities less than 0.01.

It may also be useful to consider sequences of two or three consecutive letters called *digrams* and *trigrams*, respectively. The 30 most common digrams are (in decreasing order) $TH, HE, IN, ER, AN, RE, ED, ON, ES, ST, EN, AT, TO, NT, HA, ND, OU, EA, NG, AS, OR, TI, IS, ET, IT, AR, TE, SE, HI$, and OF. The twelve most common trigrams are (in decreasing order) $THE, ING, AND, HER, ERE, ENT, THA, NTH, WAS, ETH, FOR$, and DTH.

TABLE 1.1
Probabilities of Occurrence of the 26 Letters

letter	probability	letter	probability
A	.082	N	.067
B	.015	O	.075
C	.028	P	.019
D	.043	Q	.001
E	.127	R	.060
F	.022	S	.063
G	.020	T	.091
H	.061	U	.028
I	.070	V	.010
J	.002	W	.023
K	.008	X	.001
L	.040	Y	.020
M	.024	Z	.001

1.2.1   Cryptanalysis of the Affine Cipher

As a simple illustration of how cryptanalysis can be performed using statistical data, let's look first at the **Affine Cipher**. Suppose Oscar has intercepted the following ciphertext:

Example 1.9
Ciphertext obtained from an Affine Cipher

```
FMXVEDKAPHFERBNDKRXRSREFMORUDSDKDVSHVUFEDK
APRKDLYEVLRHHRH
```

The frequency analysis of this ciphertext is given in Table 1.2.

There are only 57 characters of ciphertext, but this is sufficient to cryptanalyze an **Affine Cipher**. The most frequent ciphertext characters are: R (8 occurrences), D (7 occurrences), E, H, K (5 occurrences each), and F, S, V (4 occurrences each). As a first guess, we might hypothesize that R is the encryption of e and D is the encryption of t, since e and t are (respectively) the two most common letters. Expressed numerically, we have $e_K(4) = 17$ and $e_K(19) = 3$. Recall that $e_K(x) = ax + b$, where a and b are unknowns. So we get two linear equations in two unknowns:

$$4a + b = 17$$

$$19a + b = 3.$$

TABLE 1.2
Frequency of Occurrence of the 26 Ciphertext Letters

letter	frequency	letter	frequency
A	2	N	1
B	1	O	1
C	0	P	2
D	7	Q	0
E	5	R	8
F	4	S	3
G	0	T	0
H	5	U	2
I	0	V	4
J	0	W	0
K	5	X	2
L	2	Y	1
M	2	Z	0

This system has the unique solution $a = 6$, $b = 19$ (in $\mathbb{Z}_{26}$). But this is an illegal key, since $\gcd(a, 26) = 2 > 1$. So our hypothesis must be incorrect.

Our next guess might be that R is the encryption of e and E is the encryption of t. Proceeding as above, we obtain $a = 13$, which is again illegal. So we try the next possibility, that R is the encryption of e and H is the encryption of t. This yields $a = 8$, again impossible. Continuing, we suppose that R is the encryption of e and K is the encryption of t. This produces $a = 3$, $b = 5$, which is at least a legal key. It remains to compute the decryption function corresponding to $K = (3, 5)$, and then to decrypt the ciphertext to see if we get a meaningful string of English, or nonsense. This will confirm the validity of $(3, 5)$.

If we perform these operations, we have $d_K(y) = 9y - 19$ and the given ciphertext decrypts to yield:

```
algorithmsarequitegeneraldefinitionsofarit
hmeticprocesses
```

We conclude that we have determined the correct key. $\square$

1.2.2 Cryptanalysis of the Substitution Cipher

Here, we look at the more complicated situation, the **Substitution Cipher**. Consider the following ciphertext:

TABLE 1.3
Frequency of Occurrence of the 26 Ciphertext Letters

letter	frequency	letter	frequency
A	0	N	9
B	1	O	0
C	15	P	1
D	13	Q	4
E	7	R	10
F	11	S	3
G	1	T	2
H	4	U	5
I	5	V	5
J	11	W	8
K	1	X	6
L	0	Y	10
M	16	Z	20

Example 1.10
Ciphertext obtained from a Substitution Cipher

```
YIFQFMZRWQFYVECFMDZPCVMRZWNMDZVEJBTXCDDUMJ
NDIFEFMDZCDMQZKCEYFCJMYRNCWJCSZREXCHZUNMXZ
NZUCDRJXYYSMRTMEYIFZWDYVZVYFZUMRZCRWNZDZJJ
XZWGCHSMRNMDHNCMFQCHZJMXJZWIEJYUCFWDJNZDIR
```

The frequency analysis of this ciphertext is given in Table 1.3.

Since Z occurs significantly more often than any other ciphertext character, we might conjecture that $d_K(Z) = e$. The remaining ciphertext characters that occur at least ten times (each) are C, D, F, J, M, R, Y. We might expect that these letters are encryptions of (a subset of) t, a, o, i, n, s, h, r, but the frequencies really do not vary enough to tell us what the correspondence might be.

At this stage we might look at digrams, especially those of the form $-Z$ or $Z-$, since we conjecture that Z decrypts to e. We find that the most common digrams of this type are DZ and ZW (four times each); NZ and ZU (three times each); and $RZ, HZ, XZ, FZ, ZR, ZV, ZC, ZD$, and ZJ (twice each). Since ZW occurs four times and WZ not at all, and W occurs less often than many other characters, we might guess that $d_K(W) = d$. Since DZ occurs four times and ZD occurs twice, we would think that $D_K(D) \in \{r, s, t\}$, but it is not clear which of the three possibilities is the correct one.

If we proceed on the assumption that $d_K(Z) = e$ and $d_K(W) = d$, we might look back at the ciphertext and notice that we have ZRW and RZW both occurring near the beginning of the ciphertext, and RW occurs again later on. Since R occurs frequently in the ciphertext and nd is a common digram, we might try $d_K(R) = n$ as the most likely possibility.

At this point, we have the following:

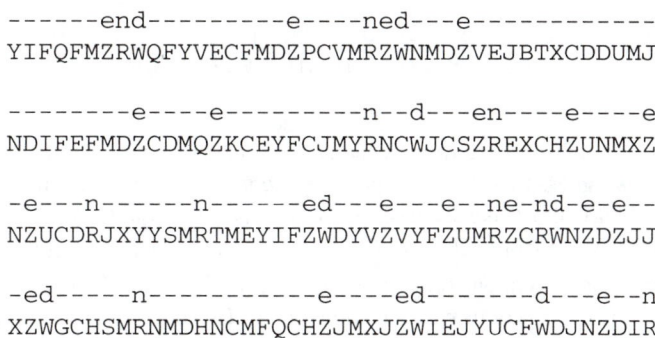

Our next step might be to try $d_K(N) = h$, since NZ is a common digram and ZN is not. If this is correct, then the segment of plaintext $ne - ndhe$ suggests that $d_K(C) = a$. Incorporating these guesses, we have:

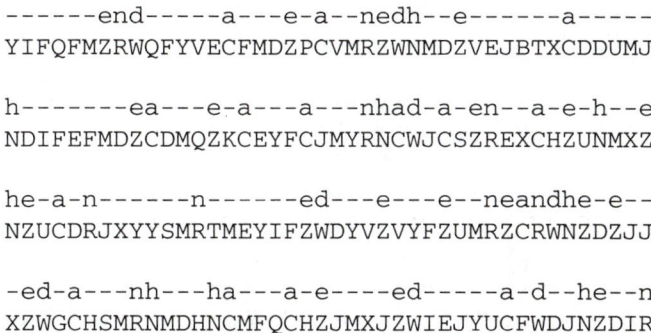

Now, we might consider M, the second most common ciphertext character. The ciphertext segment RNM, which we believe decrypts to $nh-$, suggests that $h-$ begins a word, so M probably represents a vowel. We have already accounted for a and e, so we expect that $d_K(M) = i$ or o. Since ai is a much more likely digram than ao, the ciphertext digram CM suggests that we try $d_K(M) = i$ first. Then we have:

```
-----iend-----a-i-e-a-inedhi-e------a---i-
YIFQFMZRWQFYVECFMDZPCVMRZWNMDZVEJBTXCDDUMJ

h-----i-ea-i-e-a---a-i-nhad-a-en--a-e-hi-e
NDIFEFMDZCDMQZKCEYFCJMYRNCWJCSZREXCHZUNMXZ

he-a-n-----in-i----ed---e---e-ineandhe-e--
NZUCDRJXYYSMRTMEYIFZWDYVZVYFZUMRZCRWNZDZJJ

-ed-a--inhi--hai--a-e-i--ed-----a-d--he--n
XZWGCHSMRNMDHNCMFQCHZJMXJZWIEJYUCFWDJNZDIR
```

Next, we might try to determine which letter is encrypted to o. Since o is a common letter, we guess that the corresponding ciphertext letter is one of D, F, J, Y. Y seem to be the most likely possibility, otherwise, we would get long strings of vowels, namely aoi from CFM or CJM. Hence, let's suppose $d_E(Y) = o$.

The three most frequent remaining ciphertext letters are D, F, J, which we conjecture could decrypt to r, s, t in some order. Two occurrences of the trigram NMD suggest that $d_E(D) = s$, giving the trigram his in the plaintext (this is consistent with our earlier hypothesis that $d_E(D) \in \{r, s, t\}$). The segment $HNCMF$ could be an encryption of $chair$, which would give $d_E(F) = r$ (and $d_E(H) = c$) and so we would then have $d_E(J) = t$ by process of elimination. Now, we have:

```
o-r-riend-ro--arise-a-inedhise--t---ass-it
YIFQFMZRWQFYVECFMDZPCVMRZWNMDZVEJBTXCDDUMJ

hs-r-riseasi-e-a-orationhadta-en--ace-hi-e
NDIFEFMDZCDMQZKCEYFCJMYRNCWJCSZREXCHZUNMXZ

he-asnt-oo-in-i-o-redso-e-ore-ineandhesett
NZUCDRJXYYSMRTMEYIFZWDYVZVYFZUMRZCRWNZDZJJ

-ed-ac-inhischair-aceti-ted--to-ardsthes-n
XZWGCHSMRNMDHNCMFQCHZJMXJZWIEJYUCFWDJNZDIR
```

It is now very easy to determine the plaintext and the key for Example 1.10. The complete decryption is the following:

> Our friend from Paris examined his empty glass with surprise, as
> if evaporation had taken place while he wasn't looking. I poured some

more wine and he settled back in his chair, face tilted up towards the sun.[1]

☐

1.2.3 Cryptanalysis of the Vigenere Cipher

In this section we describe some methods for cryptanalyzing the **Vigenere Cipher**. The first step is to determine the keyword length, which we denote by m. There are a couple of techniques that can be employed. The first of these is the so-called *Kasiski test* and the second uses the *index of coincidence*.

The Kasiski test was first described by Friedrich Kasiski in 1863. It is based on the observation that two identical segments of plaintext will be encrypted to the same ciphertext whenever their occurrence in the plaintext is x positions apart, where $x \equiv 0 \bmod m$. Conversely, if we observe two identical segments of ciphertext, each of length at least three, say, then there is a good chance that they do correspond to identical segments of plaintext.

The Kasiski test works as follows. We search the ciphertext for pairs of identical segments of length at least three, and record the distance between the starting positions of the two segments. If we obtain several such distances $d_1, d_2, \ldots,$ then we would conjecture that m divides the greatest common divisor of the d_i's.

Further evidence for the value of m can be obtained by the index of coincidence. This concept was defined by Wolfe Friedman in 1920, as follows.

DEFINITION 1.7 *Suppose* $\mathbf{x} = x_1 x_2 \ldots x_n$ *is a string of n alphabetic characters. The* index of coincidence *of* $\mathbf{x}$*, denoted* $I_c(\mathbf{x})$*, is defined to be the probability that two random elements of* $\mathbf{x}$ *are identical. Suppose we denote the frequencies of* $A, B, C, \ldots, Z$ *in* $\mathbf{x}$ *by* $f_0, f_1, \ldots, f_{25}$ *(respectively). We can choose two elements of* $\mathbf{x}$ *in* $\binom{n}{2}$ *ways.[2] For each i, $0 \le i \le 25$, there are* $\binom{f_i}{2}$ *ways of choosing both elements to be i. Hence, we have the formula*

$$I_c(\mathbf{x}) = \frac{\displaystyle\sum_{i=0}^{25} f_i(f_i - 1)}{n(n-1)}.$$

Now, suppose $\mathbf{x}$ is a string of English language text. Denote the expected probabilities of occurrence of the letters $A, B, \ldots, Z$ in Table 1.1 by $p_0, \ldots, p_{25}$.

---

[1] P. Mayle, A Year in Provence, A. Knopf, Inc., 1989.

[2] The *binomial coefficient* $\binom{n}{k} = n!/(k!(n-k)!)$ denotes the number of ways of choosing a subset of k objects from a set of n objects.

Then, we would expect that

$$I_c(\mathbf{x}) \approx \sum_{i=0}^{25} p_i{}^2 = 0.065,$$

since the probability that two random elements both are A is $p_0{}^2$, the probability that both are B is $p_1{}^2$, etc. The same reasoning applies if $\mathbf{x}$ is a ciphertext obtained by means of any monoalphabetic cipher. In this case, the individual probabilities will be permuted, but the quantity

$$\sum_{i=0}^{25} p_i{}^2$$

will be unchanged.

Now, suppose we start with a ciphertext $\mathbf{y} = y_1 y_2 \ldots y_n$ that has been constructed by using a **Vigenere Cipher**. Define m substrings $\mathbf{y}_1, \mathbf{y}_2, \ldots, \mathbf{y}_m$ of $\mathbf{y}$ by writing out the ciphertext, by columns, in a rectangular array of dimensions $m \times (n/m)$. The rows of this matrix are the substrings $\mathbf{y}_i$, $1 \leq i \leq m$. If this is done, and m is indeed the keyword length, then each $I_c(\mathbf{y}_i)$ should be roughly equal to 0.065. On the other hand, if m is not the keyword length, then the substrings $\mathbf{y}_i$ will look much more random, since they will have been obtained by shift encryption with different keys. Observe that a completely random string will have

$$I_c \approx 26(1/26)^2 = 1/26 = 0.038.$$

The two values 0.065 and 0.038 are sufficiently far apart that we will often be able to determine the correct keyword length (or confirm a guess that has already been made using the Kasiski test).

Let us illustrate these two techniques with an example.

Example 1.11
Ciphertext obtained from a Vigenere Cipher

```
CHREEVOAHMAERATBIAXXWTNXBEEOPHBSBQMQEQERBW
RVXUOAKXAOSXXWEAHBWGJMMQMNKGRFVGXWTRZXWIAK
LXFPSKAUTEMNDCMGTSXMXBTUIADNGMGPSRELXNJELX
VRVPRTULHDNQWTWDTYGBPHXTFALJHASVBFXNGLLCHR
ZBWELEKMSJIKNBHWRJGNMGJSGLXFEYPHAGNRBIEQJT
AMRVLCRREMNDGLXRRIMGNSNRWCHRQHAEYEVTAQEBBI
PEEWEVKAKOEWADREMXMTBHHCHRTKDNVRZCHRCLQOHP
WQAIIWXNRMGWOIIFKEE
```

First, let's try the Kasiski test. The ciphertext string CHR occurs in five places in the ciphertext, beginning at positions 1, 166, 236, 276 and 286. The distances from the first occurrence to the other three occurrences are (respectively) 165, 235, 275 and 285. The gcd of these four integers is 5, so that is very likely the keyword length.

Let's see if computation of indices of coincidence gives the same conclusion. With $m = 1$, the index of coincidence is 0.045. With $m = 2$, the two indices are 0.046 and 0.041. With $m = 3$, we get 0.043, 0.050, 0.047. With $m = 4$, we have indices 0.042, 0.039, 0.046, 0.040. Then trying $m = 5$, we obtain the values 0.063, 0.068, 0.069, 0.061 and 0.072. This also provides strong evidence that the keyword length is five. ∐

Proceeding under this assumption, how do we determine the keyword? It is useful to consider the mutual index of coincidence of two strings.

DEFINITION 1.8 *Suppose* $\mathbf{x} = x_1 x_2 \ldots x_n$ *and* $\mathbf{y} = y_1 y_2 \ldots y_{n'}$ *are strings of* n *and* n' *alphabetic characters, respectively. The* mutual index of coincidence *of* $\mathbf{x}$ *and* $\mathbf{y}$*, denoted* $MI_c(\mathbf{x}, \mathbf{y})$*, is defined to be the probability that a random element of* $\mathbf{x}$ *is identical to a random element of* $\mathbf{y}$*. If we denote the frequencies of* $A, B, C, \ldots, Z$ *in* $\mathbf{x}$ *and* $\mathbf{y}$ *by* $f_0, f_1, \ldots, f_{25}$ *and* $f'_0, f'_1, \ldots, f'_{25}$*, respectively, then* $MI_c(\mathbf{x}, \mathbf{y})$ *is seen to be*

$$MI_c(\mathbf{x}, \mathbf{y}) = \frac{\sum_{i=0}^{25} f_i f'_i}{n n'}.$$

Now, given that we have determined the value of m, the substrings $\mathbf{y}_i$ are obtained by shift encryption of the plaintext. Suppose $K = (k_1, k_2, \ldots, k_m)$ is the keyword. Let us see if we can estimate $MI_c(\mathbf{y}_i, \mathbf{y}_j)$. Consider a random character in $\mathbf{y}_i$ and a random character in $\mathbf{y}_j$. The probability that both characters are A is $p_{-k_i} p_{-k_j}$, the probability that both are B is $p_{1-k_i} p_{1-k_j}$, etc. (Note that all subscripts are reduced modulo 26.) Hence, we estimate that

$$MI_c(\mathbf{y}_i, \mathbf{y}_j) \approx \sum_{h=0}^{25} p_{h-k_i} p_{h-k_j} = \sum_{h=0}^{25} p_h p_{h+k_i-k_j}.$$

Observe that the value of this estimate depends only on the difference $k_i - k_j$ mod 26, which we call the *relative shift* of $\mathbf{y}_i$ and $\mathbf{y}_j$. Also, notice that

$$\sum_{h=0}^{25} p_h p_{h+\ell} = \sum_{h=0}^{25} p_h p_{h-\ell},$$

so a relative shift of ℓ yields the same estimate of MI_c as does a relative shift of $26 - \ell$.

We tabulate these estimates, for relative shifts ranging between 0 to 13, in Table 1.4.

TABLE 1.4
Expected Mutual Indices of Coincidence

relative shift	expected value of MI_c
0	0.065
1	0.039
2	0.032
3	0.034
4	0.044
5	0.033
6	0.036
7	0.039
8	0.034
9	0.034
10	0.038
11	0.045
12	0.039
13	0.043

The important observation is that, if the relative shift is not zero, these estimates vary between 0.031 and 0.045; whereas, a relative shift of zero yields an estimate of 0.065. We can use this observation to formulate a likely guess for $\ell = k_i - k_j$, the relative shift of $\mathbf{y}_i$ and $\mathbf{y}_j$, as follows. Suppose we fix $\mathbf{y}_i$, and consider the effect of encrypting $\mathbf{y}_j$ by $e_0, e_1, e_2, \ldots$. Denote the resulting strings by $\mathbf{y}_j^0, \mathbf{y}_j^1$, etc. It is easy to compute the indices $MI_c(\mathbf{y}_i, \mathbf{y}_j^g)$, $0 \le g \le 25$. This can be done using the formula

$$MI_c(\mathbf{x}, \mathbf{y}^g) = \frac{\displaystyle\sum_{i=0}^{25} f_i f'_{i-g}}{nn'}.$$

When $g = \ell$, the MI_c should be close to 0.065, since the relative shift of $\mathbf{y}_i$ and $\mathbf{y}_j^\ell$ is zero. However, for values of $g \ne \ell$, the MI_c should vary between 0.031 and 0.045.

By using this technique, we can obtain the relative shifts of any two of the substrings $\mathbf{y}_i$. This leaves only 26 possible keywords, which can easily be obtained by exhaustive key search, for example.

Let us illustrate by returning to Example 1.11.

Example 1.11   *(Cont.)*
We have hypothesized that the keyword length is 5. We now try to compute the relative shifts. By computer, it is not difficult to compute the 260 values $MI_c(\mathbf{y}_i, \mathbf{y}_j^g)$, where $1 \le i < j \le 5, 0 \le g \le 25$. These values are tabulated in Table 1.5. For each (i, j) pair, we look for values of $MI_c(\mathbf{y}_i, \mathbf{y}_j^g)$ that are close to 0.065. If there is a unique such value (for a given (i, j) pair), we conjecture that

TABLE 1.5
Observed Mutual Indices of Coincidence

i	j	value of $MI_c(\mathbf{y}_i, \mathbf{y}_j^g)$								
1	2	.028	.027	.028	.034	.039	.037	.026	.025	.052
		.068	.044	.026	.037	.043	.037	.043	.037	.028
		.041	.041	.034	.037	.051	.045	.042	.036	
1	3	.039	.033	.040	.034	.028	.053	.048	.033	.029
		.056	.050	.045	.039	.040	.036	.037	.032	.027
		.037	.036	.031	.037	.055	.029	.024	.037	
1	4	.034	.043	.025	.027	.038	.049	.040	.032	.029
		.034	.039	.044	.044	.034	.039	.045	.044	.037
		.055	.047	.032	.027	.039	.037	.039	.035	
1	5	.043	.033	.028	.046	.043	.044	.039	.031	.026
		.030	.036	.040	.041	.024	.019	.048	.070	.044
		.028	.038	.044	.043	.047	.033	.026	.046	
2	3	.046	.048	.041	.032	.036	.035	.036	.030	.024
		.039	.034	.029	.040	.067	.041	.033	.037	.045
		.033	.033	.027	.033	.045	.052	.042	.030	
2	4	.046	.034	.043	.044	.034	.031	.040	.045	.040
		.048	.044	.033	.024	.028	.042	.039	.026	.034
		.050	.035	.032	.040	.056	.043	.028	.028	
2	5	.033	.033	.036	.046	.026	.018	.043	.080	.050
		.029	.031	.045	.039	.037	.027	.026	.031	.039
		.040	.037	.041	.046	.045	.043	.035	.030	
3	4	.038	.036	.040	.033	.036	.060	.035	.041	.029
		.058	.035	.035	.034	.053	.030	.032	.035	.036
		.036	.028	.046	.032	.051	.032	.034	.030	
3	5	.035	.034	.034	.036	.030	.043	.043	.050	.025
		.041	.051	.050	.035	.032	.033	.033	.052	.031
		.027	.030	.072	.035	.034	.032	.043	.027	
4	5	.052	.038	.033	.038	.041	.043	.037	.048	.028
		.028	.036	.061	.033	.033	.032	.052	.034	.027
		.039	.043	.033	.027	.030	.039	.048	.035	

it is the value of the relative shift.

Six such values in Table 1.5 are boxed. They provide strong evidence that the relative shift of $\mathbf{y}_1$ and $\mathbf{y}_2$ is 9; the relative shift of $\mathbf{y}_1$ and $\mathbf{y}_5$ is 16; the relative shift of $\mathbf{y}_2$ and $\mathbf{y}_3$ is 13; the relative shift of $\mathbf{y}_2$ and $\mathbf{y}_5$ is 7; the relative shift of $\mathbf{y}_3$ and $\mathbf{y}_5$ is 20; and the relative shift of $\mathbf{y}_4$ and $\mathbf{y}_5$ is 11. This gives us the following equations in the five unknowns k_1, k_2, k_3, k_4, k_5:

$$k_1 - k_2 = 9$$

$$k_1 - k_5 = 16$$

$$k_2 - k_3 = 13$$
$$k_2 - k_5 = 7$$
$$k_3 - k_5 = 20$$
$$k_4 - k_5 = 11.$$

This allows us to express the five k_i's in terms of k_1:

$$k_2 = k_1 + 17$$
$$k_3 = k_1 + 4$$
$$k_4 = k_1 + 21$$
$$k_5 = k_1 + 10.$$

So the key is likely to be $(k_1, k_1 + 17, k_1 + 4, k_1 + 21, k_1 + 10)$ for some $k_1 \in \mathbb{Z}_{26}$. Hence, we suspect that the keyword is some cyclic shift of $AREVK$. It now does not take long to determine that the keyword is $JANET$. The complete decryption is the following:

> The almond tree was in tentative blossom. The days were longer, often ending with magnificent evenings of corrugated pink skies. The hunting season was over, with hounds and guns put away for six months. The vineyards were busy again as the well-organized farmers treated their vines and the more lackadaisical neighbors hurried to do the pruning they should have done in November.[3]

⬚

1.2.4 A Known Plaintext Attack on the Hill Cipher

The **Hill Cipher** is more difficult to break with a ciphertext-only attack, but it succumbs easily to a known plaintext attack. Let us first assume that the opponent has determined the value of m being used. Suppose he has at least m distinct pairs of m-tuples, $x_j = (x_{1,j}, x_{2,j}, \ldots, x_{m,j})$ and $y_j = (y_{1,j}, y_{2,j}, \ldots, y_{m,j})$ $(1 \leq j \leq m)$, such that $y_j = e_K(x_j)$, $1 \leq j \leq m$. If we define two $m \times m$ matrices $X = (x_{i,j})$ and $Y = (y_{i,j})$, then we have the matrix equation $Y = XK$, where the $m \times m$ matrix K is the unknown key. Provided that the matrix X is invertible, Oscar can compute $K = X^{-1}Y$ and thereby break the system. (If Y is not invertible, then it will be necessary to try other sets of m plaintext-ciphertext pairs.)

Let's look at a simple example.

---

[3] P. Mayle, A Year in Provence, A. Knopf, Inc., 1989.

Example 1.12

Suppose the plaintext $friday$ is encrypted using a **Hill Cipher** with $m = 2$, to give the ciphertext $PQCFKU$.

We have that $e_K(5, 17) = (15, 16)$, $e_K(8, 3) = (2, 5)$ and $e_K(0, 24) = (10, 20)$. From the first two plaintext-ciphertext pairs, we get the matrix equation

$$\begin{pmatrix} 15 & 16 \\ 2 & 5 \end{pmatrix} = \begin{pmatrix} 5 & 17 \\ 8 & 3 \end{pmatrix} K.$$

Using Theorem 1.3, it is easy to compute

$$\begin{pmatrix} 5 & 17 \\ 8 & 3 \end{pmatrix}^{-1} = \begin{pmatrix} 9 & 1 \\ 2 & 15 \end{pmatrix},$$

so

$$K = \begin{pmatrix} 9 & 1 \\ 2 & 15 \end{pmatrix} \begin{pmatrix} 15 & 16 \\ 2 & 5 \end{pmatrix} = \begin{pmatrix} 7 & 19 \\ 8 & 3 \end{pmatrix}.$$

This can be verified by using the third plaintext-ciphertext pair. □

What would the opponent do if he does not know m? Assuming that m is not too big, he could simply try $m = 2, 3, \ldots$, until the key is found. If a guessed value of m is incorrect, then an $m \times m$ matrix found by using the algorithm described above will not agree with further plaintext-ciphertext pairs. In this way, the value of m can be determined if it is not already known.

1.2.5 Cryptanalysis of the LFSR-based Stream Cipher

Recall that the ciphertext is the sum modulo 2 of the plaintext and the keystream, i.e., $y_i = x_i + z_i \bmod 2$. The keystream is produced from $z_1, \ldots, z_m$ using the linear recurrence relation

$$z_{m+i} = \sum_{j=0}^{m-1} c_j z_{i+j} \bmod 2,$$

where $c_0, \ldots, c_{m-1} \in \mathbb{Z}_2$ (and $c_0 = 1$).

Since all operations in this cryptosystem are linear, we might suspect that the cryptosystem is vulnerable to a known-plaintext attack, as is the case with the **Hill Cipher**. Suppose Oscar has a plaintext string $x_1 x_2 \ldots x_n$ and the corresponding ciphertext string $y_1 y_2 \ldots y_n$. Then he can compute the keystream bits $z_i = x_i + y_i \bmod 2$, $1 \le i \le n$. Let us also suppose that Oscar knows the value m. Then Oscar needs only to compute $c_0, \ldots, c_{m-1}$ in order to be able to reconstruct the entire keystream. In other words, he needs to be able to determine the values of m unknowns.

Now, for any $i \geq 1$, we have

$$z_{m+i} = \sum_{j=0}^{m-1} c_j z_{i+j} \bmod 2,$$

which is a linear equation in the m unknowns. If $n \geq 2m$, then there are m linear equations in m unknowns, which can subsequently be solved.

The system of m linear equations can be written in matrix form as follows:

$$(z_{m+1}, z_{m+2}, \ldots, z_{2m}) = (c_0, c_1, \ldots, c_{m-1}) \begin{pmatrix} z_1 & z_2 & \cdots & z_m \\ z_2 & z_3 & \cdots & z_{m+1} \\ \vdots & \vdots & & \vdots \\ z_m & z_{m+1} & \cdots & z_{2m-1} \end{pmatrix}.$$

If the coefficient matrix has an inverse (modulo 2), we obtain the solution

$$(c_0, c_1, \ldots, c_{m-1}) = (z_{m+1}, z_{m+2}, \ldots, z_{2m}) \begin{pmatrix} z_1 & z_2 & \cdots & z_m \\ z_2 & z_3 & \cdots & z_{m+1} \\ \vdots & \vdots & & \vdots \\ z_m & z_{m+1} & \cdots & z_{2m-1} \end{pmatrix}^{-1}.$$

In fact, the matrix will have an inverse if m is the degree of the recurrence used to generate the keystream (see the exercises for a proof).

Let's illustrate with an example.

Example 1.13
Suppose Oscar obtains the ciphertext string

$$101101011110010$$

corresponding to the plaintext string

$$011001111111000.$$

Then he can compute the keystream bits:

$$110100100001010.$$

Suppose also that Oscar knows that the keystream was generated using a 5-stage LFSR. Then he would solve the following matrix equation, which is obtained from the first 10 keystream bits:

$$(0,1,0,0,0) = (c_0, c_1, c_2, c_3, c_4) \begin{pmatrix} 1 & 1 & 0 & 1 & 0 \\ 1 & 0 & 1 & 0 & 0 \\ 0 & 1 & 0 & 0 & 1 \\ 1 & 0 & 0 & 1 & 0 \\ 0 & 0 & 1 & 0 & 0 \end{pmatrix}.$$

It can be checked that

$$
\begin{pmatrix}
1 & 1 & 0 & 1 & 0 \\
1 & 0 & 1 & 0 & 0 \\
0 & 1 & 0 & 0 & 1 \\
1 & 0 & 0 & 1 & 0 \\
0 & 0 & 1 & 0 & 0
\end{pmatrix}^{-1}
=
\begin{pmatrix}
0 & 1 & 0 & 0 & 1 \\
1 & 0 & 0 & 1 & 0 \\
0 & 0 & 0 & 0 & 1 \\
0 & 1 & 0 & 1 & 1 \\
1 & 0 & 1 & 1 & 0
\end{pmatrix}.
$$

This yields

$$
(c_0, c_1, c_2, c_3, c_4) = (0, 1, 0, 0, 0)
\begin{pmatrix}
0 & 1 & 0 & 0 & 1 \\
1 & 0 & 0 & 1 & 0 \\
0 & 0 & 0 & 0 & 1 \\
0 & 1 & 0 & 1 & 1 \\
1 & 0 & 1 & 1 & 0
\end{pmatrix}
$$

$$
= (1, 0, 0, 1, 0).
$$

Thus the recurrence used to generate the keystream is

$$
z_{i+5} = z_i + z_{i+3} \bmod 2.
$$

$\Box$

1.3 Notes

Much of the material on classical cryptography is covered in textbooks, for example Beker and Piper [BP82] and Denning [DE82]. The probability estimates for the 26 alphabetic characters are taken from Beker and Piper. As well, the cryptanalysis of the **Vigenere Cipher** is a modification of the description given in Beker and Piper.

A good reference for elementary number theory is Rosen [RO93]. Background in elementary linear algebra can be found in Anton [AN91].

Kahn's book "The Codebreakers" [KA67] is an entertaining and informative history of cryptography up to 1967. In it, Kahn states that the **Vigenere Cipher** is incorrectly attributed to Vigenere.

The **Hill Cipher** was first described in [HI29]. Much information on stream ciphers can be found in the book by Rueppel [RU86].

Exercises

1.1 Below are given four examples of ciphertext, one obtained from a **Substitution Cipher**, one from a **Vigenere Cipher**, one from an **Affine Cipher**, and one unspecified. In each case, the task is to determine the plaintext.

Give a clearly written description of the steps you followed to decrypt each ciphertext. This should include all statistical analysis and computations you performed.

The first two plaintexts were taken from "The Diary of Samuel Marchbanks," by Robertson Davies, Clarke Irwin, 1947; the fourth was taken from "Lake Wobegon Days," by Garrison Keillor, Viking Penguin, Inc., 1985.

(a) **Substitution Cipher**:

```
EMGLOSUDCGDNCUSWYSFHNSFCYKDPUMLWGYICOXYSIPJCK
QPKUGKMGOLICGINCGACKSNISACYKZSCKXECJCKSHYSXCG
OIDPKZCNKSHICGIWYGKKGKGOLDSILKGOIUSIGLEDSPWZU
GFZCCNDGYYSFUSZCNXEOJNCGYEOWEUPXEZGACGNFGLKNS
ACIGOIYCKXCJUCIUZCFZCCNDGYYSFEUEKUZCSOCFZCCNC
IACZEJNCSHFZEJZEGMXCYHCJUMGKUCY
```

HINT F decrypts to w.

(b) **Vigenere Cipher**:

```
KCCPKBGUFDPHQTYAVINRRTMVGRKDNBVFDETDGILTXRGUD
DKOTFMBPVGEGLTGCKQRACQCWDNAWCRXIZAKFTLEWRPTYC
QKYVXCHKFTPONCQQRHJVAJUWETMCMSPKQDYHJVDAHCTRL
SVSKCGCZQQDZXGSFRLSWCWSJTBHAFSIASPRJAHKJRJUMV
GKMITZHFPDISPZLVLGWTFPLKKEBDPGCEBSHCTJRWXBAFS
PEZQNRWXCVYCGAONWDDKACKAWBBIKFTIOVKCGGHJVLNHI
FFSQESVYCLACNVRWBBIREPBBVFEXOSCDYGZWPFDTKFQIY
CWHJVLNHIQIBTKHJVNPIST
```

(c) **Affine Cipher**:

```
KQEREJEBCPPCJCRKIEACUZBKRVPKRBCIBQCARBJCVFCUP
KRIOFKPACUZQEPBKRXPEIIEABDKPBCPFCDCCAFIEABDKP
BCPFEQPKAZBKRHAIBKAPCCIBURCCDKDCCJCIDFUIXPAFF
ERBICZDFKABICBBENEFCUPJCVKABPCYDCCDPKBCOCPERK
IVKSCPICBRKIJPKABI
```

(d) unspecified cipher:

```
BNVSNSIHQCEELSSKKYERIFJKXUMBGYKAMQLJTYAVFBKVT
DVBPVVRJYYLAOKYMPQSCGDLFSRLLPROYGESEBUUALRWXM
MASAZLGLEDFJBZAVVPXWICGJXASCBYEHOSNMULKCEAHTQ
OKMFLEBKFXLRRFDTZXCIWBJSICBGAWDVYDHAVFJXZIBKC
GJIWEAHTTOEWTUHKRQVVRGZBXYIREMMASCSPBNLHJMBLR
FFJELHWEYLWISTFVVYFJCMHYUYRUFSFMGESIGRLWALSWM
NUHSIMYYITCCQPZSICEHBCCMZFEGVJYOCDEMMPGHVAAUM
ELCMOEHVLTIPSUYILVGFLMVWDVYDBTHFRAYISYSGKVSUU
HYHGGCKTMBLRX
```

1.2   (a) How many 2×2 matrices are there that are invertible over $\mathbb{Z}_{26}$?

(b) Let p be prime. Show that the number of 2×2 matrices that are invertible over $\mathbb{Z}_p$ is $(p^2 - 1)(p^2 - p)$.

HINT   Since p is prime, $\mathbb{Z}_p$ is a field. Use the fact that a matrix over a field is invertible if and only if its rows are linearly independent vectors (i.e., there does not exist a non-zero linear combination of the rows whose sum is the vector of all 0's).

(c) For p prime, and $m \geq 2$ an integer, find a formula for the number of $m \times m$ matrices that are invertible over $\mathbb{Z}_p$.

1.3 Sometimes it is useful to choose a key such that the encryption operation is identical to the decryption operation. In the case of the **Hill Cipher**, we would be looking for matrices K such that $K = K^{-1}$ (such a matrix is called *involutory*). In fact, Hill recommended the use of involutory matrices as keys in his cipher. Determine the number of involutory matrices (over $\mathbb{Z}_{26}$) in the case $m = 2$.

HINT　Use the formula given in Theorem 1.3 and observe that $\det A = \pm 1$ for an involutory matrix over $\mathbb{Z}_{26}$.

1.4 Suppose we are told that the plaintext

$$\texttt{breathtaking}$$

yields the ciphertext

$$\texttt{RUPOTENTOSUP}$$

where the **Hill Cipher** is used (but m is not specified). Determine the encryption matrix.

1.5 An **Affine-Hill Cipher** is the following modification of a **Hill Cipher**: Let m be a positive integer, and define $\mathcal{P} = \mathcal{C} = (\mathbb{Z}_{26})^m$. In this cryptosystem, a key K consists of a pair (L, b), where L is an $m \times m$ invertible matrix over $\mathbb{Z}_{26}$, and $b \in (\mathbb{Z}_{26})^m$. For $x = (x_1, \ldots, x_m) \in \mathcal{P}$ and $K = (L, b) \in \mathcal{K}$, we compute $y = e_K(x) = (y_1, \ldots, y_m)$ by means of the formula $y = xL + b$. Hence, if $L = (\ell_{i,j})$ and $b = (b_1, \ldots, b_m)$, then

$$(y_1, \ldots, y_m) = (x_1, \ldots, x_m) \begin{pmatrix} \ell_{1,1} & \ell_{1,2} & \cdots & \ell_{1,m} \\ \ell_{2,1} & \ell_{2,2} & \cdots & \ell_{2,m} \\ \vdots & \vdots & & \vdots \\ \ell_{m,1} & \ell_{m,2} & \cdots & \ell_{m,m} \end{pmatrix} + (b_1, \ldots, b_m).$$

Suppose Oscar has learned that the plaintext

$$\texttt{adisplayedequation}$$

is encrypted to give the ciphertext

$$\texttt{DSRMSIOPLXLJBZULLM}$$

and Oscar also knows that $m = 3$. Compute the key, showing all computations.

1.6 Here is how we might cryptanalyze the **Hill Cipher** using a ciphertext-only attack. Suppose that we know that $m = 2$. Break the ciphertext into blocks of length two letters (digrams). Each such digram is the encryption of a plaintext digram using the unknown encryption matrix. Pick out the most frequent ciphertext digram and assume it is the encryption of a common digram in the list following Table 1.1 (for example, TH or ST). For each such guess, proceed as in the known-plaintext attack, until the correct encryption matrix is found.

Here is a sample of ciphertext for you to decrypt using this method:

$$\texttt{LMQETXYEAGTXCTUIEWNCTXLZEWUAISPZYVAPEWLMGQWYA}$$
$$\texttt{XFTCJMSQCADAGTXLMDXNXSNPJQSYVAPRIQSMHNOCVAXFV}$$

1.7 We describe a special case of a **Permutation Cipher**. Let m, n be positive integers. Write out the plaintext, by rows, in $m \times n$ rectangles. Then form the ciphertext by taking the columns of these rectangles. For example, if $m = 4, n = 3$, then we would encrypt the plaintext "cryptography" by forming the following rectangle:

```
cryp
togr
aphy
```

The ciphertext would be "CTAROPYGHPRY."

(a) Describe how Bob would decrypt a ciphertext (given values for m and n).

(b) Decrypt the following ciphertext, which was obtained by using this method of encryption:

MYAMRARUYIQTENCTORAHROYWDSOYEOUARRGDERNOGW

1.8 There are eight different linear recurrences over $\mathbb{Z}_2$ of degree four having $c_0 = 1$. Determine which of these recurrences give rise to a keystream of period 15 (given a non-zero initialization vector).

1.9 The purpose of this exercise is to prove the statement made in Section 1.2.5 that the $m \times m$ coefficient matrix is invertible. This is equivalent to saying that the rows of this matrix are linearly independent vectors over $\mathbb{Z}_2$.

As before, we suppose that the recurrence has the form

$$z_{m+i} = \sum_{j=0}^{m-1} c_j z_{i+j} \bmod 2.$$

$(z_1, \ldots, z_m)$ comprises the initialization vector. For $i \geq 1$, define

$$v_i = (z_i, \ldots, z_{i+m-1}).$$

Note that the coefficient matrix has the vectors $v_1, \ldots, v_m$ as its rows, so our objective is to prove that these m vectors are linearly independent.

Prove the following assertions:

(a) For any $i \geq 1$,

$$v_{m+i} = \sum_{j=0}^{m-1} c_j v_{i+j} \bmod 2.$$

(b) Choose h to be the minimum integer such that there exists a non-trivial linear combination of the vectors $v_1, \ldots, v_h$ which sums to the vector $(0, \ldots, 0)$ modulo 2. Then

$$v_h = \sum_{j=0}^{h-2} \alpha_j v_{j+1} \bmod 2,$$

and not all the α_j's are zero. Observe that $h \leq m + 1$, since any $m + 1$ vectors in an $m-$dimensional vector space are dependent.

(c) Prove that the keystream must satisfy the recurrence

$$z_{h-1+i} = \sum_{j=0}^{h-2} \alpha_j z_{j+i} \bmod 2$$

for any $i \geq 1$.

(d) Observe that if $h \leq m$, then the keystream satisfies a linear recurrence of degree less than m, a contradiction. Hence, $h = m + 1$, and the matrix must be invertible.

1.10 Decrypt the following ciphertext, obtained from the **Autokey Cipher**, by using exhaustive key search:

MALVVMAFBHBUQPTSOXALTGVWWRG

1.11 We describe a stream cipher that is a modification of the **Vigenere Cipher**. Given a keyword $(K_1, \ldots, K_m)$ of length m, construct a keystream by the rule $z_i = K_i$ $(1 \le i \le m)$, $z_{i+m} = z_i + 1 \bmod 26$ $(i \ge m + 1)$. In other words, each time we use the keyword, we replace each letter by its successor modulo 26. For example, if $SUMMER$ is the keyword, we use $SUMMER$ to encrypt the first six letters, we use $TVNNFS$ for the next six letters, and so on.

Describe how you can use the concept of index of coincidence to first determine the length of the keyword, and then actually find the keyword.

Test your method by cryptanalyzing the following ciphertext:

```
IYMYSILONRFNCQXQJEDSHBUIBCJUZBOLFQYSCHATPEQGQ
JEJNGNXZWHHGWFSUKULJQACZKKJOAAHGKEMTAFGMKVRDO
PXNEHEKZNKFSKIFRQVHHOVXINPHMRTJPYWQGJWPUUVKFP
OAWPMRKKQZWLQDYAZDRMLPBJKJOBWIWPSEPVVQMBCRYVC
RUZAAOUMBCHDAGDIEMSZFZHALIGKEMJJFPCIWKRMLMPIN
AYOFIREAOLDTHITDVRMSE
```

The plaintext was taken from "The Codebreakers," by D. Kahn, Macmillan, 1967.

2

Shannon's Theory

In 1949, Claude Shannon published a paper entitled "Communication Theory of Secrecy Systems" in the *Bell Systems Technical Journal*. This paper had a great influence on the scientific study of cryptography. In this chapter, we discuss several of Shannon's ideas.

2.1 Perfect Secrecy

There are two basic approaches to discussing the security of a cryptosystem.

computational security

This measure concerns the computational effort required to break a cryptosystem. We might define a cryptosystem to be *computationally secure* if the best algorithm for breaking it requires at least N operations, where N is some specified, very large number. The problem is that no known practical cryptosystem can be proved to be secure under this definition. In practice, people will call a cryptosystem "computationally secure" if the best known method of breaking the system requires an unreasonably large amount of computer time (but this is of course very different from a proof of security). Another approach is to provide evidence of computational security by reducing the security of the cryptosystem to some well-studied problem that is thought to be difficult. For example, it may be able to prove a statement of the type "a given cryptosystem is secure if a given integer n cannot be factored." Cryptosystems of this type are sometimes termed "provably secure," but it must be understood that this approach only provides a proof of security relative to some other problem, not an absolute proof of security. [1]

---

[1] This is a similar situation to proving that a problem is NP-complete: it proves that the given problem is at least as difficult as any other NP-complete problem, but it does not provide an absolute proof of the computational difficulty of the problem.

unconditional security

This measure concerns the security of cryptosystems when there is no bound placed on the amount of computation that Oscar is allowed to do. A cryptosystem is defined to be *unconditionally secure* if it cannot be broken, even with infinite computational resources.

When we discuss the security of a cryptosystem, we should also specify the type of attack that is being considered. In Chapter 1, we saw that neither the **Shift Cipher**, the **Substitution Cipher** nor the **Vigenere Cipher** is computationally secure against a ciphertext-only attack (given a sufficient amount of ciphertext).

What we will do in this section is to develop the theory of cryptosystems that are unconditionally secure against a ciphertext-only attack. It turns out that all three of the above ciphers are unconditionally secure if only one element of plaintext is encrypted with a given key!

The unconditional security of a cryptosystem obviously cannot be studied from the point of view of computational complexity, since we allow computation time to be infinite. The appropriate framework in which to study unconditional security is probability theory. We need only elementary facts concerning probability; the main definitions are reviewed now.

DEFINITION 2.1 *Suppose* $\mathbf{X}$ *and* $\mathbf{Y}$ *are random variables. We denote the probability that* $\mathbf{X}$ *takes on the value* x *by* $p(x)$, *and the probability that* $\mathbf{Y}$ *takes on the value* y *by* $p(y)$. *The joint probability* $p(x, y)$ *is the probability that* $\mathbf{X}$ *takes on the value* x *and* $\mathbf{Y}$ *takes on the value* y. *The conditional probability* $p(x|y)$ *denotes the probability that* $\mathbf{X}$ *takes on the value* x *given that* $\mathbf{Y}$ *takes on the value* y. *The random variables* $\mathbf{X}$ *and* $\mathbf{Y}$ *are said to be independent if* $p(x, y) = p(x)p(y)$ *for all possible values* x *of* $\mathbf{X}$ *and* y *of* $\mathbf{Y}$.

Joint probability can be related to conditional probability by the formula

$$p(x, y) = p(x|y)p(y).$$

Interchanging x and y, we have that

$$p(x, y) = p(y|x)p(x).$$

From these two expressions, we immediately obtain the following result, which is known as Bayes' Theorem.

THEOREM 2.1 *(Bayes' Theorem)*
If $p(y) > 0$, *then*

$$p(x|y) = \frac{p(x)p(y|x)}{p(y)}.$$

COROLLARY 2.2
X *and* **Y** *are independent variables if and only if* $p(x|y) = p(x)$ *for all* x, y.

Throughout this section, we assume that a particular key is used for only one encryption. Let us suppose that there is a probability distribution on the plaintext space, $\mathcal{P}$. We denote the *a priori* probability that plaintext x occurs by $p_{\mathcal{P}}(x)$. We also assume that the key K is chosen (by Alice and Bob) using some fixed probability distribution (often a key is chosen at random, so all keys will be equiprobable, but this need not be the case). Denote the probability that key K is chosen by $p_{\mathcal{K}}(K)$. Recall that the key is chosen before Alice knows what the plaintext will be. Hence, we make the reasonable assumption that the key K and the plaintext x are independent events.

The two probability distributions on $\mathcal{P}$ and $\mathcal{K}$ induce a probability distribution on $\mathcal{C}$. Indeed, it is not hard to compute the probability $p_{\mathcal{C}}(y)$ that y is the ciphertext that is transmitted. For a key $K \in \mathcal{K}$, define

$$C(K) = \{e_K(x) : x \in \mathcal{P}\}.$$

That is, $C(K)$ represents the set of possible ciphertexts if K is the key. Then, for every $y \in \mathcal{C}$, we have that

$$p_C(y) = \sum_{\{K : y \in C(K)\}} p_{\mathcal{K}}(K) p_{\mathcal{P}}(d_K(y)).$$

We also observe that, for any $y \in \mathcal{C}$ and $x \in \mathcal{P}$, we can compute the conditional probability $p_{\mathcal{C}}(y|x)$ (i.e., the probability that y is the ciphertext, given that x is the plaintext) to be

$$p_C(y|x) = \sum_{\{K : x = d_K(y)\}} p_{\mathcal{K}}(K).$$

It is now possible to compute the conditional probability $p_{\mathcal{P}}(x|y)$ (i.e., the probability that x is the plaintext, given that y is the ciphertext) using Bayes' Theorem. The following formula is obtained:

$$p_{\mathcal{P}}(x|y) = \frac{p_{\mathcal{P}}(x) \displaystyle\sum_{\{K : x = d_K(y)\}} p_{\mathcal{K}}(K)}{\displaystyle\sum_{\{K : y \in C(K)\}} p_{\mathcal{K}}(K) p_{\mathcal{P}}(d_K(y))}.$$

Observe that all these calculations can be performed by anyone who knows the probability distributions.

We present a toy example to illustrate the computation of these probability distributions.

Example 2.1
Let $\mathcal{P} = \{a, b\}$ with $p_\mathcal{P}(a) = 1/4, p_\mathcal{P}(b) = 3/4$. Let $\mathcal{K} = \{K_1, K_2, K_3\}$ with $p_\mathcal{K}(K_1) = 1/2, p_\mathcal{K}(K_2) = p_\mathcal{K}(K_3) = 1/4$. Let $\mathcal{C} = \{1, 2, 3, 4\}$, and suppose the encryption functions are defined to be $e_{K_1}(a) = 1, e_{K_1}(b) = 2$; $e_{K_2}(a) = 2, e_{K_2}(b) = 3$; and $e_{K_3}(a) = 3, e_{K_3}(b) = 4$. This cryptosystem can be represented by the following *encryption matrix*:

	a	b
K_1	1	2
K_2	2	3
K_3	3	4

We now compute the probability distribution $p_\mathcal{C}$. We obtain

$$p_\mathcal{C}(1) \ = \ \frac{1}{8}$$

$$p_\mathcal{C}(2) \ = \ \frac{3}{8} + \frac{1}{16} \ = \ \frac{7}{16}$$

$$p_\mathcal{C}(3) \ = \ \frac{3}{16} + \frac{1}{16} \ = \ \frac{1}{4}$$

$$p_\mathcal{C}(4) \ = \ \frac{3}{16}.$$

Now we can compute the conditional probability distributions on the plaintext, given that a certain ciphertext has been observed. We have:

$$p_\mathcal{P}(a|1) \ = \ 1 \qquad p_\mathcal{P}(b|1) \ = \ 0$$

$$p_\mathcal{P}(a|2) \ = \ \frac{1}{7} \qquad p_\mathcal{P}(b|2) \ = \ \frac{6}{7}$$

$$p_\mathcal{P}(a|3) \ = \ \frac{1}{4} \qquad p_\mathcal{P}(b|3) \ = \ \frac{3}{4}$$

$$p_\mathcal{P}(a|4) \ = \ 0 \qquad p_\mathcal{P}(b|4) \ = \ 1.$$

$\square$

We are now ready to define the concept of perfect secrecy. Informally, perfect secrecy means that Oscar can obtain no information about plaintext by observing the ciphertext. This idea is made precise by formulating it in terms of the probability distributions we have defined, as follows.

DEFINITION 2.2   *A cryptosystem has perfect secrecy if $p_{\mathcal{P}}(x|y) = p_{\mathcal{P}}(x)$ for all $x \in \mathcal{P}$, $y \in \mathcal{C}$. That is, the a posteriori probability that the plaintext is x, given that the ciphertext y is observed, is identical to the a priori probability that the plaintext is x.*

In Example 2.1, the perfect secrecy property is satisfied for the ciphertext 3, but not for the other three ciphertexts.

We next prove that the **Shift Cipher** provides perfect secrecy. This seems quite obvious intuitively. For, if we are given any ciphertext element $y \in \mathbb{Z}_{26}$, then any plaintext element $x \in \mathbb{Z}_{26}$ is a possible decryption of y, depending on the value of the key. The following theorem gives the formal statement and proof using probability distributions.

THEOREM 2.3
*Suppose the 26 keys in the **Shift Cipher** are used with equal probability $1/26$. Then for any plaintext probability distribution, the **Shift Cipher** has perfect secrecy.*

PROOF   Recall that $\mathcal{P} = \mathcal{C} = \mathcal{K} = \mathbb{Z}_{26}$, and for $0 \leq K \leq 25$, the encryption rule e_K is $e_K(x) = x + K \bmod 26$ ($x \in \mathbb{Z}_{26}$). First, we compute the distribution $p_{\mathcal{C}}$. Let $y \in \mathbb{Z}_{26}$; then

$$p_{\mathcal{C}}(y) = \sum_{K \in \mathbb{Z}_{26}} p_{\mathcal{K}}(K) p_{\mathcal{P}}(d_K(y))$$

$$= \sum_{K \in \mathbb{Z}_{26}} \frac{1}{26} p_{\mathcal{P}}(y - K)$$

$$= \frac{1}{26} \sum_{K \in \mathbb{Z}_{26}} p_{\mathcal{P}}(y - K).$$

Now, for fixed y, the values $y - K \bmod 26$ comprise a permutation of $\mathbb{Z}_{26}$, and $p_{\mathcal{P}}$ is a probability distribution. Hence we have that

$$\sum_{K \in \mathbb{Z}_{26}} p_{\mathcal{P}}(y - K) = \sum_{y \in \mathbb{Z}_{26}} p_{\mathcal{P}}(y)$$

$$= 1.$$

Consequently,

$$p_{\mathcal{C}}(y) = \frac{1}{26}$$

for any $y \in \mathbb{Z}_{26}$.

Next, we have that

$$p_{\mathcal{C}}(y|x) = p_{\mathcal{K}}(y - x \bmod 26)$$

$$= \frac{1}{26}$$

for every x, y, since for every x, y the unique key K such that $e_K(x) = y$ is $K = y - x \bmod 26$. Now, using Bayes' Theorem, it is trivial to compute

$$
\begin{aligned}
p_{\mathcal{P}}(x|y) &= \frac{p_{\mathcal{P}}(x)p_{\mathcal{C}}(y|x)}{p_{\mathcal{C}}(y)} \\
&= \frac{p_{\mathcal{P}}(x)\frac{1}{26}}{\frac{1}{26}} \\
&= p_{\mathcal{P}}(x),
\end{aligned}
$$

so we have perfect secrecy. ∎

So, the **Shift Cipher** is "unbreakable" provided that a new random key is used to encrypt every plaintext character.

Let us next investigate perfect secrecy in general. First, we observe that, using Bayes' Theorem, the condition that $p_{\mathcal{P}}(x|y) = p_{\mathcal{P}}(x)$ for all $x \in \mathcal{P}$, $y \in \mathcal{C}$ is equivalent to $p_{\mathcal{C}}(y|x) = p_{\mathcal{C}}(y)$ for all $x \in \mathcal{P}$, $y \in \mathcal{C}$. Now, let us make the reasonable assumption that $p_{\mathcal{C}}(y) > 0$ for all $y \in \mathcal{C}$ (if $p_{\mathcal{C}}(y) = 0$, then ciphertext y is never used and can be omitted from $\mathcal{C}$). Fix any $x \in \mathcal{P}$. For each $y \in \mathcal{C}$, we have $p_{\mathcal{C}}(y|x) = p_{\mathcal{C}}(y) > 0$. Hence, for each $y \in \mathcal{C}$, there must be at least one key K such that $e_K(x) = y$. It follows that $|\mathcal{K}| \geq |\mathcal{C}|$. In any cryptosystem, we must have $|\mathcal{C}| \geq |\mathcal{P}|$ since each encoding rule is an injection. In the boundary case $|\mathcal{K}| = |\mathcal{C}| = |\mathcal{P}|$, we can give a nice characterization of when perfect secrecy can be obtained. This characterization is originally due to Shannon.

THEOREM 2.4
Suppose $(\mathcal{P}, \mathcal{C}, \mathcal{K}, \mathcal{E}, \mathcal{D})$ is a cryptosystem where $|\mathcal{K}| = |\mathcal{C}| = |\mathcal{P}|$. Then the cryptosystem provides perfect secrecy if and only if every key is used with equal probability $1/|\mathcal{K}|$, and for every $x \in \mathcal{P}$ and every $y \in \mathcal{C}$, there is a unique key K such that $e_K(x) = y$.

PROOF Suppose the given cryptosystem provides perfect secrecy. As observed above, for each $x \in \mathcal{P}$ and $y \in \mathcal{C}$ there must be at least one key K such that $e_K(x) = y$. So we have the inequalities:

$$
\begin{aligned}
|\mathcal{C}| &= |\{e_K(x) : K \in \mathcal{K}\}| \\
&\leq |\mathcal{K}|.
\end{aligned}
$$

But we are assuming that $|\mathcal{C}| = |\mathcal{K}|$. Hence, it must be the case that

$$
|\{e_K(x) : K \in \mathcal{K}\}| = |\mathcal{K}|.
$$

That is, there do not exist two distinct keys K_1 and K_2 such that $e_{K_1}(x) = e_{K_2}(x) = y$. Hence, we have shown that for any $x \in \mathcal{P}$ and $y \in \mathcal{C}$, there is exactly one key K such that $e_K(x) = y$.

FIGURE 2.1
One-time Pad

Let $n \geq 1$ be an integer, and take $\mathcal{P} = \mathcal{C} = \mathcal{K} = (\mathbb{Z}_2)^n$. For $K \in (\mathbb{Z}_2)^n$, define $e_K(x)$ to be the vector sum modulo 2 of K and x (or, equivalently, the exclusive-or of the two associated bitstrings). So, if $x = (x_1, \ldots, x_n)$ and $K = (K_1, \ldots, K_n)$, then

$$e_K(x) = (x_1 + K_1, \ldots, x_n + K_n) \bmod 2.$$

Decryption is identical to encryption. If $y = (y_1, \ldots, y_n)$, then

$$d_K(y) = (y_1 + K_1, \ldots, y_n + K_n) \bmod 2.$$

Denote $n = |\mathcal{K}|$. Let $\mathcal{P} = \{x_i : 1 \leq i \leq n\}$ and fix a $y \in \mathcal{C}$. We can name the keys $K_1, K_2, \ldots, K_n$, in such a way that $e_{K_i}(x_i) = y$, $1 \leq i \leq n$. Using Bayes' theorem, we have

$$p_{\mathcal{P}}(x_i | y) = \frac{p_{\mathcal{C}}(y | x_i) p_{\mathcal{P}}(x_i)}{p_{\mathcal{C}}(y)}$$

$$= \frac{p_{\mathcal{K}}(K_i) p_{\mathcal{P}}(x_i)}{p_{\mathcal{C}}(y)}.$$

Consider the perfect secrecy condition $p_{\mathcal{P}}(x_i | y) = p_{\mathcal{P}}(x_i)$. From this, it follows that $p_{\mathcal{K}}(K_i) = p_{\mathcal{C}}(y)$, for $1 \leq i \leq n$. This says that the keys are used with equal probability (namely, $p_{\mathcal{C}}(y)$). But since the number of keys is $|\mathcal{K}|$, we must have that $p_{\mathcal{K}}(K) = 1/|\mathcal{K}|$ for every $K \in \mathcal{K}$.

Conversely, suppose the two hypothesized conditions are satisfied. Then the cryptosystem is easily seen to provide perfect secrecy for any plaintext probability distribution, in a similar manner as the proof of Theorem 2.3. We leave the details for the reader. ∎

One well-known realization of perfect secrecy is the **Vernam One-time Pad**, which was first described by Gilbert Vernam in 1917 for use in automatic encryption and decryption of telegraph messages. It is interesting that the **One-time Pad** was thought for many years to be an "unbreakable" cryptosystem, but there was no proof of this until Shannon developed the concept of perfect secrecy over 30 years later.

The description of the **One-time Pad** is given in Figure 2.1.

Using Theorem 2.4, it is easily seen that the **One-time Pad** provides perfect secrecy. The system is also attractive because of the ease of encryption and decryption.

Vernam patented his idea in the hope that it would have widespread commercial use. Unfortunately, there are major disadvantages to unconditionally secure cryptosystems such as the **One-time Pad**. The fact that $|\mathcal{K}| \geq |\mathcal{P}|$ means that the amount of key that must be communicated securely is at least as big as the amount of plaintext. For example, in the case of the **One-time Pad**, we require n bits of key to encrypt n bits of plaintext. This would not be a major problem if the same key could be used to encrypt different messages; however, the security of unconditionally secure cryptosystems depends on the fact that each key is used for only one encryption. (This is the reason for the term "one-time" in the **One-time Pad**.)

For example, the **One-time Pad** is vulnerable to a known-plaintext attack, since K can be computed as the exclusive-or of the bitstrings x and $e_K(x)$. Hence, a new key needs to be generated and communicated over a secure channel for every message that is going to be sent. This creates severe key management problems, which has limited the use of the **One-time Pad** in commercial applications. However, the **One-time Pad** has seen application in military and diplomatic contexts, where unconditional security may be of great importance.

The historical development of cryptography has been to try to design cryptosystems where one key can be used to encrypt a relatively long string of plaintext (i.e., one key can be used to encrypt many messages) and still maintain (at least) computational security. One such system is the **Data Encryption Standard**, which we will study in Chapter 3.

2.2 Entropy

In the previous section, we discussed the concept of perfect secrecy. We restricted our attention to the special situation where a key is used for only one encryption. We now want to look at what happens as more and more plaintexts are encrypted using the *same* key, and how likely a cryptanalyst will be able to carry out a successful ciphertext-only attack, given sufficient time.

The basic tool in studying this question is the idea of entropy, a concept from information theory introduced by Shannon in 1948. Entropy can be thought of as a mathematical measure of information or uncertainty, and is computed as a function of a probability distribution.

Suppose we have a random variable $\mathbf{X}$ which takes on a finite set of values according to a probability distribution $p(\mathbf{X})$. What is the information gained by an event which takes place according to distribution $p(\mathbf{X})$? Equivalently, if the event has not (yet) taken place, what is the uncertainty about the outcome? This quantity is called the entropy of $\mathbf{X}$ and is denoted by $H(\mathbf{X})$.

These ideas may seem rather abstract, so let's look at a more concrete example. Suppose our random variable $\mathbf{X}$ represents the toss of a coin. The probability distribution is $p(heads) = p(tails) = 1/2$. It would seem reasonable to say that the information, or entropy, of a coin toss is one bit, since we could encode *heads* by 1 and *tails* by 0, for example. In a similar fashion, the entropy of n independent coin tosses is n, since the n coin tosses can be encoded by a bit string of length n.

As a slightly more complicated example, suppose we have a random variable $\mathbf{X}$ that takes on three possible values x_1, x_2, x_3 with probabilities $1/2, 1/4, 1/4$ respectively. The most efficient "encoding" of the three possible outcomes is to encode x_1 as 0, to encode x_2 as 10 and to encode x_3 as 11. Then the average number of bits in an encoding of $\mathbf{X}$ is

$$\frac{1}{2} \times 1 + \frac{1}{4} \times 2 + \frac{1}{4} \times 2 = \frac{3}{2}.$$

The above examples suggest that an event which occurs with probability 2^{-n} can be encoded as a bit string of length n. More generally, we could imagine that an event occurring with probability p might be encoded by a bit string of length approximately $-\log_2 p$. Given an arbitrary probability distribution $p_1, p_2, \ldots, p_n$ for a random variable $\mathbf{X}$, we take the weighted average of the quantities $-\log_2 p_i$ to be our measure of information. This motivates the following formal definition.

DEFINITION 2.3 *Suppose* $\mathbf{X}$ *is a random variable which takes on a finite set of values according to a probability distribution* $p(\mathbf{X})$. *Then, the entropy of this probability distribution is defined to be the quantity*

$$H(\mathbf{X}) = -\sum_{i=1}^{n} p_i \log_2 p_i.$$

If the possible values of $\mathbf{X}$ *are* x_i, $1 \leq i \leq n$, *then we have*

$$H(\mathbf{X}) = -\sum_{i=1}^{n} p(\mathbf{X} = x_i) \log_2 p(\mathbf{X} = x_i).$$

REMARK   Observe that $\log_2 p_i$ is undefined if $p_i = 0$. Hence, entropy is sometimes defined to be the relevant sum over all the non-zero probabilities. Since $\lim_{x \to 0} x \log_2 x = 0$, there is no real difficulty with allowing $p_i = 0$ for some i. However, we will implicitly assume that, when computing the entropy of a probability distribution p_i, the sum is taken over the indices i such that $p_i \neq 0$. Also, we note that the choice of two as the base of the logarithms is arbitrary: another base would only change the value of the entropy by a constant factor. ∎

Note that if $p_i = 1/n$ for $1 \leq i \leq n$, then $H(\mathbf{X}) = \log_2 n$. Also, it is easy to see that $H(\mathbf{X}) \geq 0$, and $H(\mathbf{X}) = 0$ if and only if $p_i = 1$ for some i and $p_j = 0$ for all $j \neq i$.

Let us look at the entropy of the various components of a cryptosystem. We can think of the key as being a random variable $\mathbf{K}$ that takes on values according to the probability distribution p_K, and hence we can compute the entropy $H(\mathbf{K})$. Similarly, we can compute entropies $H(\mathbf{P})$ and $H(\mathbf{C})$ associated with plaintext and ciphertext probability distributions, respectively.

To illustrate, we compute the entropies of the cryptosystem of Example 2.1.

Example 2.1 (Cont.)
We compute as follows:

$$\begin{aligned}
H(\mathbf{P}) &= -\frac{1}{4}\log_2\frac{1}{4} - \frac{3}{4}\log_2\frac{3}{4} \\
&= -\frac{1}{4}(-2) - \frac{3}{4}(\log_2 3 - 2) \\
&= 2 - \frac{3}{4}(\log_2 3) \\
&\approx 0.81.
\end{aligned}$$

Similar calculations yield $H(\mathbf{K}) = 1.5$ and $H(\mathbf{C}) \approx 1.85$. □

2.2.1 Huffman Encodings and Entropy

In this section, we discuss briefly the connection between entropy and Huffman encodings. As the results in this section are not relevant to the cryptographic applications of entropy, it may be skipped without loss of continuity. However, this discussion may serve to further motivate the concept of entropy.

We introduced entropy in the context of encodings of random events which occur according to a specified probability distribution. We first make these ideas more precise. As before, $\mathbf{X}$ is a random variable which takes on a finite set of values, and $p(\mathbf{X})$ is the associated probability distribution.

An *encoding* of $\mathbf{X}$ is any mapping

$$f : \mathbf{X} \to \{0,1\}^*,$$

where $\{0,1\}^*$ denotes the set of all finite strings of 0's and 1's. Given a finite list (or string) of events $x_1 \ldots x_n$, we can extend the encoding f in an obvious way by defining

$$f(x_1 \ldots x_n) = f(x_1) \parallel \ldots \parallel f(x_n)$$

where $\parallel$ denotes concatenation. In this way, we can think of f as a mapping

$$f : \mathbf{X}^* \to \{0,1\}^*.$$

Now, suppose a string $x_1 \ldots x_n$ is produced by a *memoryless source*, such that each x_i occurs according to the probability distribution on $\mathbf{X}$. This means that the probability of any string $x_1 \ldots x_n$ is computed to be $p(x_1) \times \ldots \times p(x_n)$. (Notice that this string need *not* consist of distinct values, since the source is memoryless. As a simple example, consider a sequence of n tosses of a fair coin.)

Now, given that we are going to encode strings using the mapping f, it is important that we are able to decode in an unambiguous fashion. Thus it should be the case that the encoding f is injective.

Example 2.2
Suppose $\mathbf{X} = \{a, b, c, d\}$, and consider the following three encodings:

$$
\begin{array}{llllllll}
f(a) & = & 1 & f(b) & = & 10 & f(c) & = & 100 & f(d) & = & 1000 \\
g(a) & = & 0 & g(b) & = & 10 & g(c) & = & 110 & g(d) & = & 111 \\
h(a) & = & 0 & h(b) & = & 01 & h(c) & = & 10 & h(d) & = & 11
\end{array}
$$

It can be seen that f and g are injective encodings, but h is not. Any encoding using f can be decoded by starting at the end and working backwards: every time a 1 is encountered, it signals the end of the current element.

An encoding using g can be decoded by starting at the beginning and proceeding sequentially. At any point where we have a substring that is an encoding of a, b, c, or d, we decode it and chop off the substring. For example, given the string 10101110, we decode 10 to b, then 10 to b, then 111 to d, and finally 0 to a. So the decoded string is *bbda*.

To see that h is not injective, it suffices to give an example:

$$h(ac) = h(ba) = 010.$$

□

From the point of view of ease of decoding, we would prefer the encoding g to f. This is because decoding can be done sequentially from beginning to end if g is used, so no memory is required. The property that allows the simple sequential decoding of g is called the prefix-free property. (An encoding g is *prefix-free* if there do *not* exist two elements $x, y \in \mathbf{X}$, and a string $z \in \{0, 1\}^*$ such that $g(x) = g(y) \parallel z$.)

The discussion to this point has not involved entropy. Not surprisingly, entropy is related to the efficiency of an encoding. We will measure the efficiency of an encoding f as we did before: it is the weighted average length (denoted by $\ell(f)$) of an encoding of an element of $\mathbf{X}$. So we have the following definition:

$$\ell(f) = \sum_{x \in \mathbf{X}} p(x) |f(x)|,$$

where $|y|$ denotes the length of a string y.

Now, our fundamental problem is to find an injective encoding, f, that minimizes $\ell(f)$. There is a well-known algorithm, known as Huffman's algorithm, that accomplishes this goal. Moreover, the encoding f produced by Huffman's algorithm is prefix-free, and

$$H(\mathbf{X}) \leq \ell(f) < H(\mathbf{X}) + 1.$$

Thus, the value of the entropy provides a close estimate to the average length of the optimal injective encoding.

We will not prove the results stated above, but we will give a short, informal description of Huffman's algorithm. Huffman's algorithm begins with the probability distribution on the set $\mathbf{X}$, and the code of each element is initially empty. In each iteration, the two elements having lowest probability are combined into one element having as its probability the sum of the two smaller probabilities. The smaller of the two elements is assigned the value "0" and the larger of the two elements is assigned the value "1." When only one element remains, the coding for each $x \in \mathbf{X}$ can be constructed by following the sequence of elements "backwards" from the final element to the initial element x.

This is easily illustrated with an example.

Example 2.3
Suppose $\mathbf{X} = \{a, b, c, d, e\}$ has the following probability distribution: $p(a) = .05$, $p(b) = .10$, $p(c) = .12$, $p(d) = .13$ and $p(e) = .60$. Huffman's algorithm would proceed as indicated in the following table:

a	b	c	d	e
.05	.10	.12	.13	.60
0	1			
.15		.12	.13	.60
		0	1	
.15		.25		.60
0		1		
.40				.60
0				1
1.0				

This leads to the following encodings:

x	$f(x)$
a	000
b	001
c	010
d	011
e	1

Thus, the average length encoding is

$$\ell(f) = .05 \times 3 + .10 \times 3 + .12 \times 3 + .13 \times 3 + .60 \times 1$$
$$= 1.8.$$

Compare this to the entropy:

$$H(\mathbf{X}) = .2161 + .3322 + .3671 + .3842 + .4422$$
$$= 1.7402.$$

$\square$

2.3   Properties of Entropy

In this section, we prove some fundamental results concerning entropy. First, we state a fundamental result, known as Jensen's Inequality, that will be very useful to us. Jensen's Inequality involves concave functions, which we now define.

DEFINITION 2.4   *A real-valued function f is* concave *on an interval I if*

$$f\left(\frac{x+y}{2}\right) \geq \frac{f(x) + f(y)}{2}$$

for all $x, y \in I$. f is strictly concave *on an interval I if*

$$f\left(\frac{x+y}{2}\right) > \frac{f(x) + f(y)}{2}$$

for all $x, y \in I$, $x \neq y$.

Here is Jensen's Inequality, which we state without proof.

THEOREM 2.5   *(Jensen's Inequality)*
Suppose f is a continuous strictly concave function on the interval I,

$$\sum_{i=1}^{n} a_i = 1,$$

and $a_i > 0$, $1 \leq i \leq n$. Then

$$\sum_{i=1}^{n} a_i f(x_i) \leq f\left(\sum_{i=1}^{n} a_i x_i\right),$$

where $x_i \in I$, $1 \leq i \leq n$. Further, equality occurs if and only if $x_1 = \ldots = x_n$.

We now proceed to derive several results on entropy. In the next theorem, we make use of the fact that the function $\log_2 x$ is strictly concave on the interval $(0, \infty)$. (In fact, this follows easily from elementary calculus since the second deriviative of the logarithm function is negative on the interval $(0, \infty)$.)

THEOREM 2.6
Suppose $\mathbf{X}$ is a random variable having probability distribution $p_1, p_2, \ldots, p_n$, where $p_i > 0$, $1 \leq i \leq n$. Then $H(\mathbf{X}) \leq \log_2 n$, with equality if and only if $p_i = 1/n$, $1 \leq i \leq n$.

PROOF Applying Jensen's Inequality, we have the following:

$$H(\mathbf{X}) = -\sum_{i=1}^{n} p_i \log_2 p_i$$

$$= \sum_{i=1}^{n} p_i \log_2 \frac{1}{p_i}$$

$$\leq \log_2 \sum_{i=1}^{n} \left(p_i \times \frac{1}{p_i} \right)$$

$$= \log_2 n.$$

Further, equality occurs if and only if $p_i = 1/n$, $1 \leq i \leq n$. ∎

THEOREM 2.7
$H(\mathbf{X}, \mathbf{Y}) \leq H(\mathbf{X}) + H(\mathbf{Y})$, *with equality if and only if $\mathbf{X}$ and $\mathbf{Y}$ are independent events.*

PROOF Suppose $\mathbf{X}$ takes on values x_i, $1 \leq i \leq m$, and $\mathbf{Y}$ takes on values y_j, $1 \leq j \leq n$. Denote $p_i = p(\mathbf{X} = x_i)$, $1 \leq i \leq m$, and $q_j = p(\mathbf{Y} = y_j)$, $1 \leq j \leq n$. Denote $r_{ij} = p(\mathbf{X} = x_i, \mathbf{Y} = y_j)$, $1 \leq i \leq m, 1 \leq j \leq n$ (this is the joint probability distribution).
  Observe that

$$p_i = \sum_{j=1}^{n} r_{ij}$$

$(1 \leq i \leq m)$ and

$$q_j = \sum_{i=1}^{m} r_{ij}$$

$(1 \leq j \leq n)$. We compute as follows:

$$H(\mathbf{X}) + H(\mathbf{Y}) = - \left(\sum_{i=1}^{m} p_i \log_2 p_i + \sum_{j=1}^{n} q_j \log_2 q_j \right)$$

$$= -\left(\sum_{i=1}^{m}\sum_{j=1}^{n} r_{ij} \log_2 p_i + \sum_{j=1}^{n}\sum_{i=1}^{m} r_{ij} \log_2 q_j\right)$$

$$= -\sum_{i=1}^{m}\sum_{j=1}^{n} r_{ij} \log_2 p_i q_j.$$

On the other hand,

$$H(\mathbf{X}, \mathbf{Y}) = -\sum_{i=1}^{m}\sum_{j=1}^{n} r_{ij} \log_2 r_{ij}.$$

Combining, we obtain the following:

$$H(\mathbf{X}, \mathbf{Y}) - H(\mathbf{X}) - H(\mathbf{Y}) = \sum_{i=1}^{m}\sum_{j=1}^{n} r_{ij} \log_2 \frac{1}{r_{ij}} + \sum_{i=1}^{m}\sum_{j=1}^{n} r_{ij} \log_2 p_i q_j$$

$$= \sum_{i=1}^{m}\sum_{j=1}^{n} r_{ij} \log_2 \frac{p_i q_j}{r_{ij}}$$

$$\leq \log_2 \sum_{i=1}^{m}\sum_{j=1}^{n} p_i q_j$$

$$= \log_2 1$$

$$= 0.$$

(Here, we apply Jensen's Inequality, using the fact that the r_{ij}'s form a probability distribution.)

We can also say when equality occurs: it must be the case that there is a constant c such that $p_i q_j / r_{ij} = c$ for all i, j. Using the fact that

$$\sum_{j=1}^{n}\sum_{i=1}^{m} r_{ij} = \sum_{j=1}^{n}\sum_{i=1}^{m} p_i q_j = 1,$$

it follows that $c = 1$. Hence, equality occurs if and only if $r_{ij} = p_i q_j$, i.e., if and only if

$$p(\mathbf{X} = x_i, \mathbf{Y} = y_j) = p(\mathbf{X} = x_i)p(\mathbf{Y} = y_j),$$

$1 \leq i \leq m, 1 \leq j \leq n$. But this says that $\mathbf{X}$ and $\mathbf{Y}$ are independent. ∎

We next define the idea of conditional entropy.

DEFINITION 2.5 *Suppose* $\mathbf{X}$ *and* $\mathbf{Y}$ *are two random variables. Then for any fixed value y of $\mathbf{Y}$, we get a (conditional) probability distribution $p(\mathbf{X}|y)$. Clearly,*

$$H(\mathbf{X}|y) = -\sum_{x} p(x|y) \log_2 p(x|y).$$

We define the conditional entropy $H(\mathbf{X}|\mathbf{Y})$ to be the weighted average (with respect to the probabilities $p(y)$) of the entropies $H(\mathbf{X}|y)$ over all possible values y. It is computed to be

$$H(\mathbf{X}|\mathbf{Y}) = -\sum_{y}\sum_{x} p(y)p(x|y)\log_2 p(x|y).$$

The conditional entropy measures the average amount of information about $\mathbf{X}$ that is revealed by $\mathbf{Y}$.

The next two results are straightforward; we leave the proofs as exercises.

THEOREM 2.8
$H(\mathbf{X},\mathbf{Y}) = H(\mathbf{Y}) + H(\mathbf{X}|\mathbf{Y})$.

COROLLARY 2.9
$H(\mathbf{X}|\mathbf{Y}) \le H(\mathbf{X})$, *with equality if and only if $\mathbf{X}$ and $\mathbf{Y}$ are independent.*

2.4 Spurious Keys and Unicity Distance

In this section, we apply the entropy results we have proved to cryptosystems. First, we show a fundamental relationship exists among the entropies of the components of a cryptosystem. The conditional entropy $H(\mathbf{K}|\mathbf{C})$ is called the *key equivocation*, and is a measure of how much information about the key is revealed by the ciphertext.

THEOREM 2.10
Let $(\mathcal{P},\mathcal{C},\mathcal{K},\mathcal{E},\mathcal{D})$ be a cryptosystem. Then

$$H(\mathbf{K}|\mathbf{C}) = H(\mathbf{K}) + H(\mathbf{P}) - H(\mathbf{C}).$$

PROOF First, observe that $H(\mathbf{K},\mathbf{P},\mathbf{C}) = H(\mathbf{C}|\mathbf{K},\mathbf{P}) + H(\mathbf{K},\mathbf{P})$. Now, the key and plaintext determine the ciphertext uniquely, since $y = e_K(x)$. This implies that $H(\mathbf{C}|\mathbf{K},\mathbf{P}) = 0$. Hence, $H(\mathbf{K},\mathbf{P},\mathbf{C}) = H(\mathbf{K},\mathbf{P})$. But $\mathbf{K}$ and $\mathbf{P}$ are independent, so $H(\mathbf{K},\mathbf{P}) = H(\mathbf{K}) + H(\mathbf{P})$. Hence,

$$H(\mathbf{K},\mathbf{P},\mathbf{C}) = H(\mathbf{K},\mathbf{P}) = H(\mathbf{K}) + H(\mathbf{P}).$$

In a similar fashion, since the key and ciphertext determine the plaintext uniquely (i.e., $x = d_K(y)$), we have that $H(\mathbf{P}|\mathbf{K},\mathbf{C}) = 0$ and hence $H(\mathbf{K},\mathbf{P},\mathbf{C}) = H(\mathbf{K},\mathbf{C})$.

Now, we compute as follows:

$$H(\mathbf{K}|\mathbf{C}) = H(\mathbf{K}, \mathbf{C}) - H(\mathbf{C})$$
$$= H(\mathbf{K}, \mathbf{P}, \mathbf{C}) - H(\mathbf{C})$$
$$= H(\mathbf{K}) + H(\mathbf{P}) - H(\mathbf{C}),$$

giving the desired formula.   ∎

Let us return to Example 2.1 to illustrate this result.

Example 2.1   (Cont.)
We have already computed $H(\mathbf{P}) \approx 0.81$, $H(\mathbf{K}) = 1.5$ and $H(\mathbf{C}) \approx 1.85$.
Theorem 2.10 tells us that $H(\mathbf{K}|\mathbf{C}) \approx 1.5 + 0.81 - 1.85 \approx 0.46$. This can
be verified directly by applying the definition of conditional entropy, as follows.
First, we need to compute the probabilities $p(K_i|j)$, $1 \le i \le 3$, $1 \le j \le 4$. This
can be done using Bayes' Theorem, and the following values result:

$p(K_1|1)$	$=$	1	$p(K_2|1)$	$=$	0	$p(K_3|1)$	$= \quad 0$
$p(K_1|2)$	$=$	$\dfrac{6}{7}$	$p(K_2|2)$	$=$	$\dfrac{1}{7}$	$p(K_3|2)$	$= \quad 0$
$p(K_1|3)$	$=$	0	$p(K_2|3)$	$=$	$\dfrac{3}{4}$	$p(K_3|3)$	$= \quad \dfrac{1}{4}$
$p(K_1|4)$	$=$	0	$p(K_2|4)$	$=$	0	$p(K_3|4)$	$= \quad 1.$

Now we compute

$$H(\mathbf{K}|\mathbf{C}) = \frac{1}{8} \times 0 + \frac{7}{16} \times 0.59 + \frac{1}{4} \times 0.81 + \frac{3}{16} \times 0 = 0.46,$$

agreeing with the value predicted by Theorem 2.10.   ⬚

Suppose $(\mathcal{P}, \mathcal{C}, \mathcal{K}, \mathcal{E}, \mathcal{D})$ is the cryptosystem being used, and a string of plaintext

$$x_1 x_2 \ldots x_n$$

is encrypted with one key, producing a string of ciphertext

$$y_1 y_2 \ldots y_n.$$

Recall that the basic goal of the cryptanalyst is to determine the key. We are
looking at ciphertext-only attacks, and we assume that Oscar has infinite computational resources. We also assume that Oscar knows that the plaintext is a

"natural" language, such as English. In general, Oscar will be able to rule out certain keys, but many "possible" keys may remain, only one of which is the correct key. The remaining possible, but incorrect, keys are called *spurious keys*.

For example, suppose Oscar obtains the ciphertext string $WNAJW$, which has been obtained by encryption using a shift cipher. It is easy to see that there are only two "meaningful" plaintext strings, namely *river* and *arena*, corresponding respectively to the possible encryption keys F $(= 5)$ and W $(= 22)$. Of these two keys, one will be the correct key and the other will be spurious. (Actually, it is moderately difficult to find a ciphertext of length 5 for the **Shift Cipher** that has two meaningful decryptions; the reader might search for other examples.)

Our goal is to prove a bound on the expected number of spurious keys. First, we have to define what we mean by the entropy (per letter) of a natural language L, which we denote H_L. H_L should be a measure of the average information per letter in a "meaningful" string of plaintext. (Note that a random string of alphabetic characters would have entropy (per letter) equal to $\log_2 26 \approx 4.70$.) As a "first-order" approximation to H_L, we could take $H(\mathbf{P})$. In the case where L is the English language, we get $H(\mathbf{P}) \approx 4.19$ by using the probability distribution given in Table 1.1.

Of course, successive letters in a language are not independent, and correlations among successive letters reduce the entropy. For example, in English, the letter "Q" is always followed by the letter "U." For a "second-order" approximation, we would compute the entropy of the probability distribution of all digrams and then divide by 2. In general, define $\mathbf{P}^n$ to be the random variable that has as its probability distribution that of all n-grams of plaintext. We make use of the following definitions.

DEFINITION 2.6   *Suppose L is a natural language. The entropy of L is defined to be the quantity*

$$H_L = \lim_{n \to \infty} \frac{H(\mathbf{P}^n)}{n}$$

and the redundancy of L is defined to be

$$R_L = 1 - \frac{H_L}{\log_2 |\mathcal{P}|}.$$

REMARK H_L measures the entropy per letter of the language L. A random language would have entropy $\log_2 |\mathcal{P}|$. So the quantity R_L measures the fraction of "excess characters," which we think of as redundancy.   ∎

In the case of the English language, a tabulation of a large number of digrams and their frequencies would produce an estimate for $H(\mathbf{P}^2)$. $H(\mathbf{P}^2)/2 \approx 3.90$ is one estimate obtained in this way. One could continue, tabulating trigrams, etc. and thus obtain an estimate for H_L. In fact, various experiments have yielded the empirical result that $1.0 \leq H_L \leq 1.5$. That is, the average information content in English is something like one to one and a half bits per letter!

Using 1.25 as our estimate of H_L gives a redundancy of about 0.75. This means that the English language is 75% redundant! (This is not to say that one can arbitrarily remove three out of every four letters from English text and hope to still be able to read it. What it *does* mean is that it is possible to find a Huffman encoding of n-grams, for a large enough value of n, which will compress English text to about one quarter of its original length.)

Given probability distributions on $\mathcal{K}$ and $\mathcal{P}^n$, we can define the induced probability distribution on $\mathcal{C}^n$, the set of n-grams of ciphertext (we already did this in the case $n = 1$). We have defined $\mathbf{P}^n$ to be a random variable representing an n-gram of plaintext. Similarly, define $\mathbf{C}^n$ to be a random variable representing an n-gram of ciphertext.

Given $\mathbf{y} \in \mathbf{C}^n$, define

$$K(\mathbf{y}) = \{K \in \mathcal{K} : \exists \mathbf{x} \in \mathcal{P}^n, p_{\mathcal{P}^n}(\mathbf{x}) > 0, e_K(\mathbf{x}) = \mathbf{y}\}.$$

That is, $K(\mathbf{y})$ is the set of keys K for which $\mathbf{y}$ is the encryption of a meaningful string of plaintext of length n, i.e., the set of "possible" keys, given that $\mathbf{y}$ is the ciphertext. If $\mathbf{y}$ is the observed sequence of ciphertext, then the number of spurious keys is $|K(\mathbf{y})| - 1$, since only one of the "possible" keys is the correct key. The average number of spurious keys (over all possible ciphertext strings of length n) is denoted by $\bar{s}_n$. Its value is computed to be

$$\bar{s}_n = \sum_{\mathbf{y} \in \mathcal{C}^n} p(\mathbf{y})(|K(\mathbf{y})| - 1)$$

$$= \sum_{\mathbf{y} \in \mathcal{C}^n} p(\mathbf{y})|K(\mathbf{y})| - \sum_{\mathbf{y} \in \mathcal{C}^n} p(\mathbf{y})$$

$$= \sum_{\mathbf{y} \in \mathcal{C}^n} p(\mathbf{y})|K(\mathbf{y})| - 1.$$

From Theorem 2.10, we have that

$$H(\mathbf{K}|\mathbf{C}^n) = H(\mathbf{K}) + H(\mathbf{P}^n) - H(\mathbf{C}^n).$$

Also, we can use the estimate

$$H(\mathbf{P}^n) \approx nH_L = n(1 - R_L)\log_2 |\mathcal{P}|,$$

provided n is reasonably large. Certainly,

$$H(\mathbf{C}^n) \leq n\log_2 |\mathcal{C}|.$$

Then, if $|\mathcal{C}| = |\mathcal{P}|$, it follows that

$$H(\mathbf{K}|\mathbf{C}^n) \geq H(\mathbf{K}) - nR_L \log_2 |\mathcal{P}|. \tag{2.1}$$

Next, we relate the quantity $H(\mathbf{K}|\mathbf{C}^n)$ to the number of spurious keys, $\overline{s}_n$. We compute as follows:

$$H(\mathbf{K}|\mathbf{C}^n) = \sum_{\mathbf{y} \in \mathcal{C}^n} p(\mathbf{y}) H(\mathbf{K}|\mathbf{y})$$

$$\leq \sum_{\mathbf{y} \in \mathcal{C}^n} p(\mathbf{y}) \log_2 |K(\mathbf{y})|$$

$$\leq \log_2 \sum_{\mathbf{y} \in \mathcal{C}^n} p(\mathbf{y}) |K(\mathbf{y})|$$

$$= \log_2(\overline{s}_n + 1),$$

where we apply Jensen's Inequality (Theorem 2.5) with $f(x) = \log_2 x$. Thus we obtain the inequality

$$H(\mathbf{K}|\mathbf{C}^n) \leq \log_2(\overline{s}_n + 1). \tag{2.2}$$

Combining the two inequalities (2.1) and (2.2), we get that

$$\log_2(\overline{s}_n + 1) \geq H(\mathbf{K}) - nR_L \log_2 |\mathcal{P}|.$$

In the case where keys are chosen equiprobably (which maximizes $H(\mathbf{K})$), we have the following result.

THEOREM 2.11
Suppose $(\mathcal{P}, \mathcal{C}, \mathcal{K}, \mathcal{E}, \mathcal{D})$ is a cryptosystem where $|\mathcal{C}| = |\mathcal{P}|$ and keys are chosen equiprobably. Let R_L denote the redundancy of the underlying language. Then given a string of ciphertext of length n, where n is sufficiently large, the expected number of spurious keys, $\overline{s}_n$, satisfies

$$\overline{s}_n \geq \frac{|\mathcal{K}|}{|\mathcal{P}|^{nR_L}} - 1.$$

The quantity $|\mathcal{K}|/|\mathcal{P}|^{nR_L} - 1$ approaches 0 exponentially quickly as n increases. Also, note that the estimate may not be accurate for small values of n, especially since $H(\mathbf{P}^n)/n$ may not be a good estimate for H_L if n is small.

We have one more concept to define.

DEFINITION 2.7 *The unicity distance of a cryptosystem is defined to be the value of n, denoted by n_0, at which the expected number of spurious keys becomes zero; i.e., the average amount of ciphertext required for an opponent to be able to uniquely compute the key, given enough computing time.*

If we set $\overline{s}_n = 0$ in Theorem 2.11 and solve for n, we get an estimate for the unicity distance, namely

$$n_0 \approx \frac{\log_2 |\mathcal{K}|}{R_L \log_2 |\mathcal{P}|}.$$

As an example, consider the **Substitution Cipher**. In this cryptosystem, $|\mathcal{P}| = 26$ and $|\mathcal{K}| = 26!$. If we take $R_L = 0.75$, then we get an estimate for the unicity distance of

$$n_0 \approx 88.4/(0.75 \times 4.7) \approx 25.$$

This suggests that, given a ciphertext string of length at least 25, (usually) a unique decryption is possible.

2.5   Product Cryptosystems

Another innovation introduced by Shannon in his 1949 paper was the idea of combining cryptosystems by forming their "product." This idea has been of fundamental importance in the design of present-day cryptosystems such as the **Data Encryption Standard**, which we study in the next chapter.

For simplicity, we will confine our attention in this section to cryptosystems in which $\mathcal{C} = \mathcal{P}$: cryptosystems of this type are called *endomorphic*. Suppose $\mathbf{S}_1 = (\mathcal{P}, \mathcal{P}, \mathcal{K}_1, \mathcal{E}_1, \mathcal{D}_1)$ and $\mathbf{S}_2 = (\mathcal{P}, \mathcal{P}, \mathcal{K}_2, \mathcal{E}_2, \mathcal{D}_2)$ are two endomorphic cryptosystems which have the same plaintext (and ciphertext) spaces. Then the *product* of $\mathbf{S}_1$ and $\mathbf{S}_2$, denoted by $\mathbf{S}_1 \times \mathbf{S}_2$, is defined to be the cryptosystem

$$(\mathcal{P}, \mathcal{P}, \mathcal{K}_1 \times \mathcal{K}_2, \mathcal{E}, \mathcal{D}).$$

A key of the product cryptosystem has the form $K = (K_1, K_2)$, where $K_1 \in \mathcal{K}_1$ and $K_2 \in \mathcal{K}_2$. The encryption and decryption rules of the product cryptosystem are defined as follows: For each $K = (K_1, K_2)$, we have an encryption rule e_K defined by the formula

$$e_{(K_1, K_2)}(x) = e_{K_2}(e_{K_1}(x)),$$

and a decryption rule defined by the formula

$$d_{(K_1, K_2)}(y) = d_{K_1}(d_{K_2}(y)).$$

That is, we first encrypt x with e_{K_1}, and then "re-encrypt" the resulting ciphertext with e_{K_2}. Decrypting is similar, but it must be done in the reverse order:

$$\begin{aligned}
d_{(K_1, K_2)}(e_{(K_1, K_2)}(x)) &= d_{(K_1, K_2)}(e_{K_2}(e_{K_1}(x))) \\
&= d_{K_1}(d_{K_2}(e_{K_2}(e_{K_1}(x)))) \\
&= d_{K_1}(e_{K_1}(x)) \\
&= x.
\end{aligned}$$

Recall also that cryptosystems have probability distributions associated with their keyspaces. Thus we need to define the probability distribution for the keyspace

FIGURE 2.2
Multiplicative Cipher

Let $\mathcal{P} = \mathcal{C} = \mathbb{Z}_{26}$ and let

$$K = \{a \in \mathbb{Z}_{26} : \gcd(a, 26) = 1\}.$$

For $a \in K$, define

$$e_a(x) = ax \bmod 26$$

and

$$d_a(y) = a^{-1}y \bmod 26$$

$(x, y \in \mathbb{Z}_{26})$.

$\mathcal{K}$ of the product cryptosystem. We do this in a very natural way:

$$p_K(K_1, K_2) = p_{K_1}(K_1) \times p_{K_2}(K_2).$$

In other words, choose K_1 using the distribution p_{K_1}, and then independently choose K_2 using the distribution p_{K_2}.

Here is a simple example to illustrate the definition of a product cryptosystem. Suppose we define the **Multiplicative Cipher** as in Figure 2.2.

Suppose **M** is the **Multiplicative Cipher** (with keys chosen equiprobably) and **S** is the **Shift Cipher** (with keys chosen equiprobably). Then it is very easy to see that **M** × **S** is nothing more than the **Affine Cipher** (again, with keys chosen equiprobably). It is slightly more difficult to show that **S** × **M** is also the **Affine Cipher** with equiprobable keys.

Let's prove these assertions. A key in the **Shift Cipher** is an element $K \in \mathbb{Z}_{26}$, and the corresponding encryption rule is $e_K(x) = x + K \bmod 26$. A key in the **Multiplicative Cipher** is an element $a \in \mathbb{Z}_{26}$ such that $\gcd(a, 26) = 1$; the corresponding encryption rule is $e_a(x) = ax \bmod 26$. Hence, a key in the product cipher **M** × **S** has the form (a, K), where

$$e_{(a,K)}(x) = ax + K \bmod 26.$$

But this is precisely the definition of a key in the **Affine Cipher**. Further, the probability of a key in the **Affine Cipher** is $1/312 = 1/12 \times 1/26$, which is the product of the probabilities of the keys a and K, respectively. Thus **M** × **S** is the **Affine Cipher**.

Now let's consider **S** × **M**. A key in this cipher has the form (K, a), where

$$e_{(K,a)}(x) = a(x + K) = ax + aK \bmod 26.$$

Thus the key (K, a) of the product cipher $\mathbf{S} \times \mathbf{M}$ is identical to the key (a, aK) of the **Affine Cipher**. It remains to show that each key of the **Affine Cipher** arises with the same probability $1/312$ in the product cipher $\mathbf{S} \times \mathbf{M}$. Observe that $aK = K_1$ if and only if $K = a^{-1}K_1$ (recall that $\gcd(a, 26) = 1$, so a has a multiplicative inverse). In other words, the key (a, K_1) of the **Affine Cipher** is equivalent to the key $(a^{-1}K_1, a)$ of the product cipher $\mathbf{S} \times \mathbf{M}$. We thus have a bijection between the two key spaces. Since each key is equiprobable, we conclude that $\mathbf{S} \times \mathbf{M}$ is indeed the **Affine Cipher**.

We have shown that $\mathbf{M} \times \mathbf{S} = \mathbf{S} \times \mathbf{M}$. Thus we would say that the two cryptosystems *commute*. But not all pairs of cryptosystems commute; it is easy to find counterexamples. On the other hand, the product operation is always *associative*: $(\mathbf{S}_1 \times \mathbf{S}_2) \times \mathbf{S}_3 = \mathbf{S}_1 \times (\mathbf{S}_2 \times \mathbf{S}_3)$.

If we take the product of an (endomorphic) cryptosystem $\mathbf{S}$ with itself, we obtain the cryptosystem $\mathbf{S} \times \mathbf{S}$, which we denote by $\mathbf{S}^2$. If we take the n-fold product, the resulting cryptosystem is denoted by $\mathbf{S}^n$. We call $\mathbf{S}^n$ an *iterated* cryptosystem.

A cryptosystem $\mathbf{S}$ is defined to be *idempotent* if $\mathbf{S}^2 = \mathbf{S}$. Many of the cryptosystems we studied in Chapter 1 are idempotent. For example, the **Shift, Substitution, Affine, Hill, Vigenere** and **Permutation Ciphers** are all idempotent. Of course, if a cryptosystem $\mathbf{S}$ is idempotent, then there is no point in using the product system $\mathbf{S}^2$, as it requires an extra key but provides no more security.

If a cryptosystem is not idempotent, then there is a potential increase in security by iterating several times. This idea is used in the **Data Encryption Standard**, which consists of 16 iterations. But, of course, this approach requires a non-idempotent cryptosystem to start with. One way in which simple non-idempotent cryptosystems can sometimes be constructed is to take the product of two different (simple) cryptosystems.

REMARK   It is not hard to show that if $\mathbf{S}_1$ and $\mathbf{S}_2$ are both idempotent and they commute, then $\mathbf{S}_1 \times \mathbf{S}_2$ will also be idempotent. This follows from the following algebraic manipulations:

$$
\begin{aligned}
(\mathbf{S}_1 \times \mathbf{S}_2) \times (\mathbf{S}_1 \times \mathbf{S}_2) &= \mathbf{S}_1 \times (\mathbf{S}_2 \times \mathbf{S}_1) \times \mathbf{S}_2 \\
&= \mathbf{S}_1 \times (\mathbf{S}_1 \times \mathbf{S}_2) \times \mathbf{S}_2 \\
&= (\mathbf{S}_1 \times \mathbf{S}_1) \times (\mathbf{S}_2 \times \mathbf{S}_2) \\
&= \mathbf{S}_1 \times \mathbf{S}_2.
\end{aligned}
$$

(Note the use of the associative property in this proof.)

So, if $\mathbf{S}_1$ and $\mathbf{S}_2$ are both idempotent, and we want $\mathbf{S}_1 \times \mathbf{S}_2$ to be non-idempotent, then it is necessary that $\mathbf{S}_1$ and $\mathbf{S}_2$ not commute. $\blacksquare$

Fortunately, many simple cryptosystems are suitable building blocks in this type of approach. Taking the product of substitution-type ciphers with permutation-type ciphers is a commonly used technique. We will see a realization of this in the next chapter.

2.6 Notes

The idea of perfect secrecy and the use of entropy techniques in cryptography was pioneered by Shannon [SH49]. Product cryptosystems are also discussed in this paper. The concept of entropy was defined by Shannon in [SH48]. Good introductions to entropy, Huffman coding and related topics can be found in the books by Welsh [WE88] and Goldie and Pinch [GP91].

The results of Section 2.4 are due to Beauchemin and Brassard [BB88], who generalized earlier results of Shannon.

Exercises

2.1 Let n be a positive integer. A *Latin square* of order n is an $n \times n$ array L of the integers $1, \ldots, n$ such that every one of the n integers occurs exactly once in each row and each column of L. An example of a Latin square of order 3 is as follows:

1	2	3
3	1	2
2	3	1

Given any Latin square L of order n, we can define a related cryptosystem. Take $\mathcal{P} = \mathcal{C} = \mathcal{K} = \{1, ..., n\}$. For $1 \le i \le n$, the encryption rule e_i is defined to be $e_i(j) = L(i, j)$. (Hence each row of L gives rise to one encryption rule.)

Give a complete proof that this Latin square cryptosystem achieves perfect secrecy.

2.2 Prove that the **Affine Cipher** achieves perfect secrecy.

2.3 Suppose a cryptosystem achieves perfect secrecy for a particular plaintext probability distribution p_0. Prove that perfect secrecy is maintained for *any* plaintext probability distribution.

2.4 Prove that if a cryptosystem has perfect secrecy and $|\mathcal{K}| = |\mathcal{C}| = |\mathcal{P}|$, then every ciphertext is equally probable.

2.5 Suppose $\mathbf{X}$ is a set of cardinality n, where $2^k \le n < 2^{k+1}$, and $p(x) = 1/n$ for all $x \in \mathbf{X}$.

(a) Find a prefix-free encoding of $\mathbf{X}$, say f, such that $\ell(f) = k + 2 - 2^{k+1}/n$.

HINT Encode $2^{k+1} - n$ elements of $\mathbf{X}$ as strings of length k, and encode the remaining elements as strings of length $k + 1$.

(b) Illustrate your construction for $n = 6$. Compute $\ell(f)$ and $H(\mathbf{X})$ in this case.

2.6 Suppose $\mathbf{X} = \{a, b, c, d, e\}$ has the following probability distribution: $p(a) = .32$, $p(b) = .23$, $p(c) = .20$, $p(d) = .15$ and $p(e) = .10$. Use Huffman's algorithm to find the optimal prefix-free encoding of $\mathbf{X}$. Compare the length of this encoding to $H(\mathbf{X})$.

2.7 Prove that $H(\mathbf{X}, \mathbf{Y}) = H(\mathbf{Y})+H(\mathbf{X}|\mathbf{Y})$. Then show as a corollary that $H(\mathbf{X}|\mathbf{Y}) \leq H(\mathbf{X})$, with equality if and only if $\mathbf{X}$ and $\mathbf{Y}$ are independent.

2.8 Prove that a cryptosystem has perfect secrecy if and only if $H(\mathbf{P}|\mathbf{C}) = H(\mathbf{P})$.

2.9 Prove that, in any cryptosystem, $H(\mathbf{K}|\mathbf{C}) \geq H(\mathbf{P}|\mathbf{C})$. (Intuitively, this result says that, given a ciphertext, the opponent's uncertainty about the key is at least as great as his uncertainty about the plaintext.)

2.10 Consider a cryptosystem in which $\mathcal{P} = \{a, b, c\}$, $\mathcal{K} = \{K_1, K_2, K_3\}$ and $\mathcal{C} = \{1, 2, 3, 4\}$. Suppose the encryption matrix is as follows:

	a	b	c
K_1	1	2	3
K_2	2	3	4
K_3	3	4	1

Given that keys are chosen equiprobably, and the plaintext probability distribution is $p_\mathbf{P}(a) = 1/2$, $p_\mathbf{P}(b) = 1/3$, $p_\mathbf{P}(c) = 1/6$, compute $H(\mathbf{P})$, $H(\mathbf{C})$, $H(\mathbf{K})$, $H(\mathbf{K}|\mathbf{C})$ and $H(\mathbf{P}|\mathbf{C})$.

2.11 Compute $H(\mathbf{K}|\mathbf{C})$ and $H(\mathbf{K}|\mathbf{P}, \mathbf{C})$ for the **Affine Cipher**.

2.12 Consider a **Vigenere Cipher** with keyword length m. Show that the unicity distance is $1/R_L$, where R_L is the redundancy of the underlying language. (This result is interpreted as follows. If n_0 denotes the number of alphabetic characters being encrypted , then the "length" of the plaintext is n_0/m, since each plaintext element consists of m alphabetic characters. So, a unicity distance of $1/R_L$ corresponds to a plaintext consisting of m/R_L alphabetic characters.)

2.13 Show that the unicity distance of the **Hill Cipher** (with an $m \times m$ encryption matrix) is less than m/R_L (Note that the number of alphabetic characters in a plaintext of this length is m^2/R_L.)

2.14 A **Substitution Cipher** over a plaintext space of size n has $|\mathcal{K}| = n!$ Stirling's formula gives the following estimate for $n!$:

$$n! \approx \sqrt{2\pi n}\left(\frac{n}{e}\right)^n.$$

(a) Using Stirling's formula, derive an estimate of the unicity distance of the **Substitution Cipher**.

(b) Let $m \geq 1$ be an integer. The m-**gram Substitution Cipher** is the **Substitution Cipher** where the plaintext (and ciphertext) spaces consist of all 26^m m-grams. Estimate the unicity distance of the m-**gram Substitution Cipher** if $R_L = 0.75$.

2.15 Prove that the **Shift Cipher** is idempotent.

2.16 Suppose $\mathbf{S}_1$ is the **Shift Cipher** (with equiprobable keys, as usual) and $\mathbf{S}_2$ is the **Shift Cipher** where keys are chosen with respect to some probability distribution $p_\mathcal{K}$ (which need not be equiprobable). Prove that $\mathbf{S}_1 \times \mathbf{S}_2 = \mathbf{S}_1$.

2.17 Suppose $\mathbf{S}_1$ and $\mathbf{S}_2$ are **Vigenere Ciphers** with keyword lengths m_1, m_2 respectively, where $m_1 > m_2$.

(a) If $m_2 \mid m_1$, then show that $\mathbf{S}_2 \times \mathbf{S}_1 = \mathbf{S}_1$.

(b) One might try to generalize the previous result by conjecturing that $\mathbf{S}_2 \times \mathbf{S}_1 = \mathbf{S}_3$, where $\mathbf{S}_3$ is the **Vigenere Cipher** with keyword length $\operatorname{lcm}(m_1, m_2)$. Prove that this conjecture is false.

HINT If $m_1 \not\equiv 0 \bmod m_2$, then the number of keys in the product cryptosystem $\mathbf{S}_2 \times \mathbf{S}_1$ is less than the number of keys in $\mathbf{S}_3$.

3

The Data Encryption Standard

3.1 Introduction

On May 15, 1973, the National Bureau of Standards published a solicitation for cryptosystems in the Federal Register. This lead ultimately to the development of the **Data Encryption Standard**, or **DES**, which has become the most widely used cryptosystem in the world. **DES** was developed at IBM, as a modification of an earlier system known as **LUCIFER**. **DES** was first published in the Federal Register of March 17, 1975. After a considerable amount of public discussion, **DES** was adopted as a standard for "unclassified" applications on January 15, 1977. **DES** has been reviewed by the National Bureau of Standards (approximately) every five years since its adoption. Its most recent renewal was in January 1994, when it was renewed until 1998. It is anticipated that it will not remain a standard past 1998.

3.2 Description of DES

A complete description of **DES** is given in the Federal Information Processing Standards Publication 46, dated January 15, 1977. **DES** encrypts a plaintext bitstring x of length 64 using a key K which is a bitstring of length 56, obtaining a ciphertext bitstring which is again a bitstring of length 64. We first give a "high-level" description of the system.

The algorithm proceeds in three stages:

1. Given a plaintext x, a bitstring x_0 is constructed by permuting the bits of x according to a (fixed) *initial permutation* IP. We write $x_0 = \text{IP}(x) = L_0 R_0$, where L_0 comprises the first 32 bits of x_0 and R_0 the last 32 bits.

2. 16 iterations of a certain function are then computed. We compute $L_i R_i$,

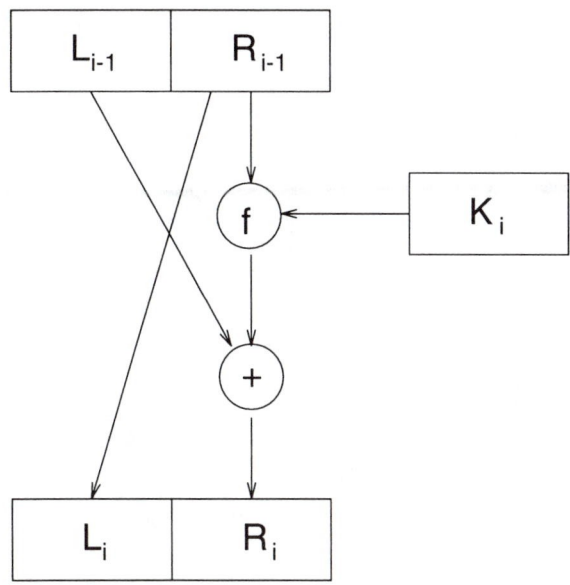

FIGURE 3.1
One round of DES encryption

$1 \leq i \leq 16$, according to the following rule:

$$L_i = R_{i-1}$$
$$R_i = L_{i-1} \oplus f(R_{i-1}, K_i),$$

where $\oplus$ denotes the exclusive-or of two bitstrings. f is a function that we will describe later, and $K_1, K_2, \ldots, K_{16}$ are each bitstrings of length 48 computed as a function of the key K. (Actually, each K_i is a permuted selection of bits from K.) $K_1, K_2, \ldots, K_{16}$ comprises the *key schedule*. One round of encryption is depicted in Figure 3.1

3. Apply the inverse permutation IP^{-1} to the bitstring $R_{16}L_{16}$, obtaining the ciphertext y. That is, $y = \text{IP}^{-1}(R_{16}L_{16})$. Note the inverted order of L_{16} and R_{16}.

The function f takes as input a first argument A, which is a bitstring of length 32, and a second argument J that is a bitstring of length 48, and produces as output a bitstring of length 32. The following steps are executed.

1. The first argument A is "expanded" to a bitstring of length 48 according to a fixed *expansion function* E. E(A) consists of the 32 bits from A, permuted in a certain way, with 16 of the bits appearing twice.

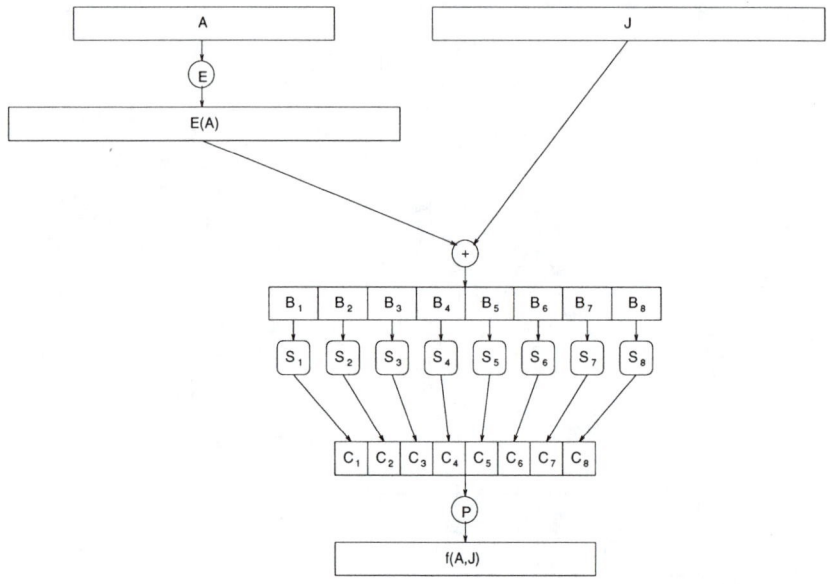

FIGURE 3.2
The DES f function

2. Compute $E(A) \oplus J$ and write the result as the concatenation of eight 6-bit strings $B = B_1 B_2 B_3 B_4 B_5 B_6 B_7 B_8$.

3. The next step uses eight *S-boxes* $S_1, \ldots, S_8$. Each S_i is a fixed 4×16 array whose entries come from the integers $0 - 15$. Given a bitstring of length six, say $B_j = b_1 b_2 b_3 b_4 b_5 b_6$, we compute $S_j(B_j)$ as follows. The two bits $b_1 b_6$ determine the binary representation of a row r of S_j ($0 \leq r \leq 3$), and the four bits $b_2 b_3 b_4 b_5$ determine the binary representation of a column c of S_j ($0 \leq c \leq 15$). Then $S_j(B_j)$ is defined to be the entry $S_j(r, c)$, written in binary as a bitstring of length four. (Hence, each S_j can be thought of as a function that accepts as input a bitstring of length two and one of length four, and produces as output a bitstring of length four.) In this fashion, we compute $C_j = S_j(B_j)$, $1 \leq j \leq 8$.

4. The bitstring $C = C_1 C_2 C_3 C_4 C_5 C_6 C_7 C_8$ of length 32 is permuted according to a fixed permutation P. The resulting bitstring $P(C)$ is defined to be $f(A, J)$.

The f function is depicted in Figure 3.2. Basically, it consists of a substitution (using an S-box) followed by the (fixed) permutation P. The 16 iterations of f comprise a product cryptosystem, as described in Section 2.5.

In the remainder of this section, we present the specific functions used in **DES**.

The initial permutation IP is as follows:

IP							
58	50	42	34	26	18	10	2
60	52	44	36	28	20	12	4
62	54	46	38	30	22	14	6
64	56	48	40	32	24	16	8
57	49	41	33	25	17	9	1
59	51	43	35	27	19	11	3
61	53	45	37	29	21	13	5
63	55	47	39	31	23	15	7

This means that the 58th bit of x is the first bit of $IP(x)$; the 50th bit of x is the second bit of $IP(x)$, etc.

The inverse permutation IP^{-1} is:

IP^{-1}							
40	8	48	16	56	24	64	32
39	7	47	15	55	23	63	31
38	6	46	14	54	22	62	30
37	5	45	13	53	21	61	29
36	4	44	12	52	20	60	28
35	3	43	11	51	19	59	27
34	2	42	10	50	18	58	26
33	1	41	9	49	17	57	25

The expansion function E is specified by the following table:

E bit-selection table					
32	1	2	3	4	5
4	5	6	7	8	9
8	9	10	11	12	13
12	13	14	15	16	17
16	17	18	19	20	21
20	21	22	23	24	25
24	25	26	27	28	29
28	29	30	31	32	1

The eight S-boxes and the permutation P are now presented:

S_1															
14	4	13	1	2	15	11	8	3	10	6	12	5	9	0	7
0	15	7	4	14	2	13	1	10	6	12	11	9	5	3	8
4	1	14	8	13	6	2	11	15	12	9	7	3	10	5	0
15	12	8	2	4	9	1	7	5	11	3	14	10	0	6	13

S_2															
15	1	8	14	6	11	3	4	9	7	2	13	12	0	5	10
3	13	4	7	15	2	8	14	12	0	1	10	6	9	11	5
0	14	7	11	10	4	13	1	5	8	12	6	9	3	2	15
13	8	10	1	3	15	4	2	11	6	7	12	0	5	14	9

S_3															
10	0	9	14	6	3	15	5	1	13	12	7	11	4	2	8
13	7	0	9	3	4	6	10	2	8	5	14	12	11	15	1
13	6	4	9	8	15	3	0	11	1	2	12	5	10	14	7
1	10	13	0	6	9	8	7	4	15	14	3	11	5	2	12

S_4															
7	13	14	3	0	6	9	10	1	2	8	5	11	12	4	15
13	8	11	5	6	15	0	3	4	7	2	12	1	10	14	9
10	6	9	0	12	11	7	13	15	1	3	14	5	2	8	4
3	15	0	6	10	1	13	8	9	4	5	11	12	7	2	14

S_5															
2	12	4	1	7	10	11	6	8	5	3	15	13	0	14	9
14	11	2	12	4	7	13	1	5	0	15	10	3	9	8	6
4	2	1	11	10	13	7	8	15	9	12	5	6	3	0	14
11	8	12	7	1	14	2	13	6	15	0	9	10	4	5	3

S_6															
12	1	10	15	9	2	6	8	0	13	3	4	14	7	5	11
10	15	4	2	7	12	9	5	6	1	13	14	0	11	3	8
9	14	15	5	2	8	12	3	7	0	4	10	1	13	11	6
4	3	2	12	9	5	15	10	11	14	1	7	6	0	8	13

S_7															
4	11	2	14	15	0	8	13	3	12	9	7	5	10	6	1
13	0	11	7	4	9	1	10	14	3	5	12	2	15	8	6
1	4	11	13	12	3	7	14	10	15	6	8	0	5	9	2
6	11	13	8	1	4	10	7	9	5	0	15	14	2	3	12

S_8															
13	2	8	4	6	15	11	1	10	9	3	14	5	0	12	7
1	15	13	8	10	3	7	4	12	5	6	11	0	14	9	2
7	11	4	1	9	12	14	2	0	6	10	13	15	3	5	8
2	1	14	7	4	10	8	13	15	12	9	0	3	5	6	11

P			
16	7	20	21
29	12	28	17
1	15	23	26
5	18	31	10
2	8	24	14
32	27	3	9
19	13	30	6
22	11	4	25

Finally, we need to describe the computation of the key schedule from the key K. Actually, K is a bitstring of length 64, of which 56 bits comprise the key and 8 bits are parity-check bits (for error-detection). The bits in positions $8, 16, \ldots, 64$ are defined so that each byte contains an odd number of 1's. Hence, a single error can be detected within each group of 8 bits. The parity-check bits are ignored in the computation of the key schedule.

1. Given a 64-bit key K, discard the parity-check bits and permute the remaining bits of K according to a (fixed) permutation PC-1. We will write PC-1$(K) = C_0 D_0$, where C_0 comprises the first 28 bits of PC-1(K) and D_0 the last 28 bits.

2. For i ranging from 1 to 16, compute

$$C_i = LS_i(C_{i-1})$$
$$D_i = LS_i(D_{i-1}),$$

and $K_i = $ PC-2$(C_i D_i)$. LS_i represents a cyclic shift (to the left) of either one or two positions, depending on the value of i: shift one position if $i = 1, 2, 9$ or 16, and shift two positions otherwise. PC-2 is another fixed permutation.

The key schedule computation is depicted in Figure 3.3.

The permutations PC-1 and PC-2 used in the key schedule computation are as follows:

PC-1						
57	49	41	33	25	17	9
1	58	50	42	34	26	18
10	2	59	51	43	35	27
19	11	3	60	52	44	36
63	55	47	39	31	23	15
7	62	54	46	38	30	22
14	6	61	53	45	37	29
21	13	5	28	20	12	4

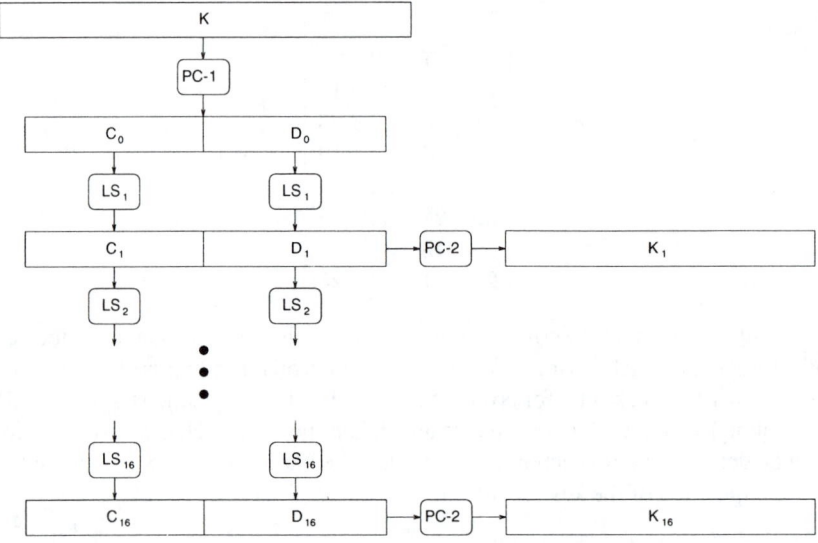

FIGURE 3.3
Computation of DES key schedule

PC-2					
14	17	11	24	1	5
3	28	15	6	21	10
23	19	12	4	26	8
16	7	27	20	13	2
41	52	31	37	47	55
30	40	51	45	33	48
44	49	39	56	34	53
46	42	50	36	29	32

We now display the resulting key schedule. As mentioned above, each round uses a 48-bit key comprised of 48 of the bits in K. The entries in the tables below refer to the bits in K that are used in the various rounds.

Round 1											
10	51	34	60	49	17	33	57	2	9	19	42
3	35	26	25	44	58	59	1	36	27	18	41
22	28	39	54	37	4	47	30	5	53	23	29
61	21	38	63	15	20	45	14	13	62	55	31

Round 2											
2	43	26	52	41	9	25	49	59	1	11	34
60	27	18	17	36	50	51	58	57	19	10	33
14	20	31	46	29	63	39	22	28	45	15	21
53	13	30	55	7	12	37	6	5	54	47	23

Round 3											
51	27	10	36	25	58	9	33	43	50	60	18
44	11	2	1	49	34	35	42	41	3	59	17
61	4	15	30	13	47	23	6	12	29	62	5
37	28	14	39	54	63	21	53	20	38	31	7

Round 4											
35	11	59	49	9	42	58	17	27	34	44	2
57	60	51	50	33	18	19	26	25	52	43	1
45	55	62	14	28	31	7	53	63	13	46	20
21	12	61	23	38	47	5	37	4	22	15	54

Round 5											
19	60	43	33	58	26	42	1	11	18	57	51
41	44	35	34	17	2	3	10	9	36	27	50
29	39	46	61	12	15	54	37	47	28	30	4
5	63	45	7	22	31	20	21	55	6	62	38

Round 6											
3	44	27	17	42	10	26	50	60	2	41	35
25	57	19	18	1	51	52	59	58	49	11	34
13	23	30	45	63	62	38	21	31	12	14	55
20	47	29	54	6	15	4	5	39	53	46	22

Round 7											
52	57	11	1	26	59	10	34	44	51	25	19
9	41	3	2	50	35	36	43	42	33	60	18
28	7	14	29	47	46	22	5	15	63	61	39
4	31	13	38	53	62	55	20	23	37	30	6

Round 8											
36	41	60	50	10	43	59	18	57	35	9	3
58	25	52	51	34	19	49	27	26	17	44	2
12	54	61	13	31	30	6	20	62	47	45	23
55	15	28	22	37	46	39	4	7	21	14	53

Round 9											
57	33	52	42	2	35	51	10	49	27	1	60
50	17	44	43	26	11	41	19	18	9	36	59
4	46	53	5	23	22	61	12	54	39	37	15
47	7	20	14	29	38	31	63	62	13	6	45

Round 10											
41	17	36	26	51	19	35	59	33	11	50	44
34	1	57	27	10	60	25	3	2	58	49	43
55	30	37	20	7	6	45	63	38	23	21	62
31	54	4	61	13	22	15	47	46	28	53	29

Round 11											
25	1	49	10	35	3	19	43	17	60	34	57
18	50	41	11	59	44	9	52	51	42	33	27
39	14	21	4	54	53	29	47	22	7	5	46
15	38	55	45	28	6	62	31	30	12	37	13

Round 12											
9	50	33	59	19	52	3	27	1	44	18	41
2	34	25	60	43	57	58	36	35	26	17	11
23	61	5	55	38	37	13	31	6	54	20	30
62	22	39	29	12	53	46	15	14	63	21	28

Round 13											
58	34	17	43	3	36	52	11	50	57	2	25
51	18	9	44	27	41	42	49	19	10	1	60
7	45	20	39	22	21	28	15	53	38	4	14
46	6	23	13	63	37	30	62	61	47	5	12

Round 14											
42	18	1	27	52	49	36	60	34	41	51	9
35	2	58	57	11	25	26	33	3	59	50	44
54	29	4	23	6	5	12	62	37	22	55	61
30	53	7	28	47	21	14	46	45	31	20	63

Round 15											
26	2	50	11	36	33	49	44	18	25	35	58
19	51	42	41	60	9	10	17	52	43	34	57
38	13	55	7	53	20	63	46	21	6	39	45
14	37	54	12	31	5	61	30	29	15	4	47

Round 16											
18	59	42	3	57	25	41	36	10	17	27	50
11	43	34	33	52	1	2	9	44	35	26	49
30	5	47	62	45	12	55	38	13	61	31	37
6	29	46	4	23	28	53	22	21	7	63	39

Decryption is done using the same algorithm as encryption, starting with y as the input, but using the key schedule $K_{16}, \ldots, K_1$ in reverse order. The output will be the plaintext x.

3.2.1 An Example of DES Encryption

Here is an example of encryption using the **DES**. Suppose we encrypt the (hexadecimal) plaintext

$$0123456789ABCDEF$$

using the (hexadecimal) key

$$133457799BBCDFF1.$$

The key, in binary, without parity-check bits, is

$$00010010011010010101101111001001101101111011011111111000.$$

Applying IP, we obtain L_0 and R_0 (in binary):

L_0	$=$	11001100000000001100110011111111
$L_1 = R_0$	$=$	11110000101010101111000010101010

The 16 rounds of encryption are then performed, as indicated.

$E(R_0)$	$=$	011110100001010101010101011110100001010101010101
K_1	$=$	000110110000001011101111111111000111000001110010
$E(R_0) \oplus K_1$	$=$	011000010001011110111010100001100110010100100111
S-box outputs		01011100100000101011010110010111
$f(R_0, K_1)$	$=$	00100011010010101010100110111011
$L_2 = R_1$	$=$	11101111010010100110010101000100

$E(R_1)$	$=$	011101011110101001010100001100001010101000001001
K_2	$=$	011110011010111011011001110110111100100111100101
$E(R_1) \oplus K_2$	$=$	000011000100010010001101111010110110001111101100
S-box outputs		11111000110100000011101010101110
$f(R_1, K_2)$	$=$	00111100101010111000011110100011
$L_3 = R_2$	$=$	11001100000000010111011100001001

$E(R_2)$	$=$	111001011000000000000101011101011101000010100111
K_3	$=$	010101011111110010001010010000101100111110011001
$E(R_2) \oplus K_3$	$=$	101100001111100100010001111100000100011111001010
S-box outputs		00100111000100001110000101101111
$f(R_2, K_3)$	$=$	01001101000101100110111010110000
$L_4 = R_3$	$=$	10100010010111000000101111110100

$E(R_3)$	$=$	010100001000010111110000000010101111111110101001
K_4	$=$	011100101010110111010110101101101100110101010010110
$E(R_3) \oplus K_4$	$=$	001000101110111100101110110111100100101010110100
S-box outputs		00100001111011011001111100111010
$f(R_3, K_4)$	$=$	10111011001000110111011101001100
$L_5 = R_4$	$=$	01110111001000100000000001000101

$E(R_4)$	$=$	1011101011101001000001000000000000000001000001010
K_5	$=$	011111001110110000000111111010110101001110101000
$E(R_4) \oplus K_5$	$=$	110001100000010100000011111101011010100011010010
S-box outputs		01010000110010000011000111101011
$f(R_4, K_5)$	$=$	00101000000100111010110111000011
$L_6 = R_5$	$=$	10001010010010011110100110001101 11

$E(R_5)$	$=$	110001010100001001011111110100001100000110101111
K_6	$=$	011000111010010100111110010100000111101100101111
$E(R_5) \oplus K_6$	$=$	101001101110011101100001100000001011101010000000
S-box outputs		01000001111100110100110000111101
$f(R_5, K_6)$	$=$	10011110010001011100110100101100
$L_7 = R_6$	$=$	11101001011001111100110101101001

$E(R_6)$	$=$	111101010010101100001111111001011010101101010011
K_7	$=$	111011001000010010110111111011000011000101111100
$E(R_6) \oplus K_7$	$=$	000110011010111101111000000100111011001111101111
S-box outputs		00010000011101010100000010101101
$f(R_6, K_7)$	$=$	10001100000000101000111000010011 1
$L_8 = R_7$	$=$	00000110010010101011101000010000

$E(R_7)$	$=$	000000001100001001010101010111110100000010100000
K_8	$=$	111101111000101000111010110000010011101111111011
$E(R_7) \oplus K_8$	$=$	111101110100100001101111100111100111101101011011
S-box outputs		01101100000110000111110010101110
$f(R_7, K_8)$	$=$	00111100000011101000011011111001
$L_9 = R_8$	$=$	11010101011010010100101110010000

$E(R_8)$	$=$	011010101010101101010010101001010111110010100001
K_9	$=$	111000001101101111101011110110111100111100000001
$E(R_8) \oplus K_9$	$=$	100010100111000010111001010010001001101100100000
S-box outputs		00010001000011000101011101110111
$f(R_8, K_9)$	$=$	00100010001101100111110001101010
$L_{10} = R_9$	$=$	00100100011110011000110011111010

$E(R_9)$	$=$	000100001000001111111100101100000110000111110100
K_{10}	$=$	101100011110011010001111011101001000110010011111
$E(R_9) \oplus K_{10}$	$=$	101000010111000010111101101101010000101101111011
S-box outputs		11011010000010001010010011101101 01
$f(R_9, K_{10})$	$=$	01100010101111001001110000100010
$L_{11} = R_{10}$	$=$	10110111110101011101011110110010

$E(R_{10})$	$=$	010110101111111010101011111010101111110110100101
K_{11}	$=$	001000010101111110100111101111011101001110000110
$E(R_{10}) \oplus K_{11}$	$=$	011110111010000101111000011010000101110001000011
S-box outputs		01110011000001011101000100000001
$f(R_{10}, K_{11})$	$=$	11100001000001001111101000000010
$L_{12} = R_{11}$	$=$	11000101011110000011110001111000

$E(R_{11})$	$=$	01100000101010111111000000011111110000011111110001
K_{12}	$=$	011101010111000111111010110010100011001111111101001
$E(R_{11}) \oplus K_{12}$	$=$	0001010111011010000001011000101111100100000011000
S-box outputs		011110111000101100100110001110101
$f(R_{11}, K_{12})$	$=$	11000010011010001100111111101010
$L_{13} = R_{12}$	$=$	01110101101111010001100001011000

$E(R_{12})$	$=$	001110101011110111111010100011110000001011110000
K_{13}	$=$	100101111100010111010001111110101011101001000001
$E(R_{12}) \oplus K_{13}$	$=$	101011010111000001010110110101101111000010110001
S-box outputs		100110101101000110001011010011111
$f(R_{12}, K_{13})$	$=$	11011101101110110010100100100010
$L_{14} = R_{13}$	$=$	00011000110000110001010101011010

$E(R_{13})$	$=$	000011110001011000000110100010101010101011110100
K_{14}	$=$	010111110100001110110111111100101110011100111010
$E(R_{13}) \oplus K_{14}$	$=$	010100000101010110110001011110000100110111001110
S-box outputs		011001000111100110011010101110001
$f(R_{13}, K_{14})$	$=$	10110111001100011000111001010101
$L_{15} = R_{14}$	$=$	11000010100011001001011000001101

$E(R_{14})$	$=$	111000000101010001011001010010101100000001011011
K_{15}	$=$	101111111001000110001101001111010011111100001010
$E(R_{14}) \oplus K_{15}$	$=$	010111111100010111010100011101111111111101010001
S-box outputs		101100101110100010001101001111000
$f(R_{14}, K_{15})$	$=$	01011011100000010010011101101110
$L_{16} = R_{15}$	$=$	01000011010000100011001000110100

$E(R_{15})$	$=$	001000000110101000000100000110100100000110101000
K_{16}	$=$	110010110011110110001011000011100001011111110101
$E(R_{15}) \oplus K_{16}$	$=$	111010110101011110001111000101000101011001011101
S-box outputs		101001110000011001001000001010001
$f(R_{15}, K_{16})$	$=$	11001000110000000100111110011000
R_{16}	$=$	00001010010011001101100110010101

Finally, applying IP^{-1} to $R_{16}L_{16}$, we obtain the ciphertext, which (in hexadecimal form) is:

$$85\text{E}813540\text{F}0\text{AB}405.$$

3.3 The DES Controversy

When **DES** was proposed as a standard, there was considerable criticism. One objection to **DES** concerned the S-boxes. All computations in **DES**, with the exception of the S-boxes, are *linear*, e.g., computing the exclusive-or of two outputs is the same as forming the exclusive-or of two inputs and then computing the output. The S-boxes, being the non-linear component of the cryptosystem, are vital to its security (We saw in Chapter 1 how linear cryptosystems, such as the

Hill Cipher, could easily be cryptanalyzed by a known plaintext attack.) However, the design criteria of the S-boxes are not completely known. Several people have suggested that the S-boxes might contain hidden "trapdoors" which would allow the National Security Agency to decrypt messages while maintaining that **DES** is "secure." It is, of course, impossible to disprove such an assertion, but no evidence has come to light that indicates that trap-doors in **DES** do in fact exist.

In 1976, the National Security Agency (NSA) asserted that the following properties of the S-boxes are design criteria:

P0 Each row of each S-box is a permutation of the integers $0, \ldots, 15$.

P1 No S-box is a linear or affine function of its inputs.

P2 Changing one input bit to an S-box causes at least two output bits to change.

P3 For any S-box and any input x, $S(x)$ and $S(x \oplus 001100)$ differ in at least two bits (here x is a bitstring of length 6).

Two other properties of the S-boxes were designated as "caused by design criteria" by NSA.

P4 For any S-box, for any input x, and for $e, f \in \{0, 1\}$, $S(x) \neq S(x \oplus 11ef00)$.

P5 For any S-box, if one input bit is fixed, and we look at the value of one fixed output bit, the number of inputs for which this output bit equals 0 will be "close to" the number of inputs for which the output bit equals 1. (Note that if we fix the value of either the first or sixth input bit, then 16 inputs will cause a particular output bit to equal 0 and 16 inputs will cause the output to equal 1. For the second through fifth input bits, this will not be true, but the resulting distribution will be "close to" uniform. More precisely, for any S-box, if the value of any input bit is fixed, then the number of inputs for which any fixed output bit has the value 0 (or 1) is always between 13 and 19.)

It is not publicly known if further design criteria were used in the construction of the S-boxes.

The most pertinent criticism of **DES** is that the size of the keyspace, 2^{56}, is too small to be really secure. Various special-purpose machines have been proposed for a known plaintext attack, which would essentially perform an exhaustive search for the key. That is given a 64-bit plaintext x and corresponding ciphertext y, every possible key would be tested until a key K such that $e_K(x) = y$ is found (and note that there may be more than one such key K).

As early as 1977, Diffie and Hellman suggested that one could build a VLSI chip which could test 10^6 keys per second. A machine with 10^6 chips could search the entire key space in about a day. They estimated that such a machine could be built for about $20,000,000.

At the CRYPTO '93 Rump Session, Michael Wiener gave a very detailed design of a key search machine. The machine is based on a key search chip which is pipelined, so that 16 encryptions take place simultaneously. This chip can test

5×10^7 keys per second, and can be built using current technology for $10.50 per chip. A frame consisting of 5760 chips can be built for $100,000. This would allow a **DES** key to be found in about 1.5 days on average. A machine using 10 frames would cost $1,000,000, but would reduce the average search time to about 3.5 hours.

3.4 DES in Practice

Even though the description of **DES** is quite lengthy, it can be implemented very efficiently, either in hardware or in software. The only arithmetic operations to be performed are exclusive-ors of bitstrings. The expansion function E, the S-boxes, the permutations IP and P, and the computation of $K_1, K_2, \ldots, K_{16}$ can all be done in constant time by table look-up (in software) or by hard-wiring them into a circuit.

Current hardware implementations can attain extremely fast encryption rates. Digital Equipment Corporation announced at CRYPTO '92 that they have fabricated a chip with 50K transistors that can encrypt at the rate of 1 Gbit/second using a clock rate of 250 MHz! The cost of this chip is about $300. As of 1991, there were 45 hardware and firmware implementations of **DES** that had been validated by the National Bureau of Standards.

One very important application of **DES** is in banking transactions, using standards developed by the American Bankers Association. **DES** is used to encrypt personal identification numbers (PINs) and account transactions carried out by automated teller machines (ATMs). **DES** is also used by the Clearing House Interbank Payments System (CHIPS) to authenticate transactions involving over 1.5×10^{12} per week.

DES is also widely used in government organizations, such as the Department of Energy, the Justice Department, and the Federal Reserve System.

3.4.1 DES Modes of Operation

Four modes of operation have been developed for **DES**: *electronic codebook mode* (ECB), *cipher feedback mode* (CFB), *cipher block chaining mode* (CBC) and *output feedback mode* (OFB).

ECB mode corresponds to the usual use of a block cipher: given a sequence $x_1 x_2 \ldots$ of 64-bit plaintext blocks, each x_i is encrypted with the same key K, producing a string of ciphertext blocks, $y_1 y_2 \ldots$.

In CBC mode, each ciphertext block y_i is x-ored with the next plaintext block x_{i+1} before being encrypted with the key K. More formally, we start with a 64-bit initialization vector IV, and define $y_0 = \text{IV}$. Then we construct $y_1, y_2, \ldots$ from the rule $y_i = e_K(y_{i-1} \oplus x_i)$, $i \geq 1$. The use of CBC mode is depicted in Figure 3.4.

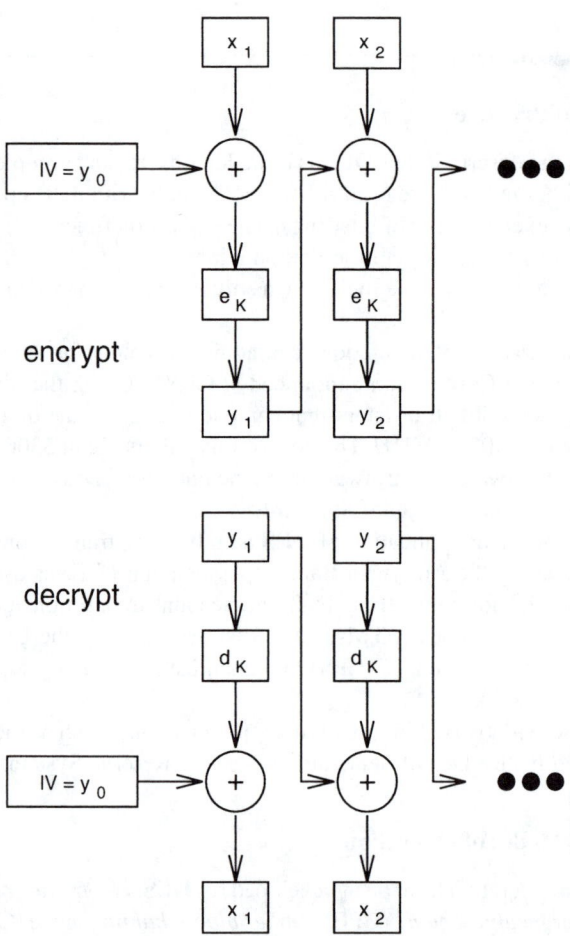

FIGURE 3.4
CBC mode

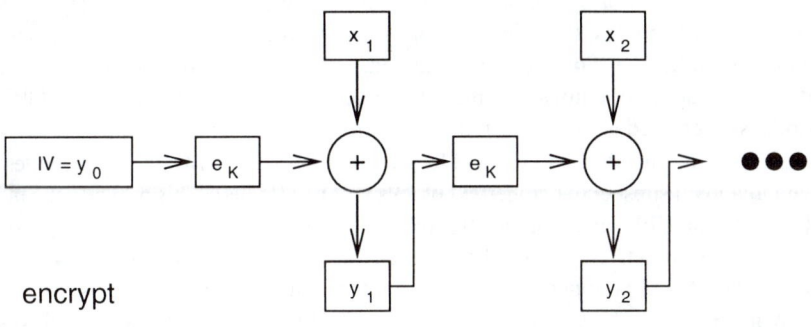

encrypt

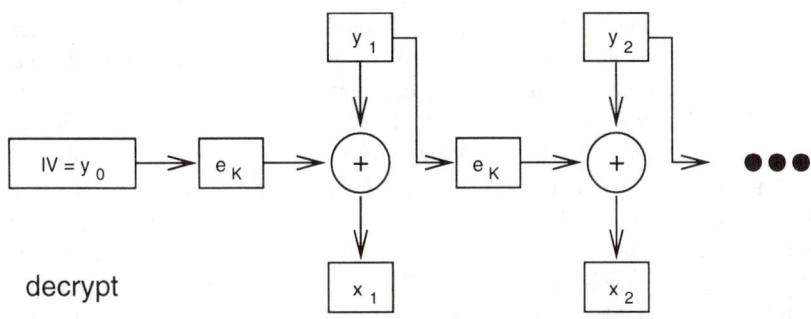

decrypt

FIGURE 3.5
CFB mode

In OFB and CFB modes, a keystream is generated which is then x-ored with the plaintext (i.e., it operates as a stream cipher, cf. Section 1.1.7). OFB is actually a synchronous stream cipher: the keystream is produced by repeatedly encrypting a 64-bit initialization vector, IV. We define $z_0 = \text{IV}$, and then compute the keystream $z_1 z_2 \ldots$ from the rule $z_i = e_K(z_{i-1})$, $i \geq 1$. The plaintext sequence $x_1 x_2 \ldots$ is then encrypted by computing $y_i = x_i \oplus z_i$, $i \geq 1$.

In CFB mode, we start with $y_0 = \text{IV}$ (a 64-bit initialization vector) and we produce the keystream element z_i by encrypting the previous ciphertext block. That is, $z_i = e_K(y_{i-1})$, $i \geq 1$. As in OFB mode, $y_i = x_i \oplus z_i$, $i \geq 1$. The use of CFB is depicted in Figure 3.5 (note that the **DES** encryption function e_K is used for both encryption and decryption in CFB and OFB modes).

There are also variations of OFB and CFB mode called k-bit feedback modes ($1 \leq k \leq 64$). We have described the 64-bit feedback modes here. 1-bit and 8-bit feedback modes are often used in practice for encrypting data one bit (or byte) at a time.

The four modes of operation have different advantages and disadvantages. In ECB and OFB modes, changing one 64-bit plaintext block, x_i, causes the corresponding ciphertext block, y_i, to be altered, but other ciphertext blocks are not affected. In some situations this might be a desirable property. For example, OFB mode is often used to encrypt satellite transmissions.

On the other hand, if a plaintext block x_i is changed in CBC and CFB modes, then y_i and all subsequent ciphertext blocks will be affected. This property means that CBC and CFB modes are useful for purposes of authentication. More specifically, these modes can be used to produce a *message authentication code*, or MAC. The MAC is appended to a sequence of plaintext blocks, and is used to convince Bob that the given sequence of plaintext originated with Alice and was not tampered with by Oscar. Thus the MAC guarantees the integrity (or authenticity) of a message (but it does not provide secrecy, of course).

We will describe how CBC mode is used to produce a MAC. We begin with the initialization vector IV consisting of all zeroes. Then construct the ciphertext blocks $y_1, \ldots, y_n$ with key K, using CBC mode. Finally, define the MAC to be y_n. Then Alice transmits the sequence of plaintext blocks, $x_1 \ldots x_n$, along with the MAC. When Bob receives $x_1 \ldots x_n$, he can reconstruct $y_1, \ldots, y_n$ using the (secret) key K, and verify that y_n is the same as the MAC that he received.

Note that Oscar cannot produce a valid MAC since he does not know the key K being used by Alice and Bob. Further, if Oscar intercepts a sequence of plaintext blocks $x_1 \ldots x_n$, and changes one or more of them, then it is highly unlikely that Oscar can change the MAC so that it will be accepted by Bob.

It is often desirable to combine authenticity and secrecy. This could be done as follows: Alice first uses key K_1 to produce a MAC for $x_1 \ldots x_n$. Then she defines x_{n+1} to be the MAC, and she encrypts the sequence $x_1 \ldots x_{n+1}$ using a second key, K_2, yielding $y_1 \ldots y_{n+1}$. When Bob receives $y_1 \ldots y_{n+1}$, he first decrypts (using K_2) and then checks that x_{n+1} is the MAC for $x_1 \ldots x_n$ using K_1.

Alternatively, Alice could use K_1 to encrypt $x_1 \ldots x_n$, obtaining $y_1 \ldots y_n$, and then use K_2 to produce a MAC y_{n+1} for $y_1 \ldots y_n$. Bob would use K_2 to verify the MAC, and then use K_1 to decrypt $y_1 \ldots y_n$.

3.5   A Time-memory Trade-off

In this section, we describe an interesting time-memory tradeoff for a chosen plaintext attack. Recall that in a chosen plaintext attack, Oscar obtains a plaintext-ciphertext pair produced using the (unknown) key K. So Oscar has x and y, where $y = e_K(x)$, and he wants to determine K.

A feature of this time-memory trade-off is that it does not depend on the "structure" of **DES** in any way. The only aspects of **DES** that are relevant to the attack

FIGURE 3.6
Computation of $X(i,j)$

$$
\begin{array}{ccccccc}
X(1,0) & \xrightarrow{g} & X(1,1) & \xrightarrow{g} & \ldots & \xrightarrow{g} & X(1,t) \\
X(2,0) & \xrightarrow{g} & X(2,1) & \xrightarrow{g} & \ldots & \xrightarrow{g} & X(2,t) \\
\vdots & & & & & & \vdots \\
X(m,0) & \xrightarrow{g} & X(m,1) & \xrightarrow{g} & \ldots & \xrightarrow{g} & X(m,t)
\end{array}
$$

are that plaintexts and ciphertexts have 64 bits, while keys have 56 bits.

We have already discussed the idea of exhaustive search: given a plaintext-ciphertext pair, try all 2^{56} possible keys. This requires no memory but, on average, 2^{55} keys will be tried before the correct one is found. On the other hand, for a given plaintext x, Oscar could precompute $y_K = e_K(x)$ for all 2^{56} keys K, and construct a table of ordered pairs (y_K, K), sorted by their first coordinates. At a later time, when Oscar obtains the ciphertext y which is an encryption of plaintext x, he looks up the value y in the table, immediately obtaining the key K. Now the actual determination of the key requires only constant time, but we have a large memory requirement and a large precomputation time. (Note that this approach would yield no advantage in total computation time if only one key is to be found, since constructing the table takes at least as much time as an exhaustive search. The advantage occurs when several keys are to be found over a period of time, since the same table can be used in each case.)

The time-memory trade-off combines a smaller computation time than exhaustive search with a smaller memory requirement than table look-up. The algorithm can be described in terms of two parameters m and t, which are positive integers. The algorithm requires a *reduction function* R which reduces a bitstring of length 64 to one of length 56. (R might just discard eight of the 64 bits, for example.) Let x be a fixed plaintext string of length 64. Define the function $g(K_0) = R(e_{K_0}(x))$ for a bitstring K_0 of length 56. Note that g is a function that maps 56 bits to 56 bits.

In the pre-processing stage, Oscar chooses m random bitstrings of length 56, denoted $X(i,0)$, $1 \leq i \leq m$. Oscar computes $X(i,j)$ for $1 \leq j \leq t$ according to the recurrence relation $X(i,j) = g(X(i,j-1))$, $1 \leq i \leq m$, $1 \leq j \leq t$, as indicated in Figure 3.6.

Then Oscar constructs a table of ordered pairs $T = (X(i,t), X(i,0))$, sorted by their first coordinate (i.e., only the first and last columns of X are stored).

At a later time, Oscar obtains a ciphertext y which is an encryption of the chosen plaintext x (as before). He again wants to determine K. He is going to determine if K is in the first t columns of the array X, but he will do this by looking only at the table T.

FIGURE 3.7
DES time-memory trade-off

---

1.   compute $y_1 = R(y)$
2.   **for** $j = 1$ **to** t **do**
3.   **if** $y_j = X(i, t)$ for some i **then**
4.   compute $X(i, t - j)$ from $X(i, 0)$ by iterating the g
          function $t - j$ times
5.   **if** $y = e_{X(i,t-j)}(x)$ **then**
              set $K = X(i, t - j)$ and **QUIT**
7.   compute $y_{j+1} = g(y_j)$

Suppose that $K = X(i, t - j)$ for some j, $1 \leq j \leq t$ (i.e., suppose that K is in the first t columns of X). Then it is clear that $g^j(K) = X(i, t)$, where g^j denotes the function obtained by iterating g, j times. Now, observe that

$$g^j(K) = g^{j-1}(g(K))$$
$$= g^{j-1}(R(e_K(x)))$$
$$= g^{j-1}(R(y)).$$

Suppose we compute y_j, $1 \leq j \leq t$, from the recurrence relation

$$y_j = \begin{cases} R(y) & \text{if } j = 1 \\ g(y_{j-1}) & \text{if } 2 \leq j \leq t. \end{cases}$$

Then it follows that $y_j = X(i, t)$ if $K = X(i, t - j)$. However, note that $y_j = X(i, t)$ is not sufficient to ensure that $K = X(i, t - j)$. This is because the reduction function R is not an injection: The domain of R has cardinality 2^{64} and the range of R has cardinality 2^{56}, so, on average, there are $2^8 = 256$ pre-images of any given bitstring of length 56. So we need to check whether $y = e_{X(i,t-j)}(x)$, to see if $X(i, t - j)$ is indeed the key. We did not store the value $X(i, t-j)$, but we can easily re-compute it from $X(i, 0)$ by iterating the g function $t - j$ times.

Oscar proceeds according to the algorithm presented in Figure 3.7.

By analyzing the probability of success for the algorithm, it can be shown that if $mt^2 \approx N = 2^{56}$, then the probability that $K = X(i, t-j)$ for some i, j is about $0.8mt/N$. The factor 0.8 accounts for the fact that the numbers $X(i, t)$ may not all be distinct. It is suggested that one should take $m \approx t \approx N^{1/3}$ and construct about $N^{1/3}$ tables, each using a different reduction function R. If this is done, the memory requirement is $112 \times N^{2/3}$ bits (since we need to store $2 \times N^{2/3}$ integers, each of which has 56 bits). The precomputation time is easily seen to be $O(N)$.

The running time is a bit more dificult to analyze. First, note that step 3 can be implemented to run in (expected) constant time (using hash coding) or (worst-case) time $O(\log m)$ using a binary search. If step 3 is never satisfied (i.e., the search fails), then the running time is $O(N^{2/3})$. A more detailed analysis shows that even when the running time of steps 4 and 5 is taken into account, the expected running time increases by only a constant factor.

3.6 Differential Cryptanalysis

One very well-known attack on **DES** is the method of "differential cryptanalysis" introduced by Biham and Shamir. This is a chosen-plaintext attack. Although it does not provide a practical method of breaking the usual 16-round **DES**, it does succeed in breaking **DES** if the number of rounds of encryption is reduced. For instance, 8-round **DES** can be broken in only a couple of minutes on a small personal computer.

We will now describe the basic ideas used in this technique. For the purposes of this attack, we can ignore the initial permutation IP and its inverse (it has no effect on cryptanalysis). As mentioned above, we consider **DES** restricted to n rounds, for various values of $n \leq 16$. So, in this setting, we will regard $L_0 R_0$ as the plaintext, and $L_n R_n$ as the ciphertext, in an n-round **DES**. (Note also that we are not inverting $L_n R_n$.)

Differential cryptanalysis involves comparing the x-or (exclusive-or) of two plaintexts to the x-or of the corresponding two ciphertexts. In general, we will be looking at two plaintexts $L_0 R_0$ and $L_0^* R_0^*$ with a specified x-or value $L_0' R_0' = L_0 R_0 \oplus L_0^* R_0^*$. Throughout this discussion, we will use prime markings ($'$) to indicate the x-or of two bitstrings.

DEFINITION 3.1 *Let S_j be a particular S-box ($1 \leq j \leq 8$). Consider an (ordered) pair of bitstrings of length six, say (B_j, B_j^*). We say that the input x-or (of S_j) is $B_j \oplus B_j^*$ and the output x-or (of S_j) is $S_j(B_j) \oplus S_j(B_j^*)$.*

Note that an input x-or is a bitstring of length six and an output x-or is a bitstring of length four.

DEFINITION 3.2 *For any $B_j' \in (\mathbb{Z}_2)^6$, define the set $\Delta(B_j')$ to consist of the ordered pairs (B_j, B_j^*) having input x-or B_j'.*

It is easy to see that any set $\Delta(B_j')$ contains $2^6 = 64$ pairs, and that

$$\Delta(B_j') = \{(B_j, B_j \oplus B_j') : B_j \in (\mathbb{Z}_2)^6\}.$$

For each pair in $\Delta(B_j')$, we can compute the output x-or of S_j and tabulate the resulting distribution. There are 64 output x-ors, which are distributed among $2^4 = 16$ possible values. The non-uniformity of these distributions will be the basis for the attack.

Example 3.1
Suppose we consider the first S-box, S_1, and the input x-or 110100. Then

$$\Delta(110100) = \{(000000, 110100), (000001, 110101), \ldots, (111111, 001011)\}.$$

For each ordered pair in the set $\Delta(110100)$, we compute output x-or of S_1. For example, $S_1(000000) = E_{16} = 1110$ and $S_1(110100) = 9_{16} = 1001$, so the output x-or for the pair $(000000, 110100)$ is 0111.

If this is done for all 64 pairs in $\Delta(110100)$, then the following distribution of output x-ors is obtained:

0000	0001	0010	0011	0100	0101	0110	0111
0	8	16	6	2	0	0	12

1000	1001	1010	1011	1100	1101	1110	1111
6	0	0	0	0	8	0	6

$\square$

In Example 3.1, only eight of the 16 possible output x-ors actually occur. This particular example has a very non-uniform distribution. In general, if we fix an S-box S_j and an input x-or B_j', then on average, it turns out that about $75 - 80\%$ of the possible output x-ors actually occur.

It will be convenient to have some notation to describe these distributions and how they arise, so we make the following definitions.

DEFINITION 3.3   *For $1 \leq j \leq 8$, and for bitstrings B_j' of length six and C_j' of length four, define*

$$IN_j(B_j', C_j') = \{B_j \in (\mathbb{Z}_2)^6 : S_j(B_j) \oplus S_j(B_j \oplus B_j') = C_j'\}$$

and

$$N_j(B_j', C_j') = |IN_j(B_j', C_j')|.$$

FIGURE 3.8
Possible inputs with input x-or 110100

output x-or	possible inputs
0000	
0001	$000011, 001111, 011110, 011111$ $101010, 101011, 110111, 111011$
0010	$000100, 000101, 001110, 010001$ $010010, 010100, 011010, 011011$ $100000, 100101, 010110, 101110$ $101111, 110000, 110001, 111010$
0011	$000001, 000010, 010101, 100001$ $110101, 110110$
0100	$010011, 100111$
0101	
0110	
0111	$000000, 001000, 001101, 010111$ $011000, 011101, 100011, 101001$ $101100, 110100, 111001, 111100$
1000	$001001, 001100, 011001, 101101$ $111000, 111101$
1001	
1010	
1011	
1100	
1101	$000110, 010000, 010110, 011100$ $100010, 100100, 101000, 110010$
1110	
1111	$000111, 001010, 001011, 110011$ $111110, 111111$

$N_j(B'_j, C'_j)$ counts the number of pairs with input x-or equal to B'_j which have output x-or equal to C'_j for the S-box S_j. The actual pairs having the specified input x-ors and giving rise to the specified output x-ors can be obtained from the set $IN_j(B'_j, C'_j)$. Observe that this set can be partitioned into $N_j(B'_j, C'_j)/2$ pairs, each of which has (input) x-or equal to B'_j.

Observe that the distribution tabulated in Example 3.1 consists of the values $N_1(110100, C'_1)$, $C'_1 \in (\mathbb{Z}_2)^4$. The sets $IN_1(110100, C'_1)$ are listed in Figure 3.8.

For each of the eight S-boxes, there are 64 possible input x-ors. Thus, there are 512 distributions which can be computed. These could easily be tabulated by computer.

Recall that the input to the S-boxes in round i is formed as $B = E \oplus J$, where $E = \mathrm{E}(R_{i-1})$ is the expansion of R_{i-1} and $J = K_i$ consists of the key bits for round i. Now, the input x-or (for all eight S-boxes) can be computed as follows:

$$B \oplus B^* = (E \oplus J) \oplus (E^* \oplus J)$$
$$= E \oplus E^*.$$

It is very important to observe that the input x-or does not depend on the key bits J. (However, the output x-or certainly does depend on these key bits.)

We will write each of B, E and J as the concatenation of eight 6-bit strings:

$$B = B_1 B_2 B_3 B_4 B_5 B_6 B_7 B_8$$
$$E = E_1 E_2 E_3 E_4 E_5 E_6 E_7 E_8$$
$$J = J_1 J_2 J_3 J_4 J_5 J_6 J_7 J_8,$$

and we write B^* and E^* in a similar way. Let us suppose for the moment that we know the values E_j and E_j^* for some j, $1 \le j \le 8$, and the value of the output x-or for S_j, $C_j' = S_j(B_j) \oplus S_j(B_j^*)$. Then it must be the case that

$$E_j \oplus J_j \in IN_j(E_j', C_j'),$$

where $E_j' = E_j \oplus E_j^*$.

Suppose we define a set $test_j$ as follows:

DEFINITION 3.4 *Suppose E_j and E_j^* are bitstrings of length six, and C_j' is a bitstring of length four. Define*

$$test_j(E_j, E_j^*, C_j') = \{B_j \oplus E_j : B_j \in IN_j(E_j', C_j')\},$$

where $E_j' = E_j \oplus E_j^*$.

That is, we take the x-or of E_j with every element of the set $IN_j(E_j', C_j')$.

The following result is an immediate consequence of the discussion above.

THEOREM 3.1
Suppose E_j and E_j^* are two inputs to the S-box S_j, and the output x-or for S_j is C_j'. Denote $E_j' = E_j \oplus E_j^*$. Then the key bits J_j occur in the set $test_j(E_j, E_j^*, C_j')$.

Observe that there will be exactly $N_j(E_j', C_j')$ bitstrings of length six in the set $test_j(E_j, E_j^*, C_j')$; the correct value of J_j must be one of these possibilities.

Example 3.2

Suppose $E_1 = 000001$, $E_1^* = 110101$ and $C_1' = 1101$. Since $N_1(110100, 1101) = 8$, there will be exactly eight bitstrings in the set $test_1(000001, 110101, 1101)$. From Figure 3.8, we see that

$$IN_1(110100, 1101) =$$

$$\{000110, 010000, 010110, 011100, 100010, 100100, 101000, 110010\}.$$

Hence,

$$test_1(000001, 110101, 1101) =$$

$$\{000111, 010001, 010111, 011101, 100011, 100101, 101001, 110011\}.$$

$\square$

If we have a second such triple E_1, E_1^*, C_1', then we can obtain a second set $test_1$ of possible values for the keybits in J_1. The true value of J_1 must be in the intersection of both sets. If we have several such triples, then we can quickly determine the key bits in J_1. One straightforward way to do this is to maintain an array of 64 counters, representing the 64 possibilities for the six key bits in J_1. A counter is incremented every time the corresponding key bits occur in a set $test_1$ for a particular triple. Given t triples, we hope to find a unique counter which has the value t; this will correspond to the true value of the keybits in J_1.

3.6.1 An Attack on a 3-round DES

Let's now see how the ideas of the previous section can be applied in a chosen plaintext attack of a 3-round **DES**. We will begin with a pair of plaintexts and corresponding ciphertexts: L_0R_0, $L_0^*R_0^*$, L_3R_3 and $L_3^*R_3^*$. We can express R_3 as follows:

$$R_3 = L_2 \oplus f(R_2, K_3)$$
$$= R_1 \oplus f(R_2, K_3)$$
$$= L_0 \oplus f(R_0, K_1) \oplus f(R_2, K_3).$$

R_3^* can be expressed in a similar way, and hence

$$R_3' = L_0' \oplus f(R_0, K_1) \oplus f(R_0^*, K_1) \oplus f(R_2, K_3) \oplus f(R_2^*, K_3).$$

Now, suppose we have chosen the plaintexts so that $R_0 = R_0^*$, i.e., so that

$$R_0' = 00\ldots0.$$

FIGURE 3.9
Differential attack on 3-round DES

Input: $L_0 R_0, L_0^* R_0^*, L_3 R_3$ and $L_3^* R_3^*$, where $R_0 = R_0^*$
1.   compute $C' = \mathrm{P}^{-1}(R_3' \oplus L_0')$
2.   compute $E = \mathrm{E}(L_3)$ and $E^* = \mathrm{E}(L_3^*)$
3.   **for** $j = 1$ **to** 8 **do**
         compute $test_j(E_j, E_j^*, C_j')$

Then $f(R_0, K_1) = f(R_0^*, K_1)$ and so

$$R_3' = L_0' \oplus f(R_2, K_3) \oplus f(R_2^*, K_3).$$

At this point, R_3' is known since it can be computed from the two ciphertexts, and L_0' is known since it can be computed from the two plaintexts. This means that we can compute $f(R_2, K_3) \oplus f(R_2^*, K_3)$ from the equation

$$f(R_2, K_3) \oplus f(R_2^*, K_3) = R_3' \oplus L_0'.$$

Now, $f(R_2, K_3) = \mathrm{P}(C)$ and $f(R_2^*, K_3) = \mathrm{P}(C^*)$, where C and C^*, respectively, denote the two outputs of the eight S-boxes (recall that P is a fixed, publicly known permutation). Hence,

$$\mathrm{P}(C) \oplus \mathrm{P}(C^*) = R_3' \oplus L_0',$$

and consequently

$$C' = C \oplus C^* = \mathrm{P}^{-1}(R_3' \oplus L_0'). \tag{3.1}$$

This is the output x-or for the eight S-boxes in round three.

Now, $R_2 = L_3$ and $R_2^* = L_3^*$ are also known (they are part of the ciphertexts). Hence, we can compute

$$E = \mathrm{E}(L_3) \tag{3.2}$$

and

$$E^* = \mathrm{E}(L_3^*) \tag{3.3}$$

using the publicly known expansion function E. These are the inputs to the S-boxes for round three. So, we now know E, E^*, and C' for the third round, and we can proceed, as in the previous section, to construct the sets $test_1, \ldots, test_8$ of possible values for the key bits in $J_1, \ldots, J_8$.

A pseudo-code description of this algorithm is given in Figure 3.9. The attack will use several such triples E, E^*, C'. We set up eight arrays of counters, and thereby determine the 48 bits in K_3, the key for the third round. The 56 bits in the key can then be computed by an exhaustive search of the $2^8 = 256$ possibilities for the remaining eight key bits.

Let's look at an example to illustrate.

Example 3.3
Suppose we have the following three pairs of plaintexts and ciphertexts, where the plaintexts have the specified x-ors, that are encrypted using the same key. We use a hexadecimal representation, for brevity:

plaintext	ciphertext
748502CD38451097	03C70306D8A09F10
3874756438451097	78560A0960E6D4CB
486911026ACDFF31	45FA285BE5ADC730
375BD31F6ACDFF31	134F7915AC253457
357418DA013FEC86	D8A31B2F28BBC5CF
12549847013FEC86	0F317AC2B23CB944

From the first pair, we compute the S-box inputs (for round 3) from Equations (3.2) and (3.3). They are:

$$E = 000000000111111000001110100000000110100000001100$$

$$E^* = 101111110000001010101100000000101010000001010010.$$

The S-box output x-or is calculated using Equation (3.1) to be:

$$C' = 100101100101110101011011101100111.$$

From the second pair, we compute the S-box inputs to be

$$E = 101000001011111111101000001010100000001011110110$$

$$E^* = 100010100110101001011110101111110010100010101010$$

and the S-box output x-or is

$$C' = 100111001001110000011111101010110.$$

From the third pair, the S-box inputs are

$$E = 111011110001010100000110100011110110100101011111$$

$$E^* = 000001011110100110100010101111110101011000000100$$

and the S-box output x-or is

$$C' = 110101010111010111011011100101011.$$

Next, we tabulate the values in the eight counter arrays for each of the three pairs. We illustrate the procedure with the counter array for J_1 from the first pair. In this pair, we have $E'_1 = 101111$ and $C'_1 = 1001$. The set

$$IN_1(101111, 1001) = \{000000, 000111, 101000, 101111\}.$$

Since $E_1 = 000000$, we have that

$$J_1 \in test_1(000000, 101111, 1001) = \{000000, 000111, 101000, 101111\}.$$

Hence, we increment the values $0, 7, 40$, and 47 in the counter array for J_1.

The final tabulations are now presented. If we think of a bit-string of length six as being the binary representation of an integer between 0 and 63, then the 64 values correspond to the counts of $0, 1, \dots, 63$. The counter arrays are as follows:

J_1															
1	0	0	0	0	1	0	1	0	0	0	0	0	0	0	0
0	0	0	0	0	1	1	0	0	0	0	1	1	0	0	0
0	1	0	0	0	1	0	0	1	0	0	0	0	0	0	3
0	0	0	0	0	0	0	0	0	0	0	0	0	0	0	1

J_2															
0	0	0	1	0	3	0	0	1	0	0	1	0	0	0	0
0	1	0	0	0	2	0	0	0	0	0	0	1	0	0	0
0	0	0	0	0	1	0	0	1	0	1	0	0	0	1	0
0	0	1	1	0	0	0	0	1	0	1	0	2	0	0	0

J_3															
0	0	0	0	1	1	0	0	0	0	0	0	0	0	1	0
0	0	0	3	0	0	0	0	0	0	0	0	0	0	1	1
0	2	0	0	0	0	0	0	0	0	0	0	1	1	0	0
0	0	0	0	0	0	1	0	0	0	0	0	1	0	0	0

J_4															
3	1	0	0	0	0	0	0	0	0	2	2	0	0	0	0
0	0	0	0	1	1	0	0	0	0	0	0	1	0	1	1
1	1	1	0	1	0	0	0	0	1	1	1	0	0	1	0
0	0	0	0	1	1	0	0	0	0	0	0	0	0	2	1

J_5															
0	0	0	0	0	0	1	0	0	0	1	0	0	0	0	0
0	0	0	0	2	0	0	0	3	0	0	0	0	0	0	0
0	0	0	0	0	0	0	0	0	0	0	0	0	0	0	0
0	0	2	0	0	0	0	0	0	1	0	0	0	0	2	0

J_6															
1	0	0	1	1	0	0	3	0	0	0	0	1	0	0	1
0	0	0	0	1	1	0	0	0	0	0	0	0	0	0	0
0	0	0	0	1	1	0	1	0	0	0	0	0	0	0	0
1	0	0	1	1	0	1	1	0	0	0	0	0	0	0	0

J_7															
0	0	2	1	0	1	0	3	0	0	0	1	1	0	0	0
0	1	0	0	0	0	0	0	0	0	0	1	0	0	0	1
0	0	2	0	0	0	2	0	0	0	0	1	2	1	1	0
0	0	0	0	0	0	0	0	0	0	1	0	0	0	1	1

J_8															
0	0	0	0	0	0	0	0	0	0	0	0	0	0	0	0
0	0	0	0	0	0	0	0	0	0	0	0	0	0	0	0
0	0	0	0	0	0	0	0	1	0	1	0	0	1	0	1
0	3	0	0	0	0	1	0	0	0	0	0	0	0	0	0

In each of the eight counter arrays, there is a unique counter having the value 3. The positions of these counters determine the key bits in $J_1, \ldots, J_8$. These positions are (respectively): $47, 5, 19, 0, 24, 7, 7, 49$. Converting these integers to binary, we obtain $J_1, \ldots, J_8$:

$$J_1 = 101111$$

$$J_2 = 000101$$

$$J_3 = 010011$$

$$J_4 = 000000$$

$$J_5 = 011000$$

$$J_6 = 000111$$

$$J_7 = 000111$$

$$J_8 = 110001.$$

We can now construct 48 bits of the key, by looking at the key schedule for round 3. It follows that K has the form

```
0001101  0110001  01?01?0  1?00100
0101001  0000??0  111?11?  ?100011
```

where parity bits are omitted and "?" denotes an unknown key bit. The complete key (in hexadecimal, including parity bits), is:

$$1A624C89520DEC46.$$

□

3.6.2   An Attack on a 6-round DES

We now describe an extension of these ideas to a probabilistic attack on a 6-round **DES**. The idea is to carefully choose a pair of plaintexts with a specified x-or, and then to determine the probabilities of a specified sequence of x-ors through the rounds of encryption. We need to define an important concept now.

DEFINITION 3.5   *Let $n \geq 1$ be an integer. An n-round characteristic is a list of the form*

$$L_0', R_0', L_1', R_1', p_1, \ldots, L_n', R_n', p_n,$$

which satisfies the following properties:

1. $L_i' = R_{i-1}'$ *for* $1 \leq i \leq n$.
2. *Let* $1 \leq i \leq n$, *and let* L_{i-1}, R_{i-1} *and* L_{i-1}^*, R_{i-1}^* *be chosen such that* $L_{i-1} \oplus L_{i-1}^* = L_{i-1}'$ *and* $R_{i-1} \oplus R_{i-1}^* = R_{i-1}'$. *Suppose* L_i, R_i *and* L_i^*, R_i^* *are computed by applying one round of* **DES** *encryption. Then the probability that* $L_i \oplus L_i^* = L_i'$ *and* $R_i \oplus R_i^* = R_i'$ *is precisely* p_i. *(Note that this probability is computed over all possible 48-tuples* $J = J_1 \ldots J_8$.)

The probability of the characteristic is defined to be the product $p = p_1 \times \ldots \times p_n$.

REMARK   Suppose we choose L_0, R_0 and L_0^*, R_0^* so that $L_0 \oplus L_0^* = L_0'$ and $R_0 \oplus R_0^* = R_0'$ and we apply n rounds of **DES** encryption, obtaining $L_1, \ldots, L_n$ and $R_1, \ldots, R_n$. Then we cannot claim that the probability that $L_i \oplus L_i^* = L_i'$ and $R_i \oplus R_i^* = R_i'$ for all i ($1 \leq i \leq n$) is $p_1 \times \ldots \times p_n$. This is because the 48-tuples in the key schedule $K_1, \ldots, K_n$, are not mutually independent. (If these n 48-tuples were chosen independently at random, then the assertion would be true.) But we nevertheless expect $p_1 \times \ldots \times p_n$ to be a fairly accurate estimate of this probability.

We also need to recognize that the probabilities p_i in a characteristic are defined with respect to an arbitrary (but fixed) pair of plaintexts having a specified x-or, where the 48 key bits for one round of **DES** encryption vary over all 2^{48} possibilities. However, a cryptanalyst is attempting to determine a fixed (but unknown) key. He is going to choose plaintexts at random (such that they have specified x-ors), hoping that the probabilities that the x-ors during the n rounds of encryption agree with the x-ors specified in the characteristic are fairly close to $p_1, \ldots, p_n$, respectively.   ∎

As a simple example, we present in Figure 3.10 a 1-round characteristic which was the basis of the attack on the 3-round **DES** (as before, we use hexadecimal representations). We depict another 1-round characteristic in Figure 3.11.

Let's look at the characteristic in Figure 3.11 in more detail. When $f(R_0, K_1)$ and $f(R_0^*, K_1)$ are computed, the first step is to expand R_0 and R_0^*. The resulting

FIGURE 3.10
A 1-round characteristic

L_0'	$=$	anything	R_0'	$=$	00000000_{16}	
L_1'	$=$	00000000_{16}	R_1'	$=$	L_0'	$p = 1$

FIGURE 3.11
Another 1-round characteristic

L_0'	$=$	00000000_{16}	R_0'	$=$	60000000_{16}	
L_1'	$=$	60000000_{16}	R_1'	$=$	00808200_{16}	$p = 14/64$

x-or of the two expansions is

$$001100\ldots0.$$

So the input x-or to S_1 is 001100 and the input x-ors for the other seven S-boxes are all 000000. The output x-ors for S_2 through S_8 will all be 0000. The output x-or for S_1 will be 1110 with probability $14/64$ (since it can be computed that $N_1(001100, 1110) = 14$). So we obtain

$$C' = 11100000000000000000000000000000$$

with probability $14/64$. Applying P, we get

$$P(C) \oplus P(C^*) = 00000000100000001000001000000000,$$

which in hexadecimal is 00808200_{16}. When this is x-ored with L_0', we get the specified R_1' with probability $14/64$. Of course $L_1' = R_0'$ always.

The attack on the 6-round **DES** is based on the 3-round characteristic given in Figure 3.12. In the 6-round attack, we will start with $L_0 R_0$, $L_0^* R_0^*$, $L_6 R_6$ and $L_6^* R_6^*$, where we have chosen the plaintexts so that $L_0' = 40080000_{16}$ and

FIGURE 3.12
A 3-round characteristic

L_0'	$=$	40080000_{16}	R_0'	$=$	04000000_{16}	
L_1'	$=$	04000000_{16}	R_1'	$=$	00000000_{16}	$p = 1/4$
L_2'	$=$	00000000_{16}	R_2'	$=$	04000000_{16}	$p = 1$
L_3'	$=$	04000000_{16}	R_3'	$=$	40080000_{16}	$p = 1/4$

$R_0' = 04000000_{16}$. We can express R_6 as follows:

$$
\begin{aligned}
R_6 &= L_5 \oplus f(R_5, K_6) \\
&= R_4 \oplus f(R_5, K_6) \\
&= L_3 \oplus f(R_3, K_4) \oplus f(R_5, K_6).
\end{aligned}
$$

R_6^* can be expressed in a similar way, and hence we get

$$
R_6' = L_3' \oplus f(R_3, K_4) \oplus f(R_3^*, K_4) \oplus f(R_5, K_6) \oplus f(R_5^*, K_6). \tag{3.4}
$$

(Note the similarity with the 3-round attack.)

R_6' is known. From the characteristic, we estimate that $L_3' = 04000000_{16}$ and $R_3' = 40080000_{16}$ with probability $1/16$. If this is in fact the case, then the input x-or for the S-boxes in round 4 can be computed by the expansion function to be:

$$
0010000000000000001010000 \ldots 0.
$$

The input x-ors for S_2, S_5, S_6, S_7 and S_8 are all 000000, and hence the output x-ors are 0000 for these five S-boxes in round 4. This means that we can compute the output x-ors of these five S-boxes in round 6 from Equation (3.4). So, suppose we compute

$$
C_1' C_2' C_3' C_4' C_5' C_6' C_7' C_8' = \mathrm{P}^{-1}(R_6' \oplus 04000000_{16})
$$

where each C_i is a bitstring of length four. Then with probability $1/16$, it will be the case that C_2', C_5', C_6', C_7' and C_8' are respectively the output x-ors of S_2, S_5, S_6, S_7 and S_8 in round 6. The inputs to these S-boxes in round 6 can be computed to be E_2, E_5, E_6, E_7 and E_8, and E_2^*, E_5^*, E_6^*, E_7^* and E_8^*, where

$$
E_1 E_2 E_3 E_4 E_5 E_6 E_7 E_8 = \mathrm{E}(R_5) = \mathrm{E}(L_6)
$$

and

$$
E_1^* E_2^* E_3^* E_4^* E_5^* E_6^* E_7^* E_8^* = \mathrm{E}(R_5^*) = \mathrm{E}(L_6^*)
$$

can be computed from the ciphertexts, as indicated in Figure 3.13.

We would like to determine the 30 key bits in J_2, J_5, J_6, J_7 and J_8 as we did in the 3-round attack. The problem is that the hypothesized output x-or for round 6 is correct only with probability $1/16$. So $15/16$ of the time we will obtain random garbage rather than possible key bits. We somehow need to be able to determine the correct key from the given data, $15/16$ of which is incorrect. This might not seem very promising, but fortunately our prospects are not as bleak as they initially appear.

DEFINITION 3.6   *Suppose $L_0 \oplus L_0^* = L_0'$ and $R_0 \oplus R_0^* = R_0'$. We say that the pair of plaintexts $L_0 R_0$ and $L_0^* R_0^*$ is right pair with respect to a characteristic if $L_i \oplus L_i^* = L_i'$ and $R_i \oplus R_i^* = R_i'$ for all i, $1 \leq i \leq n$. The pair is defined to be a wrong pair, otherwise.*

FIGURE 3.13
Differential attack on 6-round DES

> Input: $L_0 R_0$, $L_0^* R_0^*$, $L_6 R_6$ and $L_6^* R_6^*$, where $L_0' = 40080000_{16}$ and $R_0' = 04000000_{16}$
>
> 1. compute $C' = P^{-1}(R_6' \oplus 04000000_{16})$
> 2. compute $E = E(L_6)$ and $E^* = E(L_6^*)$
> 3. **for** $j \in \{2, 5, 6, 7, 8\}$ **do**
>   compute $test_j(E_j, E_j^*, C_j')$.

We expect that about $1/16$ of our pairs are right pairs and the rest are wrong pairs with respect to our 3-round characteristic.

Our strategy is to compute E_j, E_j^*, and C_j', as described above, and then to determine $test_j(E_j, E_j^*, C_j')$, for $j = 2, 5, 6, 7, 8$. If we start with a right pair, then the correct key bits for each J_j will be included in the set $test_j$. If the pair is a wrong pair, then the value of C_j' will be incorrect, and it seems reasonable to hypothesize that each set $test_j$ will be essentially random.

We can often identify a wrong pair by this method: If $|test_j| = 0$, for any $j \in \{2, 5, 6, 7, 8\}$, then we necessarily have a wrong pair. Now, given a wrong pair, we might expect that the probability that $|test_j| = 0$ for a particular j is approximately $1/5$. This is a reasonable assumption since $N_j(E_j', C_j') = |test_j|$ and, as mentioned earlier, the probability that $N_j(E_j', C_j') = 0$ is approximately $1/5$. The probability that all five $test_j$'s have positive cardinality is estimated to be $.8^5 \approx .33$, so the probability that at least one $test_j$ has zero cardinality is about $.67$. So we expect to eliminate about $2/3$ of the wrong pairs by this simple observation, which we call the *filtering operation*. The proportion of right pairs that remain after filtering is approximately

$$\frac{\frac{1}{16}}{\frac{1}{16} + \frac{15}{16} \times \frac{1}{3}} = \frac{1}{6}.$$

Example 3.4
Suppose we have the following plaintext-ciphertext pair:

plaintext	ciphertext
86FA1C2B1F51D3BE	1E23ED7F2F553971
C6F21C2B1B51D3BE	296DE2B687AC6340

Observe that $L_0' = 40080000_{16}$ and $R_0' = 04000000_{16}$. The S-box inputs and outputs for round 6 are computed to be the following:

j	E_j	E_j^*	C_j'
2	111100	010010	1101
5	111101	111100	0001
6	011010	000101	0010
7	101111	010110	1100
8	111110	101100	1101

Then, the sets $test_j$ are as follows:

j	$test_j$
2	$14, 15, 26, 30, 32, 33, 48, 52$
5	
6	$7, 24, 36, 41, 54, 59$
7	
8	$34, 35, 48, 49$

We see that both $test_5$ and $test_7$ are empty sets, so this pair is a wrong pair and is discarded by the filtering operation. $\square$

Now suppose that we have a pair such that $|test_j| > 0$ for $j = 2, 5, 6, 7, 8$, so that it survives the filtering operation. (Of course, we do not know if the pair is a right pair or a wrong pair.) We say that the bitstring $J_2 J_5 J_6 J_7 J_8$ of length 30 is *suggested* by the pair if $J_j \in test_j$ for $j = 2, 5, 6, 7, 8$. The number of suggested bitstrings is

$$\prod_{j \in \{2,5,6,7,8\}} |test_j|.$$

It is not unusual for the number of suggested bitstrings to be quite large (for example, greater than 80000).

Suppose we were to tabulate all the suggested bitstrings obtained from the N pairs that were not discarded by the filtering operation. For every right pair, the correct bitstring $J_2 J_5 J_6 J_7 J_8$ will be a suggested bitstring. This correct bitstring will be counted about $3N/16$ times. Incorrect bitstrings should occur much less often, since they will occur essentially at random and there are 2^{30} possibilities (a very large number).

It would get extremely unwieldy to tabulate all the suggested bitstrings, so we use an algorithm that requires less space and time. We can encode any $test_j$ as a vector T_j of length 64, where the ith coordinate of T_j is set to 1 (for $0 \le i \le 63$) if the bitstring of length six that is the binary representation of i is in the set $test_j$; and the ith coordinate is set to 0 otherwise (this is essentially the same as the counter array representation that we used in the 3-round attack).

For each remaining pair, construct these vectors as described above, and name them T_j^i, $j = 2, 5, 6, 7, 8$, $1 \leq i \leq N$. For $I \subseteq \{1, \ldots, N\}$, we say that I is *allowable* if for each $j \in \{2, 5, 6, 7, 8\}$, there is at least one coordinate equal to $|I|$ in the vector

$$\sum_{i \in I} T_j^i.$$

If the ith pair is a right pair for every $i \in I$, then the set I is allowable. Hence, we expect there to be an allowable set of size (approximately) $3N/16$, which we hope will suggest the correct key bits and no other. It is a simple matter to construct all the allowable sets I by means of a recursive algorithm.

Example 3.5

We did some computer runs to test this approach. A random sample of 120 pairs of plaintexts with the specified x-ors was generated, and these were encrypted using the same (random) key. We present the 120 pairs of ciphertexts and corresponding plaintexts in hexadecimal form in Table 3.1.

When we compute the allowable sets, we obtain n_i allowable sets of cardinality i, for the following values:

i	n_i
2	111
3	180
4	231
5	255
6	210
7	120
8	45
9	10
10	1

The unique allowable set of size 10 is

$$\{24, 29, 30, 48, 50, 52, 55, 83, 92, 118\}.$$

In fact, it does arise from the 10 right pairs. This allowable set suggests the correct key bits for J_2, J_5, J_6, J_7 and J_8 and no others. They are as follows:

$$J_2 = 011001$$

$$J_5 = 110000$$

$$J_6 = 001001$$

$$J_7 = 101010$$

$$J_8 = 100011$$

FIGURE 3.14
Another 3-round characteristic

L'_0	$=$	00200008_{16}	R'_0	$=$	00000400_{16}	
L'_1	$=$	00000400_{16}	R'_1	$=$	00000000_{16}	$p = 1/4$
L'_2	$=$	00000000_{16}	R'_2	$=$	00000400_{16}	$p = 1$
L'_3	$=$	00000400_{16}	R'_3	$=$	00200008_{16}	$p = 1/4$

Note that all the allowable sets of cardinality at least 6, and all but three of the allowable sets of cardinality 5, arise from right pairs, since $\binom{10}{5} = 252$ and $\binom{10}{i} = n_i$ for $6 \le i \le 10$.

This method yields 30 of the 56 key bits. By means of a different 3-round characteristic, presented in Figure 3.14, it is possible to compute 12 further key bits, namely those in J_1 and J_4. Now only 14 key bits remain unknown. Since $2^{14} = 16384$ is quite small, an exhaustive search can be used to determine the remaining 14 key bits.

The entire key (in hexadecimal, including parity-check bits) is:

$$34E9F71A20756231.$$

As mentioned above, the 120 pairs are given in Table 3.1. In the second column, a * denotes that a pair is a right pair, while a ** denotes that the pair is an identifiable wrong pair and is discarded by the filtering operation. Of the 120 pairs, 73 are identified as being wrong pairs by the filtering process, so 47 pairs remain as "possible" right pairs. $\quad\Box$

3.6.3 Other examples of Differential Cryptanalysis

Differential cryptanalysis techniques can be used to attack **DES** with more than six rounds. An 8-round **DES** requires 2^{14} chosen plaintexts, and 10-, 12-, 14- and 16-round **DES**s can be broken with 2^{24}, 2^{31}, 2^{39} and 2^{47} chosen plaintexts, respectively. The attacks on more than 10 rounds are probably not practical at this time.

Several substitution-permutation product ciphers other than **DES** are also susceptible (to varying degrees) to differential cryptanalysis. These cryptosystems include several substitution-premutation cryptosystems that have been proposed in recent years, such as FEAL, REDOC-II, and LOKI.

TABLE 3.1
Cryptanalysis of 6-round DES

pair	right pair?	plaintext	ciphertext
1	**	86FA1C2B1F51D3BE	1E23ED7F2F553971
		C6F21C2B1B51D3BE	296DE2B687AC6340
2	**	EDC439EC935E1ACD	0F847EFE90466588
		ADCC39EC975E1ACD	93E84839F374440B
3	**	9468A0BE00166155	3D6A906A6566D0BF
		D460A0BE04166155	3BC3B236398379E1
4	**	D4FF2B18A5A8AAC8	26B14738C2556BA4
		94F72B18A1A8AAC8	15753FDE86575A8F
5		09D0F2CF277AF54F	15751F4F11308114
		49D8F2CF237AF54F	6046A7C863F066AF
6		CBC7157240D415DF	7FCDC300FB9698E5
		8BCF157244D415DF	522185DD7E47D43A
7		0D4A1E84890981C1	E7C0B01E32557558
		4D421E848D0981C1	912C6341A69DF295
8	**	6CE6B2A9B8194835	75D52E028A5C48A3
		2CEEB2A9BC194835	6C88603B48E5A8CE
9	**	799F63C3C9322C1A	A6DA322B8F2444B5
		399763C3CD322C1A	6634AA9DF18307F4
10	**	1B36645E381EDF48	1F91E295D559091B
		5B3E645E3C1EDF48	D094FC12C02C17CA
11		85CA13F50B4ADBB9	ED108EE7397DDE0A
		C5C213F50F4ADBB9	3F405F4A3E254714
12	**	7963A8EFD15BC4A1	8C714399715A33BA
		396BA8EFD55BC4A1	C344C73CC97E4AC4
13		7BCFF7BCA455E65E	475A2D0459BCCE62
		3BC7F7BCA055E65E	8E94334AEF359EF8
14		0C505CEDB499218C	D3C66239E89CC076
		4C585CEDB099218C	9A316E801EE18EB1
15		6C5EA056CDC91A14	BC7EBA159BCA94E6
		2C56A056C9C91A14	67DB935C21FF1A8D
16	**	6622A441A0D32415	35F8616FEBA62883
		262AA441A4D32415	4313E1925F5B64BC
17		C0333C994AFF1C99	D46A4CF1C0221B11
		803B3C994EFF1C99	D22B42DB150E2CE8
18		9E7B2974F00E1A6E	172D286D9606E6FE
		DE732974F40E1A6E	2217A91F8C427D27
19	**	CF592897BFD70C7E	FB892B59E7DCE7EC
		8F512897BBD70C7E	C328B765E1CC6653
20		E976CF19124A9FA1	905BF24188509FA6
		A97ECF19164A9FA1	9ADDBA0C23DD724F
21	**	5C09696E7363675D	92D60E5C71801A99
		1C01696E7763675D	DD90908A4FE8168F
22	**	A8145AB3C1B2C7DE	F68FC9F80564847B
		E81C5AB3C5B2C7DE	51C041B5711B8132
23		47DF6A0BB1787159	52E36C4CA22EA5A2
		07D76A0BB5787159	373EAFD503F68DE4
24	*	7CE65464329B4E6D	832A9D7032015D9F
		3CEE5464369B4E6D	85E2CE665571E99C

pair	right pair?	plaintext	ciphertext
25	**	421FB6AD95791BA7	D1E730BA1DB565E7
		0217B6AD91791BA7	188E61735FA4F3CE
26	**	C58E9A361368FFD6	795EB9D30CAE6879
		85869A361768FFD6	26D37AC4867ACC61
27	**	DD86B6C74C8EA4E2	CC3B6915C9A348DF
		9D8EB6C7488EA4E2	104C2394555645F0
28	**	43DB9D2F483CA585	E3E4DA503D1B9396
		03D39D2F4C3CA585	4EA02C0061332443
29	*	855A309F96FEA5EA	85AD6E9E352AFAFA
		C552309F92FEA5EA	929D22370ACAB80D
30	*	AB3CA25B02BD18C8	0F7D768E9203F786
		EB34A25B06BD18C8	A1313BC26A99D353
31	**	A9F7A6F4A7C00E06	F26B385E6BA057FD
		E9FFA6F4A3C00E06	203D8384F8F54D19
32	**	688B9ACD856D1312	C41D99C107B4EF76
		28839ACD816D1312	6CC817CA025A7DAC
33	**	76BF0621C03D4CD9	BBE1F95AFC1E052A
		36B70621C43D4CD9	561F4801F2EB0C63
34	**	014CF8D1F981B8EE	D27091C4314CBFE8
		4144F8D1FD81B8EE	B7976D6A80E3DB61
35	**	487D66EDE0405F8C	8136325C0AEB84CE
		087566EDE4405F8C	8C638BC4495B69A0
36	**	DDCA47093A362521	51040CF16B600FAA
		9DC247093E362521	7FC75515AC3CAAF9
37	**	45A9D34A3996F6D9	F2004B854AE6C46C
		05A1D34A3D96F6D9	546825016B03D193
38	**	295D2FBFB00875EA	A309DF027E69C265
		69552FBFB40875EA	4F633FFB95A0C11E
39		964C8B98D590D524	1FF1D0271D6F6C18
		D6448B98D190D524	8CF2D8D401EBFC0F
40		60383D2BAF0836BC	10A82D55FC480640
		20303D2BAB0836BC	602346173581EF79
41	**	5CF8D539A22A1CAD	92685D806FBE8738
		1CF0D539A62A1CAD	17006DAB2D28081C
42		F95167CAB6565609	C52E2EB27446054E
		B95967CAB2565609	0C219F686840E57A
43		49F1C83615874122	2680C8ECDF5E51CD
		09F9C83611874122	5022A7B69B4E75EF
44	**	ACB2EC1941B03765	D6B593460098DEC5
		ECBAEC1945B03765	D3190A0200FC6B9B
45		CCCC129D5CB55EC0	3AD22B7EF59E0D5E
		8CC4129D58B55EC0	A48C92CBEC17E430
46	**	917FF8E2EE6B78D5	EF847E058DB71724
		D177F8E2EA6B78D5	F243F0554A00E4C5
47	**	51DBCF028E96DE00	574897CA1EE73885
		11D3CF028A96DE00	9F0FD0A5B2C2B5FD
48	*	2094942E093463CE	59F6A018C6A0D820
		609C942E0D3463CE	799FE001432346C0

pair	right pair?	plaintext	ciphertext
49	**	50FB0723D7CD1081	16AF758395EA3A7D
		10F30723D3CD1081	CDCB23392D144BED
50	*	740815A4F6CDCABB	4A84D2ED4D9351AB
		340015A4F2CDCABB	5923D04CE94D6111
51	**	EDA46A1AE93735DC	0B302A51B7E5476A
		ADAC6A1AED3735DC	5F817F0ABC770E75
52	*	08BC39B766B2C128	DFB5F3F500BC0100
		48B439B762B2C128	B7B9FED8AC93EBFA
53	**	A74E29BBA98F2312	A2B352B7F922E8DA
		E74629BBAD8F2312	D6BC4B89CED2DEAC
54	**	D6F50D31EE4E68AB	4D464847065C0938
		96FD0D31EA4E68AB	7554D87AEDCE5634
55	*	06191AA594891CF5	649C1D084F920F9E
		46111AA590891CF5	BE12A10384365E19
56		5EA7EFD557946962	15E664293F4D77EE
		1EAFEFD553946962	E23396A758DC9CE6
57	**	41FB7704781CC88A	8ABD385C441FD6CE
		01F377047C1CC88A	06DE8D55777AB65C
58	**	9689B9123F7C5431	E1E63120742099BB
		D681B9123B7C5431	1AF88A2CF6649A4A
59		6F25032B4A309BFE	48FE50DE774288D7
		2F2D032B4E309BFE	47950691260D5E10
60	**	D8C4B02D8E8BF1E9	F34D565E6AE85683
		98CCB02D8A8BF1E9	A4D2DB548622A8E8
61	**	F663E8CCEE86805B	51BD62C9D5D0F0BB
		B66BE8CCEA86805B	D2ABB03CF9D26C0A
62	**	428B29BFDFA838DB	006D62A65761089F
		028329BFDBA838DB	9FD73EF6124B0C11
63	**	04BE2D22D81EDC66	26D99536D99B5707
		44B62D22DC1EDC66	94144EBDA0CDEB55
64	**	667B779123A3EF80	5D09CBF2CE7E5A69
		2673779127A3EF80	5EFF8BFCA7BAA152
65	**	BC86D401D6572438	E05572AAA5F6C377
		FC8ED401D2572438	3C670BC455144F61
66	**	6FE5E9547659E401	2C465BF6F52F864C
		2FEDE9547259E401	B71D106444F95F31
67	**	27D3BAC6453BE3DE	8F160E29000461CD
		67DBBAC6413BE3DE	2A6660F46487F885
68	**	1D864E7642A7023A	65F91EEBFD8A9C05
		5D8E4E7646A7023A	84761791B3C36661
69	**	5256CA6894707CBA	91527F9349ABCF15
		125ECA6890707CBA	30F28F06A7B0A35A
70	**	C05383B8EFCD2BD7	710B6EC61BF63E9C
		805B83B8EBCD2BD7	53AC029D8E0179D5
71		50EB21CA13F9A96E	26D95BA4DE4C85CF
		10E321CA17F9A96E	8F01A90F638AFFF6
72	**	60EB1229ACD90EDC	3890EE8567782F96
		20E31229A8D90EDC	EE404DF7BE537589

pair	right pair?	plaintext	ciphertext
73		8E9A17D17B173B99	885C3933627EDEF0
		CE9217D17F173B99	B7ABB6DF5835E962
74		6EC5CD0802C98817	A985ADFB1FEE013C
		2ECDCD0806C98817	0428DE024B7E4604
75	**	1E81712FF1145C06	417E667A99B3CFA5
		5E89712FF5145C06	5C24AA056EB1ADBA
76	**	DF3C5C13311AEC7C	BF01675096F1C48A
		9F345C13351AEC7C	243D99BCE12DB864
77	**	7C34472994127C2D	713915DA311A7CF4
		3C3C472990127C2D	E9733D11D787E20B
78	**	37304DABA75EAFB3	EFB5C37FA0238ADF
		77384DABA35EAFB3	A728F7407AF958B3
79		D03A16E4C2D8B54B	423FC0AC24CEFEDD
		903216E4C6D8B54B	047D8595DB4D372E
80	**	8CED882B5D91832E	0006E2DE3AF5C2B5
		CCE5882B5991832E	00F6AA9ED614001B
81	**	1BB0E6C79EFBEC41	E9AED4363915775A
		5BB8E6C79AFBEC41	655BC48F1FFB5165
82		D41B8346DA9E2252	34F5E0BCC5B042EA
		94138346DE9E2252	702D2C48CDBE5173
83	*	02A9D0A0A91F6304	E2F1C10E59AF07C5
		42A1D0A0AD1F6304	BDEE6AA00F25F840
84	**	841B3E27C8F0A561	2B288E554D712C92
		C4133E27CCF0A561	FF8609C9E7301162
85	**	CDF0A8D6EE909185	5D661834D1C76324
		8DF8A8D6EA909185	22034D57D21FFB56
86	**	4C31AC854F44EA34	BD016309AEDB9BB1
		0C39AC854B44EA34	C72EEDC4FA1D9312
87		DB3FC0703C972930	296ABCFBF01DF991
		9B37C07038972930	CA4700686F9F83A2
88		E4B362BFD6A7CFD1	20FDAF335F25B1DA
		A4BB62BFD2A7CFD1	008C24D75E14ACBD
89		F234232A0E0A4A28	90CFD699F2DEC5BD
		B23C232A0A0A4A28	2918D3DE0C1B689C
90	**	71265345A5874004	3052CE3CE88710AE
		312E5345A1874004	38F0FC685DF30564
91	**	3E6364548C857110	0E8581E42C9FEC6F
		7E6B645488857110	4DD1751861EC5529
92	*	464FBEDBD78900A7	90F5F9ADEDED627A
		0647BEDBD38900A7	2EF4C540425E339B
93	**	373B75F847480BB0	5408B964F8442D16
		773375F843480BB0	805287D52599E9F0
94	**	D714E87810DE97AC	4EC4D623108FA909
		971CE87814DE97AC	0AA0725CED10D6A3
95		B9B5932EF54B2C60	4B438B3CCF36DEC9
		F9BD932EF14B2C60	054C6A337709280D
96	**	2F283C38D2E4E1DD	83515FB6DFEA90B8
		6F203C38D6E4E1DD	09BCC4FF38C78C23

pair	right pair?	plaintext	ciphertext
97	**	1EB8ADAA43BBD575	21A1E04813616E42
		5EB0ADAA47BBD575	D044BA3F25DFD02A
98	**	3164AA5454D9F991	9382C6C1883F1038
		716CAA5450D9F991	5CDFED4FF2117DEC
99		D78C1C5C6F2243D2	1CCEB091E030E6A6
		97841C5C6B2243D2	4DA2CD67CC449B21
100		BBE212A7D3CE3D14	2917C207B4D93E0D
		FBEA12A7D7CE3D14	A01D50E5A2B902D8
101	**	104917795E98D0FB	40916A71385C2803
		504117795A98D0FB	413FD26EF671F46D
102	**	4DDA114D6EFEEEB4	2E2C65E1D5CBAC31
		0DD2114D6AFEEEB4	A16FF03BC0913ED6
103		E0BED7B285BF0A77	5D9EFEFF0AD10490
		A0B6D7B281BF0A77	4C6CA1FAC36A8E5B
104	**	0AE1555FA1716214	378400BCED39EB81
		4AE9555FA5716214	A1E0C758BD8912C2
105	**	4657C26790FCB354	588BA079B2E7ED20
		065FC26794FCB354	DA90827AEED7A41F
106	**	32BD719B0DC1B091	F3477C7552BCB05D
		72B5719B09C1B091	EFF444449D66BE9E
107	**	0992F8C8C73A9BFE	9F3FFD0F158295F6
		499AF8C8C33A9BFE	C138358DCECC8FC7
108		02C3F061A237BBEB	AC28B0307127EA7C
		42CBF061A637BBEB	3FF1DAED9E0FCBC5
109	**	80E529E69EDE6827	1DF1DB7B66BA1AF1
		C0ED29E69ADE6827	15700151A5804549
110		B55E84630067B8D5	88321611FF9DA421
		F55684630467B8D5	90649D7EACF91F9A
111		2749C2EBC603BFF2	A62B23A7348E2C3A
		6741C2EBC203BFF2	EB760A09C7FF5153
112	**	C4C5E14D4C5D9FF5	ABC2312FBFD94DF5
		84CDE14D485D9FF5	D2BB5954E5062D53
113	**	1566BA21F2647E18	A247ED988457CB78
		556EBA21F6647E18	5E99F231005F5249
114	**	2D093D426D922F92	5DF62030B9F23AE9
		6D013D4269922F92	5D92DA1FA3D07BA1
115		004518468E0C96C3	F28D85FF7E84F38F
		404D18468A0C96C3	52541B0443053C57
116	**	437B70A98AE03344	04B3FBF9823B4CF7
		037370A98EE03344	14EBEC79DAD3093E
117		2D01F1073D3E375B	F10B3E1EE356226C
		6D09F107393E375B	6FF26DA5E3525B62
118	*	66573DD7E0D7F110	F2F26204C29FE51E
		265F3DD7E4D7F110	083A4ECE57E429AC
119		0846DB9538155201	F120D0D2AE788057
		484EDB953C155201	00CC914A33034782
120		ABB34FC195C820D1	5F17AE066B50FC81
		EBBB4FC191C820D1	2858DD63A2FA4B53

3.7 Notes and References

A nice article on the history **DES** is by Smid and Branstad [SB92]. Federal Information Processing Standards (FIPS) publications concerning **DES** include the following: description of **DES** [NBS77]; implementing and using **DES** [NBS81]; modes of operation of **DES** [NBS80]; and authentication using **DES** [NBS85].

Some properties of the S-boxes are studied by Brickell, Moore, and Purtill [BMP87].

The DEC **DES** chip is described in [EB93]. Wiener's key search machine was described at CRYPTO '93 [WI94].

The time-memory trade-off for **DES** is due to Hellman [HE80]. A more general time-memory trade-off is presented by Fiat and Naor in [FN91].

The technique of differential cryptanalysis was developed by Biham and Shamir [BS91] (see also [BS93A] and their book [BS93], where cryptanalysis of other cryptosystems is also discussed). Our treatment of differential cryptanalysis is based largely on [BS93].

Another new method of cryptanalysis that can be used to attack **DES** and other similar cryptosystems is the linear cryptanalysis of Matsui [MA94, MA94A].

Descriptions of other substitution-permutation cryptosystems can be found in the following sources: LUCIFER [FE73]; FEAL [MI91]; REDOC-II [CW91]; and LOKI [BKPS90].

Exercises

3.1 Prove that **DES** decryption can be done by applying the **DES** encryption algorithm to the ciphertext with the key schedule reversed.

3.2 Let $\mathbf{DES}(x, K)$ represent the encryption of plaintext x with key K using the **DES** cryptosystem. Suppose $y = \mathbf{DES}(x, K)$ and $y' = \mathbf{DES}(c(x), c(K))$, where $c(\cdot)$ denotes the bitwise complement of its argument. Prove that $y' = c(y)$ (i.e., if we complement the plaintext and the key, then the ciphertext is also complemented). Note that this can be proved using only the "high-level" description of **DES** — the actual structure of S-boxes and other components of the system are irrelevant.

3.3 One way to strengthen **DES** is by *double encryption*: Given two keys, K_1 and K_2, define $y = e_{K_2}(e_{K_1}(x))$ (of course, this is just the product of **DES** with itself). If it happened that the encryption function e_{K_2} was the same as the decryption function d_{K_1}, then K_1 and K_2 are said to be *dual keys*. (This is very undesirable for double encryption, since the resulting ciphertext is identical to the plaintext.) A key is *self-dual* if it is its own dual key.

    (a) Prove that if C_0 is either all 0's or all 1's and D_0 is either all 0's or all 1's, then K is self-dual.

    (b) Prove that the following keys (given in hexadecimal notation) are self-dual:
```
0101010101010101
```

$$\text{FEFEFEFEFEFEFEFE}$$
$$\text{1F1F1F1F0E0E0E0E}$$
$$\text{E0E0E0E0F1F1F1F1}$$

(c) Prove that if $C_0 = 0101\ldots01$ or $1010\ldots10$ (in binary), then the x-or of the bitstrings C_i and C_{17-i} is $1111\ldots11$, for $1 \le i \le 16$ (a similar statement holds for the D_i's).

(d) Prove that the following pairs of keys (given in hexadecimal notation) are dual:

E001E001F101F101	01E001E001F101F1
FE1FFE1FFE0EFE0E	1FFE1FFE0EFE0EFE
E01FE01FF10EF10E	1FE01FE00EF10EF1

3.4 A message authentication code (MAC) can be produced by using CFB mode, as well as by using CBC mode. Given a sequence of plaintext blocks $x_1 \ldots x_n$, suppose we define the initialization vector IV to be x_1. Then encrypt $x_2 \ldots x_n$ using key K in CFB mode, obtaining $y_1 \ldots y_{n-1}$ (note that there are only $n - 1$ ciphertext blocks). Finally, define the MAC to be $e_K(y_{n-1})$. Prove that this MAC is identical to the MAC produced in Section 3.4.1 using CBC mode.

3.5 Suppose a sequence of plaintext blocks, $x_1 \ldots x_n$, is encrypted using **DES**, producing ciphertext blocks $y_1 \ldots y_n$. Suppose that one ciphertext block, say y_i, is transmitted incorrectly (i.e., some 1's are changed to 0's and vice versa). Show that the number of plaintext blocks that will be decrypted incorrectly is equal to one if ECB or OFB modes were used for encryption; and equal to two if CBC or CFB modes were used.

3.6 The purpose of this question is to investigate a simplified time-memory trade-off for a chosen plaintext attack. Suppose we have a cryptosystem in which $\mathcal{P} = \mathcal{C} = \mathcal{K}$, which attains perfect secrecy. Then it must be the case that $e_K(x) = e_{K_1}(x)$ implies $K = K_1$. Denote $\mathcal{P} = Y = \{y_1, \ldots, y_N\}$. Let x be a fixed plaintext. Define the function $g : Y \to Y$ by the rule $g(y) = e_y(x)$. Define a directed graph G having vertex set Y, in which the edge set consists of all the directed edges of the form $(y_i, g(y_i))$, $1 \le i \le N$.

(a) Prove that G consists of the union of disjoint directed cycles.

(b) Let T be a desired time parameter. Suppose we have a set of elements $Z = \{z_1, \ldots, z_m\} \subseteq Y$ such that, for every element $y_i \in Y$, either y_i is contained in a cycle of length at most T, or there exists an element $z_j \ne y_i$ such that the distance from y_i to z_j (in G) is at most T. Prove that there exists such a set Z such that

$$|Z| \le \frac{2N}{T},$$

so $|Z|$ is $O(N/T)$.

(c) For each $z_j \in Z$, define $g^{-T}(z_j)$ to be the element y_i such that $g^T(y_i) = z_j$, where g^T is the function that consists of T iterations of g. Construct a table X consisting of the ordered pairs $(z_j, g^{-T}(z_j))$, sorted with respect to their first coordinates.

A pseudo-code description of an algorithm to find K, given $y = e_K(x)$, is presented in Figure 3.15. Prove that this algorithm finds K in at most T steps. (Hence the time-memory trade-off is $O(N)$.)

FIGURE 3.15
Time-memory trade-off

1.	$y_{start} = y$
2.	$backup = $ **false**
3.	**while** $g(y) \neq y_{start}$ **do**
4.	**if** $y = z_j$ for some j **and not** $backup$ **then**
5.	$y = g^{-T}(z_j)$
6.	$backup = $ **true**
	else
7.	$y = g(y)$
8.	$K = y$

FIGURE 3.16
Differential attack on 4-round DES

Input: $L_0 R_0$, $L_0^* R_0^*$, $L_4 R_4$ and $L_4^* R_4^*$, where $L_0' = 20000000_{16}$ and $R_0' = 00000000_{16}$
1. compute $C' = P^{-1}(R_4')$
2. compute $E = E(L_4)$ and $E^* = E(L_4^*)$
3. **for** $j = 2$ **to** 8 **do**
    compute $test_j(E_j, E_j^*, C_j')$

    (d) Describe a pseudo-code algorithm to construct the desired set Z in time $O(NT)$ without using an array of size N.

3.7 Compute the probabilities of the following 3-round characteristic:

L_0'	$=$	00200008_{16}	R_0'	$=$	00000400_{16}	
L_1'	$=$	00000400_{16}	R_1'	$=$	00000000_{16}	$p = ?$
L_2'	$=$	00000000_{16}	R_2'	$=$	00000400_{16}	$p = ?$
L_3'	$=$	00000400_{16}	R_3'	$=$	00200008_{16}	$p = ?$

3.8 Here is a differential attack on a 4-round **DES**. It uses the following characteristic, which is a special case of the characteristic presented in Figure 3.10:

L_0'	$=$	20000000_{16}	R_0'	$=$	00000000_{16}	
L_1'	$=$	00000000_{16}	R_1'	$=$	20000000_{16}	$p = 1$

    (a) Suppose that the following algorithm presented in Figure 3.16 is used to compute sets $test_2, \ldots test_8$. Show that $J_j \in test_j$ for $2 \leq j \leq 8$.

(b) Given the following plaintext-ciphertext pairs, find the key bits $J_2, \ldots, J_8$.

plaintext	ciphertext
18493AC485B8D9A0	E332151312A18B4F
38493AC485B8D9A0	87391C27E5282161
482765DDD7009123	B5DDD8339D82D1D1
682765DDD7009123	81F4B92BD94B6FD8
ABCD098733731FF1	93A4B42F62EA59E4
8BCD098733731FF1	ABA494072BF411E5
13578642AAFFEDCB	FDEB526275FB9D94
33578642AAFFEDCB	CC8F72AAE685FDB1

(c) Compute the entire key (14 key bits remain to be determined, which can be done by exhaustive search).

4

The RSA System and Factoring

4.1 Introduction to Public-key Cryptography

In the classical model of cryptography that we have been studying up until now, Alice and Bob secretly choose the key K. K then gives rise to an encryption rule e_K and a decryption rule d_K. In the cryptosystems we have seen so far, d_K is either the same as e_K, or easily derived from it (for example, **DES** decryption is identical to encryption, but the key schedule is reversed). Cryptosystems of this type are known as *private-key* systems, since exposure of e_K renders the system insecure.

One drawback of a private-key system is that it requires the prior communication of the key K between Alice and Bob, using a secure channel, before any ciphertext is transmitted. In practice, this may be very difficult to achieve. For example, suppose Alice and Bob live far away from each other and they decide that they want to communicate electronically, using e-mail. In a situation such as this, Alice and Bob may not have access to a reasonable secure channel.

The idea behind a *public-key* system is that it might be possible to find a cryptosystem where it is computationally infeasible to determine d_K given e_K. If so, then the encryption rule e_K could be made public by publishing it in a directory (hence the term public-key system). The advantage of a public-key system is that Alice (or anyone else) can send an encrypted message to Bob (without the prior communication of a secret key) by using the public encryption rule e_K. Bob will be the only person that can decrypt the ciphertext, using his secret decryption rule d_K.

Consider the following analogy: Alice places an object in a metal box, and then locks it with a combination lock left there by Bob. Bob is the only person who can open the box since only he knows the combination.

The idea of a public-key system was due to Diffie and Hellman in 1976. The first realization of a public-key system came in 1977 by Rivest, Shamir, and Adleman, who invented the well-known **RSA Cryptosystem** which we study in this chapter. Since then, several public-key systems have been proposed, whose security rests on different computational problems. Of these, the most important are the following:

RSA

> The security of **RSA** is based on the difficulty of factoring large integers. This system is described in Section 4.3.

Merkle-Hellman Knapsack

> This and related systems are based on the difficulty of the subset sum problem (which is **NP**-complete[1]); however, all of the various knapsack systems have been shown to be insecure (with the exception of the **Chor-Rivest Cryptosystem** mentioned below). See Chapter 5 for a discussion of this cryptosystem.

McEliece

> The **McEliece Cryptosystem** is based on algebraic coding theory and is still regarded as being secure. It is based on the problem of decoding a linear code (which is also **NP**-complete). (See Chapter 5.)

ElGamal

> The **ElGamal Cryptosystem** is based on the difficulty of the discrete logarithm problem for finite fields. (See Chapter 5.)

Chor-Rivest

> This is also referred to as a "knapsack" type system, but it is still regarded as being secure.

Elliptic Curve

> The **Elliptic Curve Cryptosystems** are modifications of other systems (such as the ElGamal Cryptosystem, for example) that work in the domain of elliptic curves rather than finite fields. The **Elliptic Curve Cryptosystems** appear to remain secure for smaller keys than other public-key cryptosystems. (See Chapter 5.)

One very important observation is that a public-key cryptosystem can never provide unconditional security. This is because an opponent, on observing a ciphertext y, can encrypt each possible plaintext in turn using the public encryption rule e_K until he finds the unique x such that $y = e_K(x)$. This x is the decryption of y. Consequently, we study the computational security of public-key systems.

It is helpful conceptually to think of a public-key system in terms of an abstraction called a trapdoor one-way function. We informally define this notion now.

Bob's public encryption function, e_K, should be easy to compute. We have just noted that computing the inverse function (i.e., decrypting) should be hard (for

---

[1] The **NP**-complete problems are a large class of problems for which no polynomial-time algorithms are known.

anyone other than Bob). This property of being easy to compute but hard to invert is often called the *one-way* property . Thus, we desire that e_K be an (injective) one-way function.

One-way functions play a central role in cryptography; they are important for constructing public-key cryptosystems and in various other contexts. Unfortunately, although there are many functions that are believed to be one-way, there currently do not exist functions that can be proved to be one-way.

Here is an example of a function which is believed to be one-way. Suppose n is the product of two large primes p and q, and let b be a positive integer. Then define $f : \mathbb{Z}_n \to \mathbb{Z}_n$ to be

$$f(x) = x^b \bmod n.$$

(For a suitable choice of b and n, this is in fact the **RSA** encryption function; we will have much more to say about it later.)

If we are to construct a public-key cryptosystem, then it is not sufficient to find a one-way function. We do not want e_K to be a one-way function from Bob's point of view, since he wants to be able to decrypt messages that he receives in an efficient way. Thus, it is necessary that Bob possesses a *trapdoor*, which consists of secret information that permits easy inversion of e_K. That is, Bob can decrypt efficiently because he has some extra secret knowledge about K. So, we say that a function is a *trapdoor one-way* function if it is a one-way function, but it becomes easy to invert with the knowledge of a certain trapdoor.

We will see in Section 4.3 how to find a trapdoor for the function f defined above. This will lead to the **RSA Cryptosystem.**

4.2   More Number Theory

Before describing how **RSA** works, we need to discuss some more facts concerning modular arithmetic and number theory. Two fundamental results that we require are the Euclidean algorithm and the Chinese remainder theorem.

4.2.1   The Euclidean Algorithm

We already observed in Chapter 1 that $\mathbb{Z}_n$ is a ring for any positive integer n. We also proved there that $b \in \mathbb{Z}_n$ has a multiplicative inverse if and only if $\gcd(b, n) = 1$, and that the number of positive integers less than n and relatively prime to n is $\phi(n)$.

The set of residues modulo n that are relatively prime to n is denoted $\mathbb{Z}_n^*$. It is not hard to see that $\mathbb{Z}_n^*$ forms an abelian group under multiplication. We already have stated that multiplication modulo n is associative and commutative, and that 1 is the multiplicative identity. Any element in $\mathbb{Z}_n^*$ will have a multiplicative inverse (which is also in $\mathbb{Z}_n^*$). Finally, $\mathbb{Z}_n^*$ is closed under multiplication since

xy is relatively prime to n whenever x and y are relatively prime to n (prove this!).

At this point, we know that any $b \in \mathbb{Z}_n^*$ has a multiplicative inverse, b^{-1}, but we do not yet have an efficient algorithm to compute b^{-1}. Such an algorithm exists; it is called the extended Euclidean algorithm.

First, we describe the Euclidean algorithm, in its basic form, which is used to compute the greatest common divisor of two positive integers, say r_0 and r_1, where $r_0 > r_1$. The Euclidean algorithm consists of performing the following sequence of divisions:

$$
\begin{aligned}
r_0 &= q_1 r_1 + r_2, & 0 < r_2 < r_1 \\
r_1 &= q_2 r_2 + r_3, & 0 < r_3 < r_2 \\
&\;\;\vdots \\
r_{m-2} &= q_{m-1} r_{m-1} + r_m, & 0 < r_m < r_{m-1} \\
r_{m-1} &= q_m r_m.
\end{aligned}
$$

Then it is not hard to show that

$$
\gcd(r_0, r_1) = \gcd(r_1, r_2) = \ldots = \gcd(r_{m-1}, r_m) = r_m.
$$

Hence, it follows that $\gcd(r_0, r_1) = r_m$.

Since the Euclidean algorithm computes greatest common divisors, it can be used to determine if a positive integer $b < n$ has a multiplicative inverse modulo n, by starting with $r_0 = n$ and $r_1 = b$. However, it does not compute the value of the multiplicative inverse (if it exists).

Now, suppose we define a sequence of numbers $t_0, t_1, \ldots, t_m$ according to the following recurrence (where the q_j's are defined as above):

$$
\begin{aligned}
t_0 &= 0 \\
t_1 &= 1 \\
t_j &= t_{j-2} - q_{j-1} t_{j-1} \bmod r_0, \quad \text{if } j \geq 2.
\end{aligned}
$$

Then we have the following useful result.

THEOREM 4.1
For $0 \leq j \leq m$, we have that $r_j \equiv t_j r_1 \pmod{r_0}$, where the q_j's and r_j's are defined as in the Euclidean algorithm, and the t_j's are defined in the above recurrence.

PROOF The proof is by induction on j. The assertion is trivially true for $j = 0$ and $j = 1$. Assume the assertion is true for $j = i - 1$ and $i - 2$, where $i \geq 2$; we will prove the assertion is true for $j = i$. By induction, we have that

$$
r_{i-2} \equiv t_{i-2} r_1 \pmod{r_0}
$$

and

$$
r_{i-1} \equiv t_{i-1} r_1 \pmod{r_0}.
$$

Now, we compute:

$$r_i = r_{i-2} - q_{i-1}r_{i-1}$$
$$\equiv t_{i-2}r_1 - q_{i-1}t_{i-1}r_1 \pmod{r_0}$$
$$\equiv (t_{i-2} - q_{i-1}t_{i-1})r_1 \pmod{r_0}$$
$$\equiv t_i r_1 \pmod{r_0}.$$

Hence, the result is true by induction. ∎

The next corollary is an immediate consequence.

COROLLARY 4.2
Suppose $\gcd(r_0, r_1) = 1$. *Then* $t_m = r_1^{-1} \bmod r_0$.

Now, the sequence of numbers $t_0, t_1, \ldots t_m$ can be calculated in the Euclidean algorithm at the same time as the q_j's and the r_j's. In Figure 4.1, we present the extended Euclidean algorithm to compute the inverse of b modulo n, if it exists. In this version of the algorithm, we do not use an array to keep track of the q_j's, r_j's and t_j's, since it suffices to remember only the "last" two terms in each of these sequences at any point in the algorithm.

In step 10 of the algorithm, we have written the expression for *temp* in such a way that the reduction modulo n is done with a positive argument. (We mentioned earlier that modular reductions of negative numbers yield negative results in many computer languages; of course, we want to end up with a positive result here.) We also mention that at step 12, it is always the case that $tb \equiv r \pmod{n}$ (this is the result proved in Theorem 4.1).

Here is a small example to illustrate:

Example 4.1
Suppose we wish to compute $28^{-1} \bmod 75$. The Extended Euclidean algorithm proceeds as follows:

$75 = 2 \times 28 + 19$	step 6
$73 \times 28 \bmod 75 = 19$	step 12
$28 = 1 \times 19 + 9$	step 16
$3 \times 28 \bmod 75 = 9$	step 12
$19 = 2 \times 9 + 1$	step 16
$67 \times 28 \bmod 75 = 1$	step 12
$9 = 9 \times 1$	step 16

Hence, $28^{-1} \bmod 75 = 67$. □

FIGURE 4.1
Extended Euclidean algorithm

1. $n_0 = n$
2. $b_0 = b$
3. $t_0 = 0$
4. $t = 1$
5. $q = \lfloor \frac{n_0}{b_0} \rfloor$
6. $r = n_0 - q \times b_0$
7. **while** $r > 0$ **do**
8. $temp = t_0 - q \times t$
9.   **if** $temp \geq 0$ **then** $temp = temp \bmod n$
10.   **if** $temp < 0$ **then** $temp = n - ((-temp) \bmod n)$
11. $t_0 = t$
12. $t = temp$
13. $n_0 = b_0$
14. $b_0 = r$
15. $q = \lfloor \frac{n_0}{b_0} \rfloor$
16. $r = n_0 - q \times b_0$
17. **if** $b_0 \neq 1$ **then**

     b has no inverse modulo n

else

     $b^{-1} = t \bmod n$

4.2.2 The Chinese Remainder Theorem

The Chinese remainder theorem is really a method of solving certain systems of congruences. Suppose $m_1, \ldots, m_r$ are pairwise relatively prime positive integers (that is, $\gcd(m_i, m_j) = 1$ if $i \neq j$). Suppose $a_1, \ldots, a_r$ are integers, and consider the following system of congruences:

$$x \equiv a_1 \pmod{m_1}$$

$$x \equiv a_2 \pmod{m_2}$$

$$\vdots$$

$$x \equiv a_r \pmod{m_r}.$$

The Chinese remainder theorem asserts that this system has a unique solution modulo $M = m_1 \times m_2 \times \ldots \times m_r$. We will prove this result in this section, and also describe an efficient algorithm for solving systems of congruences of this type.

It is convenient to study the function $\pi : \mathbb{Z}_M \to \mathbb{Z}_{m_1} \times \ldots \times \mathbb{Z}_{m_r}$, which we define as follows:

$$\pi(x) = (x \bmod m_1, \ldots, x \bmod m_r).$$

Example 4.2
Suppose $r = 2$, $m_1 = 5$ and $m_2 = 3$, so $M = 15$. Then the function π has the following values:

$\pi(0)$	$=$	$(0,0)$	$\pi(1)$	$=$	$(1,1)$	$\pi(2)$	$=$	$(2,2)$
$\pi(3)$	$=$	$(3,0)$	$\pi(4)$	$=$	$(4,1)$	$\pi(5)$	$=$	$(0,2)$
$\pi(6)$	$=$	$(1,0)$	$\pi(7)$	$=$	$(2,1)$	$\pi(8)$	$=$	$(3,2)$
$\pi(9)$	$=$	$(4,0)$	$\pi(10)$	$=$	$(0,1)$	$\pi(11)$	$=$	$(1,2)$
$\pi(12)$	$=$	$(2,0)$	$\pi(13)$	$=$	$(3,1)$	$\pi(14)$	$=$	$(4,2)$.

$\square$

Proving the Chinese remainder theorem amounts to proving that this function π we have defined is a bijection. In Example 4.2 this is easily seen to be the case. In fact, we will be able to give an explicit general formula for the inverse function π^{-1}.

For $1 \leq i \leq r$, define

$$M_i = \frac{M}{m_i}.$$

Then it is not difficult to see that

$$\gcd(M_i, m_i) = 1$$

for $1 \leq i \leq r$. Next, for $1 \leq i \leq r$, define

$$y_i = M_i^{-1} \bmod m_i.$$

(This inverse exists since $\gcd(M_i, m_i) = 1$, and it can be found using the Euclidean algorithm .) Note that

$$M_i y_i \equiv 1 \pmod{m_i}$$

for $1 \leq i \leq r$.

Now, define a function $\rho : \mathbb{Z}_{m_1} \times \ldots \times \mathbb{Z}_{m_r} \to \mathbb{Z}_M$ as follows:

$$\rho(a_1, \ldots, a_r) = \sum_{i=1}^{r} a_i M_i y_i \bmod M.$$

We will show that the function $\rho = \pi^{-1}$, i.e., it provides an explicit formula for solving the original system of congruences.

Denote $X = \rho(a_1, \ldots, a_r)$, and let $1 \leq j \leq r$. Consider a term $a_i M_i y_i$ in the above summation, reduced modulo m_j: If $i = j$, then

$$a_i M_i y_i \equiv a_i \pmod{m_i}$$

since

$$M_i y_i \equiv 1 \pmod{m_i}.$$

On the other hand, if $i \neq j$, then

$$a_i M_i y_i \equiv 0 \pmod{m_j}$$

since $m_j \mid M_i$ in this case. Thus, we have that

$$X \equiv \sum_{i=1}^{r} a_i M_i y_i \pmod{m_j}$$

$$\equiv a_j \pmod{m_j}.$$

Since this is true for all j, $1 \leq j \leq r$, X is a solution to the system of congruences.

At this point, we need to show that the solution X is unique modulo M. But this can be done by simple counting. The function π is a function from a domain of cardinality M to a range of cardinality M. We have just proved that π is a surjective (i.e., onto) function. Hence, π must also be injective (i.e., one-to-one), since the domain and range have the same cardinality. It follows that π is a bijection and $\pi^{-1} = \rho$. Note also that π^{-1} is a linear function of its arguments $a_1, \ldots, a_r$.

Here is a bigger example to illustrate.

Example 4.3
Suppose $r = 3$, $m_1 = 7$, $m_2 = 11$ and $m_3 = 13$. Then $M = 1001$. We compute $M_1 = 143$, $M_2 = 91$ and $M_3 = 77$, and then $y_1 = 5$, $y_2 = 4$ and $y_3 = 12$. Then the function $\pi^{-1} : \mathbb{Z}_7 \times \mathbb{Z}_{11} \times \mathbb{Z}_{13} \to \mathbb{Z}_{1001}$ is the following:

$$\pi^{-1}(a_1, a_2, a_3) = 715a_1 + 364a_2 + 924a_3 \bmod 1001.$$

For example, if $x \equiv 5 \pmod 7$, $x \equiv 3 \pmod{11}$ and $x \equiv 10 \pmod{13}$, then this formula tells us that

$$x = 715 \times 5 + 364 \times 3 + 924 \times 10 \bmod 1001$$

$$= 13907 \bmod 1001$$

$$= 894 \bmod 1001.$$

This can be verified by reducing 894 modulo 7, 11 and 13.   ⬚

For future reference, we record the results of this section as a theorem.

THEOREM 4.3   *(Chinese Remainder Theorem)*
Suppose $m_1, \ldots, m_r$ are pairwise relatively prime positive integers, and suppose $a_1, \ldots, a_r$ are integers. Then, the system of r congruences $x \equiv a_i \pmod{m_i}$ $(1 \leq i \leq r)$ has a unique solution modulo $M = m_1 \times \ldots \times m_r$, which is given by

$$x = \sum_{i=1}^{r} a_i M_i y_i \bmod M,$$

where $M_i = M/m_i$ and $y_i = M_i^{-1} \bmod m_i$, for $1 \leq i \leq r$.

4.2.3   Other Useful Facts

We next mention another result from elementary group theory, called Lagrange's Theorem, that will be relevant in our treatment of the **RSA Cryptosystem**. For a (finite) multiplicative group G, define the *order* of an element $g \in G$ to be the smallest positive integer m such that $g^m = 1$. The following result is fairly simple, but we will not prove it here.

THEOREM 4.4   *(Lagrange)*
Suppose G is a multiplicative group of order n, and $g \in G$. Then the order of g divides n.

For our purposes, the following corollaries are essential.

COROLLARY 4.5
If $b \in \mathbb{Z}_n^*$, then $b^{\phi(n)} \equiv 1 \pmod{n}$.

PROOF $\mathbb{Z}_n^*$ is a multiplicative group of order $\phi(n)$.   ∎

COROLLARY 4.6   *(Fermat)*
Suppose p is prime and $b \in \mathbb{Z}_p$. Then $b^p \equiv b \pmod{p}$.

PROOF   If p is prime, then $\phi(p) = p - 1$. So, for $b \not\equiv 0 \pmod{p}$, the result follows from Corollary 4.5. For $b \equiv 0 \pmod{p}$, the result is also true since $0^p \equiv 0 \pmod{p}$.   ∎

At this point, we know that if p is prime, then $\mathbb{Z}_p^*$ is a group of order $p - 1$, and any element in $\mathbb{Z}_p^*$ has order dividing $p - 1$. However, if p is prime, then the group $\mathbb{Z}_p^*$ is in fact *cyclic*: there exists an element $\alpha \in \mathbb{Z}_p^*$ having order equal to $p - 1$. We will not prove this very important fact, but we do record it for future reference:

THEOREM 4.7
If p is prime, then $\mathbb{Z}_p^*$ is a cyclic group.

An element α having order $p - 1$ is called a *primitive* element modulo p. Observe that α is a primitive element if and only if

$$\{\alpha^i : 0 \le i \le p - 2\} = \mathbb{Z}_p^*.$$

Now, suppose p is prime and α is a primitive element modulo p. Any element $\beta \in \mathbb{Z}_p^*$ can be written as $\beta = \alpha^i$, where $0 \le i \le p - 2$, in a unique way. It is not difficult to prove that the order of $\beta = \alpha^i$ is

$$\frac{p - 1}{\gcd(p - 1, i)}.$$

Thus β is itself a primitive element if and only if $\gcd(p - 1, i) = 1$. It follows that the number of primitive elements modulo p is $\phi(p - 1)$.

Example 4.4
Suppose $p = 13$. By computing successive powers of 2, we can verify that 2 is a primitive element modulo 13:

$$2^0 \bmod 13 = 1$$
$$2^1 \bmod 13 = 2$$
$$2^2 \bmod 13 = 4$$
$$2^3 \bmod 13 = 8$$
$$2^4 \bmod 13 = 3$$
$$2^5 \bmod 13 = 6$$
$$2^6 \bmod 13 = 12$$
$$2^7 \bmod 13 = 11$$
$$2^8 \bmod 13 = 9$$
$$2^9 \bmod 13 = 5$$
$$2^{10} \bmod 13 = 10$$
$$2^{11} \bmod 13 = 7.$$

The element 2^i is primitive if and only if $\gcd(i, 12) = 1$; i.e., if and only if $i = 1, 5, 7$ or 11. Hence, the primitive elements modulo 13 are 2, 6, 7 and 11. $\Box$

FIGURE 4.2
RSA Cryptosystem

Let $n = pq$, where p and q are primes. Let $\mathcal{P} = \mathcal{C} = \mathbb{Z}_n$, and define

$$K = \{(n, p, q, a, b) : n = pq, p, q \text{ prime}, ab \equiv 1 \pmod{\phi(n)}\}.$$

For $K = (n, p, q, a, b)$, define

$$e_K(x) = x^b \bmod n$$

and

$$d_K(y) = y^a \bmod n$$

$(x, y \in \mathbb{Z}_n)$. The values n and b are public, and the values p, q, a are secret.

4.3 The RSA Cryptosystem

We can now describe the **RSA Cryptosystem**. This cryptosystem uses computations in $\mathbb{Z}_n$, where n is the product of two distinct odd primes p and q. For such n, note that $\phi(n) = (p-1)(q-1)$.

The formal description of the cryptosystem is given in Figure 4.2. Let's verify that encryption and decryption are inverse operations. Since

$$ab \equiv 1 \pmod{\phi(n)},$$

we have that

$$ab = t\phi(n) + 1$$

for some integer $t \geq 1$. Suppose that $x \in \mathbb{Z}_n{}^*$; then we have

$$
\begin{aligned}
(x^b)^a &\equiv x^{t\phi(n)+1} \pmod{n} \\
&\equiv (x^{\phi(n)})^t x \pmod{n} \\
&\equiv 1^t x \pmod{n} \\
&\equiv x \pmod{n},
\end{aligned}
$$

as desired. We leave it as an exercise for the reader to show that $(x^b)^a \equiv x \pmod{n}$ if $x \in \mathbb{Z}_n \backslash \mathbb{Z}_n{}^*$.

Here is a small (insecure) example of the **RSA Cryptosystem**.

Example 4.5

Suppose Bob chooses $p = 101$ and $q = 113$. Then $n = 11413$ and $\phi(n) = 100 \times 112 = 11200$. Since $11200 = 2^6 5^2 7$, an integer b can be used as an encryption exponent if and only if b is not divisible by 2, 5 or 7. (In practice, however, Bob will not factor $\phi(n)$. He will verify that $\gcd(\phi(n), b) = 1$ using the Euclidean algorithm.) Suppose Bob chooses $b = 3533$. Then the Extended Euclidean algorithm will yield

$$b^{-1} = 6597 \bmod 11200.$$

Hence, Bob's secret decryption exponent is $a = 6597$.

Bob publishes $n = 11413$ and $b = 3533$ in a directory. Now, suppose Alice wants to send the plaintext 9726 to Bob. She will compute

$$9726^{3533} \bmod 11413 = 5761$$

and send the ciphertext 5761 over the channel. When Bob receives the ciphertext 5761, he uses his secret decryption exponent to compute

$$5761^{6597} \bmod 11413 = 9726.$$

(At this point, the encryption and decryption operations might appear to be very complicated, but we will discuss efficient algorithms for these operations in the next section.) ☐

The security of **RSA** is based on the hope that the encryption function $e_K(x) = x^b \bmod n$ is one-way, so it will be computationally infeasible for an opponent to decrypt a ciphertext. The trapdoor that allows Bob to decrypt is the knowledge of the factorization $n = pq$. Since Bob knows this factorization, he can compute $\phi(n) = (p-1)(q-1)$ and then compute the decryption exponent a using the Extended Euclidean algorithm. We will say more about the security of **RSA** later on.

4.4 Implementing RSA

There are many aspects of the **RSA Cryptosystem** to discuss, including the details of setting up the cryptosystem, the efficiency of encrypting and decrypting, and security issues. In order to set up the system, Bob follows the steps indicated in Figure 4.3. How Bob carries out these steps will be discussed later in this chapter.

One obvious attack on the cryptosystem is for a cryptanalyst to attempt to factor n. If this can be done, it is a simple manner to compute $\phi(n) = (p-1)(q-1)$ and then compute the decryption exponent a from b exactly as Bob did. (It has

FIGURE 4.3
Setting up RSA

---

1. Bob generates two large primes, p and q
2. Bob computes $n = pq$ and $\phi(n) = (p-1)(q-1)$
3. Bob chooses a random b ($1 < b < \phi(n)$) such that $\gcd(b, \phi(n)) = 1$
4. Bob computes $a = b^{-1} \bmod \phi(n)$ using the Euclidean algorithm
5. Bob publishes n and b in a directory as his public key.

---

been conjectured that breaking **RSA** is polynomially equivalent[2] to factoring n, but this remains unproved.)

Hence, if the **RSA Cryptosystem** is to be secure, it is certainly necessary that $n = pq$ must be large enough that factoring it will be computationally infeasible. Current factoring algorithms are able to factor numbers having up to 130 decimal digits (for more information on factoring, see Section 4.8). Hence, it is recommended that, to be on the safe side, one should choose p and q to each be primes having about 100 digits; then n will have 200 digits. Several hardware implementations of **RSA** use a modulus which is 512 bits in length. However, a 512-bit modulus corresponds to about 154 decimal digits (since the number of bits in the binary representation of an integer is $\log_2 10$ times the number of decimal digits), and hence it does not offer good long-term security.

Leaving aside for the moment the question of how to find 100 digit primes, let us look now at the arithmetic operations of encryption and decryption. An encryption (or decryption) involves performing one exponentiation modulo n. Since n is very large, we must use multiprecision arithmetic to perform computations in $\mathbb{Z}_n$, and the time required will depend on the number of bits in the binary representation of n.

Suppose n has k bits in its binary representation; i.e., $k = \lfloor \log_2 n \rfloor + 1$. Using standard "grade-school" arithmetic techniques, it is not difficult to see that an addition of two k-bit integers can be done in time $O(k)$, and a multiplication can be done in time $O(k^2)$. Also, a reduction modulo n of an integer having at most $2k$ bits can be performed in time $O(k^2)$ (this amounts to doing long division and retaining the remainder). Now, suppose that $x, y \in \mathbb{Z}_n$ (where we are assuming that $0 \leq x, y \leq n - 1$). Then $xy \bmod n$ can be computed by first calculating the product xy (which is a $2k$-bit integer), and then reducing it modulo n. These two steps can be peformed in time $O(k^2)$. We call this computation *modular multiplication*.

---

[2]Two problems are said to be *polynomially equivalent* if the existence of a polynomial-time algorithm for either problem implies the existence of a polynomial-time algorithm for the other problem.

FIGURE 4.4
The square-and-multiply algorithm to compute x^b mod n

1.	$z = 1$
2.	**for** $i = \ell - 1$ **downto** 0 **do**
3.	$z = z^2$ mod n
4.	**if** $b_i = 1$ **then** $z = z \times x$ mod n

We now consider *modular exponentiation*, i.e., computation of a function of the form x^c mod n. As noted above, both the encryption and the decryption operations in **RSA** are modular exponentiations. Computation of x^c mod n can be done using $c - 1$ modular multiplications; however, this is very inefficient if c is large. Note that c might be as big as $\phi(n) - 1$, which is exponentially large compared to k.

The well-known "square-and-multiply" approach reduces the number of modular multiplications required to compute x^c mod n to at most 2ℓ, where ℓ is the number of bits in the binary representation of c. Since $\ell \leq k$, it follows that x^c mod n can be computed in time $O(k^3)$. Hence, **RSA** encryption and decryption can both be done in polynomial time (as a function of k, which is the number of bits in one plaintext (or ciphertext) character).

Square-and-multiply assumes that the exponent, b say, is represented in binary notation, say

$$b = \sum_{i=0}^{\ell-1} b_i 2^i,$$

where $b_i = 0$ or 1, $0 \leq i \leq \ell - 1$. The algorithm to compute $z = x^b$ mod n is presented in Figure 4.4. It is easy to count the number of modular multiplications performed by the square-and-multiply algorithm. There are always ℓ squarings performed (step 3). The number of modular multiplications in step 4 is equal to the number of 1's in the binary representation of b, which is an integer between 0 and ℓ. Thus, the total number of modular multiplications is at least ℓ and at most 2ℓ.

We will illustrate the use of square-and-multiply by returning to Example 4.5.

Example 4.5 *(Cont.)*

Recall that $n = 11413$, and the public encryption exponent is $b = 3533$. Alice encrypts the plaintext 9726 by computing 9726^{3533} mod 11413, using the square-

and multiply algorithm, as follows:

i	b_i	z
11	1	$1^2 \times 9726 = 9726$
10	1	$9726^2 \times 9726 = 2659$
9	0	$2659^2 = 5634$
8	1	$5634^2 \times 9726 = 9167$
7	1	$9167^2 \times 9726 = 4958$
6	1	$4958^2 \times 9726 = 7783$
5	0	$7783^2 = 6298$
4	0	$6298^2 = 4629$
3	1	$4629^2 \times 9726 = 10185$
2	1	$10185^2 \times 9726 = 105$
1	0	$105^2 = 11025$
0	1	$11025^2 \times 9726 = 5761$

Hence, as stated earlier, the ciphertext is 5761.   □

It should be emphasized that the most efficient current hardware implementations of **RSA** achieve encryption rates of about 600 Kbits per second (using a 512 bit modulus n), as compared to 1 Gbit per second for **DES**. Stated another way, **RSA** is roughly 1500 times slower than **DES**.

At this point we have discussed the encryption and decryption operations for **RSA**. In terms of setting up **RSA**, the generation of the primes p and q (Step 1) will be discussed in the next section. Step 2 is straightforward and can be done in time $O((\log n)^2)$. Steps 3 and 4 involve the Euclidean algorithm, so let's briefly consider its complexity.

Suppose we compute the greatest common divisor of r_0 and r_1, where $r_0 > r_1$. In each iteration of the algorithm, we compute a quotient and remainder, which can be done in time $O((\log r_0)^2)$. If we can obtain an upper bound on the number of iterations, then we will have a bound on the complexity of the algorithm. There is a well-known result, known as Lamé's Theorem, that provides such a bound. It asserts that if s is the number of iterations, then $f_{s+2} \leq r_0$, where f_i denotes the ith Fibonacci number. Since

$$f_i \approx \left(\frac{1 + \sqrt{5}}{2} \right)^i,$$

it follows that s is $O(\log r_0)$.

This shows that the running time of the Euclidean algorithm is $O((\log n)^3)$. (Actually, a more careful analysis can be used to show that the running time is, in fact, $O((\log n)^2)$.)

4.5 Probabilistic Primality Testing

In setting up the **RSA Cryptosystem**, it is necessary to generate large (e.g., 80 digit) "random primes." In practice, the way this is done is to generate large random numbers, and then test them for primality using a probabilistic polynomial-time Monte Carlo algorithm such as the Solovay-Strassen or Miller-Rabin algorithm, both of which we will present in this section. These algorithms are fast (i.e., an integer n can be tested in time that is polynomial in $\log_2 n$, the number of bits in the binary representation of n), but there is a possibility that the algorithm may claim that n is prime when it is not. However, by running the algorithm enough times, the error probability can be reduced below any desired threshold. (We will discuss this in more detail a bit later.)

The other pertinent question is how many random integers (of a specified size) will need to be tested until we find one that is prime. A famous result in number theory, called the Prime number theorem, states that the number of primes not exceeding N is approximately $N/\ln N$. Hence, if p is chosen at random, the probability that it is prime is about $1/\ln p$. For a 512 bit modulus, we have $1/\ln p \approx 1/177$. That is, on average, of 177 random integers p of the appropriate size, one will be prime (of course, if we restrict our attention to odd integers, the probability doubles, to about $2/177$). So it is indeed practical to generate sufficiently large random numbers that are "probably prime," and hence it is practical to set up the **RSA Cryptosystem**. We proceed to describe how this is done.

A *decision problem* is a problem in which a question is to be answered "yes" or "no." A *probabilistic* algorithm is any algorithm that uses random numbers (in contrast, an algorithm that does not use random numbers is called a *deterministic* algorithm). The following definitions pertain to probabilistic algorithms for decision problems.

DEFINITION 4.1 *A yes-biased Monte Carlo algorithm is a probabilistic algorithm for a decision problem in which a "yes" answer is (always) correct, but a "no" answer may be incorrect. A no-biased Monte Carlo algorithm is defined in the obvious way. We say that a yes-biased Monte Carlo algorithm has error probability equal to ϵ if, for any instance in which the answer is "yes," the algorithm will give the (incorrect) answer "no" with probability at most ϵ. (This probability is computed over all possible random choices made by the algorithm when it is run with a given input.)*

The decision problem called **Composites** is described in Figure 4.5.

Note that an algorithm for a decision problem only has to answer "yes" or "no." In particular, in the case of the problem **Composites**, we do not require the algorithm to find a factorization in the case that n is composite.

We will first describe the Solovay-Strassen algorithm, which is a yes-biased Monte Carlo algorithm for **Composites** with error probability $1/2$. Hence, if the

FIGURE 4.5
Composites

Problem Instance A positive integer $n \geq 2$.

Question Is n composite?

FIGURE 4.6
Quadratic Residues

Problem Instance An odd prime p, and an integer x such that $0 \leq x \leq p - 1$.

Question Is x a quadratic residue modulo p?

---

algorithm answers "yes," then n is composite; conversely, if n is composite, then the algorithm answers "yes" with probability at least $1/2$.

Although the Miller-Rabin algorithm (which we will discuss later) is faster than Solovay-Strassen, we begin by looking at the Solovay-Strassen algorithm because it is easier to understand conceptually and because it involves some number-theoretic concepts that will be useful in later chapters of the book. We begin by developing some further background from number theory before describing the algorithm.

DEFINITION 4.2 *Suppose p is an odd prime and x is an integer, $1 \leq x \leq p - 1$. x is defined to be a quadratic residue modulo p if the congruence $y^2 \equiv x \pmod{p}$ has a solution $y \in \mathbb{Z}_p$. x is defined to be a **quadratic non-residue** modulo p if $x \not\equiv 0 \pmod{p}$ and x is not a quadratic residue modulo p.*

Example 4.6
The quadratic residues modulo 11 are $1, 3, 4, 5$ and 9. Note that $(\pm 1)^2 = 1$, $(\pm 5)^2 = 3$, $(\pm 2)^2 = 4$, $(\pm 4)^2 = 5$ and $(\pm 3)^2 = 9$ (where all arithmetic is in $\mathbb{Z}_{11}$). □

The decision problem **Quadratic Residues** is defined in Figure 4.6 in the obvious way.

We prove a result, known as *Euler's criterion*, that will give rise to a polynomial-time deterministic algorithm for **Quadratic Residues**.

THEOREM 4.8 *(Euler's Criterion)*
Let p be an odd prime. Then x is a quadratic residue modulo p if and only if

$$x^{(p-1)/2} \equiv 1 \pmod{p}.$$

PROOF First, suppose $x \equiv y^2 \pmod{p}$. Recall from Corollary 4.6 that if p is prime, then $x^{p-1} \equiv 1 \pmod{p}$ for any $x \not\equiv 0 \pmod{p}$. Thus we have

$$x^{(p-1)/2} \equiv (y^2)^{(p-1)/2} \pmod{p}$$

$$\equiv y^{p-1} \pmod{p}$$

$$\equiv 1 \pmod{p}.$$

Conversely, suppose $x^{(p-1)/2} \equiv 1 \pmod{p}$. Let b be a primitive element modulo p. Then $x \equiv b^i \pmod{p}$ for some i. Then we have

$$x^{(p-1)/2} \equiv (b^i)^{(p-1)/2} \pmod{p}$$

$$\equiv b^{i(p-1)/2} \pmod{p}.$$

Since b has order $p - 1$, it must be the case that $p - 1$ divides $i(p - 1)/2$. Hence, i is even, and then the square roots of x are $\pm b^{i/2}$.   ∎

Theorem 4.8 yields a polynomial-time algorithm for **Quadratic Residues**, by using the "square-and-multiply" technique for exponentiation modulo p. The complexity of the algorithm will be $O((\log p)^3)$.

We now need to give some further definitions from number theory.

DEFINITION 4.3 *Suppose p is an odd prime. For any integer $a \geq 0$, we define the Legendre symbol $\left(\frac{a}{p}\right)$ as follows:*

$$\left(\frac{a}{p}\right) = \begin{cases} 0 & \text{if } a \equiv 0 \pmod{p} \\ 1 & \text{if } a \text{ is a quadratic residue modulo } p \\ -1 & \text{if } a \text{ is a quadratic non-residue modulo } p. \end{cases}$$

We have already seen that $a^{(p-1)/2} \equiv 1 \pmod{p}$ if and only if a is a quadratic residue modulo p. If a is a multiple of p, then it is clear that $a^{(p-1)/2} \equiv 0 \pmod{p}$. Finally, if a is a quadratic non-residue modulo p, then $a^{(p-1)/2} \equiv -1 \pmod{p}$ since $a^{p-1} \equiv 1 \pmod{p}$. Hence, we have the following result, which provides an efficient algorithm to evaluate Legendre symbols:

THEOREM 4.9
Suppose p is an odd prime. Then

$$\left(\frac{a}{p}\right) \equiv a^{(p-1)/2} \pmod{p}.$$

Next, we define a generalization of the Legendre symbol.

DEFINITION 4.4   *Suppose n is an odd positive integer, and the prime power factorization of n is $p_1^{e_1} \ldots p_k^{e_k}$. Let $a \geq 0$ be an integer. The Jacobi symbol $\left(\frac{a}{n}\right)$ is defined to be*

$$\left(\frac{a}{n}\right) = \prod_{i=1}^{k} \left(\frac{a}{p_i}\right)^{e_i}.$$

Example 4.7
Consider the Jacobi symbol $\left(\frac{6278}{9975}\right)$. The prime power factorization of 9975 is $9975 = 3 \times 5^2 \times 7 \times 19$. Thus we have

$$\left(\frac{6278}{9975}\right) = \left(\frac{6278}{3}\right) \left(\frac{6278}{5}\right)^2 \left(\frac{6278}{7}\right) \left(\frac{6278}{19}\right)$$

$$= \left(\frac{2}{3}\right) \left(\frac{3}{5}\right)^2 \left(\frac{6}{7}\right) \left(\frac{8}{19}\right)$$

$$= (-1)(-1)^2(-1)(-1)$$

$$= -1.$$

$\square$

Suppose $n > 1$ is odd. If n is prime then $\left(\frac{a}{n}\right) \equiv a^{(n-1)/2} \pmod{n}$ for any a. On the other hand, if n is composite, it may or may not be the case that $\left(\frac{a}{n}\right) \equiv a^{(n-1)/2} \pmod{n}$. If this equation holds, then n is called an *Euler pseudo-prime* to the base a. For example, 91 is an Euler pseudo-prime to the base 10, since

$$\left(\frac{10}{91}\right) = -1 = 10^{45} \bmod 91.$$

However, it can be shown that, for any odd composite n, n is an Euler pseudo-prime to the base a for at most half of the integers a such that $1 \leq a \leq n-1$ (see the exercises). This fact shows that the Solovay-Strassen primality test, which we present in Figure 4.7, is a yes-biased Monte Carlo algorithm with error probability at most $1/2$. At this point it is not clear that the algorithm is a polynomial-time algorithm. We already know how to evaluate $a^{(n-1)/2} \bmod n$ in time $O((\log n)^3)$, but how do we compute Jacobi symbols efficiently? It might appear to be necessary to first factor n, since the Jacobi symbol $\left(\frac{a}{n}\right)$ is defined in terms of the factorization of n. But, if we could factor n, we would already know if it is prime, so this approach ends up in a vicious circle.

Fortunately, we can evaluate a Jacobi symbol without factoring n by using some results from number theory, the most important of which is a generalization of the law of quadratic reciprocity (property 4 below). We now enumerate these properties without proof:

FIGURE 4.7
The Solovay-Strassen primality test for an odd integer n

1. choose a random integer a, $1 \leq a \leq n - 1$
2. **if** $\left(\frac{a}{n}\right) \equiv a^{(n-1)/2} \pmod{n}$ **then**
   answer "n is prime"
   else
   answer "n is composite"

1. If n is an odd integer and $m_1 \equiv m_2 \pmod{n}$, then

$$\left(\frac{m_1}{n}\right) = \left(\frac{m_2}{n}\right).$$

2. If n is an odd integer, then

$$\left(\frac{2}{n}\right) = \begin{cases} 1 & \text{if } n \equiv \pm 1 \pmod{8} \\ -1 & \text{if } n \equiv \pm 3 \pmod{8}. \end{cases}$$

3. If n is an odd integer then

$$\left(\frac{m_1 m_2}{n}\right) = \left(\frac{m_1}{n}\right)\left(\frac{m_2}{n}\right).$$

In particular, if $m = 2^k t$, where t is odd, then

$$\left(\frac{m}{n}\right) = \left(\frac{2}{n}\right)^k \left(\frac{t}{n}\right).$$

4. Suppose m and n are odd integers. Then

$$\left(\frac{m}{n}\right) = \begin{cases} -\left(\frac{n}{m}\right) & \text{if } m \equiv n \equiv 3 \pmod{4} \\ \left(\frac{n}{m}\right) & \text{otherwise.} \end{cases}$$

Example 4.8

As an illustration of the application of these properties, we evaluate the Jacobi symbol $\left(\frac{7411}{9283}\right)$ as follows:

$$
\begin{aligned}
\left(\frac{7411}{9283}\right) &= -\left(\frac{9283}{7411}\right) && \text{by property 4} \\
&= -\left(\frac{1872}{7411}\right) && \text{by property 1} \\
&= -\left(\frac{2}{7411}\right)^4\left(\frac{117}{7411}\right) && \text{by property 3} \\
&= -\left(\frac{117}{7411}\right) && \text{by property 2} \\
&= -\left(\frac{7411}{117}\right) && \text{by property 4} \\
&= -\left(\frac{40}{117}\right) && \text{by property 1} \\
&= -\left(\frac{2}{117}\right)^3\left(\frac{5}{117}\right) && \text{by property 3} \\
&= \left(\frac{5}{117}\right) && \text{by property 2} \\
&= \left(\frac{117}{5}\right) && \text{by property 4} \\
&= \left(\frac{2}{5}\right) && \text{by property 1} \\
&= -1 && \text{by property 2}
\end{aligned}
$$

Notice that we successively apply properties 4, 1, 3, and 2 in this computation.
$\square$

   In general, by applying these four properties, it is possible to compute a Jacobi symbol $\left(\frac{m}{n}\right)$ in polynomial time. The only arithmetic operations that are required are modular reductions and factoring out powers of two. Note that if an integer is represented in binary notation, then factoring out powers of two amounts to determining the number of trailing zeroes. So, the complexity of the algorithm is determined by the number of modular reductions that must be done. It is not difficult to show that at most $O(\log n)$ modular reductions are performed, each of which can be done in time $O((\log n)^2)$. This shows that the complexity is $O((\log n)^3)$, which is polynomial in $\log n$. (In fact, the complexity can be shown to be $O((\log n)^2)$ by more precise analysis.)

Suppose that we have generated a random number n and tested it for primality using the Solovay-Strassen algorithm. If we have run the algorithm m times, what is our confidence that n is prime? It is tempting to conclude that the probability that such an integer n is prime is $1 - 2^{-m}$. This conclusion is often stated in both textbooks and technical articles, but it cannot be inferred from the given data.

We need to be careful about our use of probabilities. We will define the following random variables: **a** denotes the event

"a random odd integer n of a specified size is composite,"

and **b** denotes the event

"the algorithm answers `n is prime' m times in succession."

It is certainly the case that $prob(\mathbf{b}|\mathbf{a}) \leq 2^{-m}$. However, the probability that we are really interested is $prob(\mathbf{a}|\mathbf{b})$, which is usually not the same as $prob(\mathbf{b}|\mathbf{a})$.

We can compute $prob(\mathbf{a}|\mathbf{b})$ using Bayes' theorem (Theorem 2.1). In order to do this, we need to know $prob(\mathbf{a})$. Suppose $N \leq n \leq 2N$. Applying the Prime number theorem, the number of (odd) primes between N and $2N$ is approximately

$$\frac{2N}{\ln 2N} - \frac{N}{\ln N} \approx \frac{N}{\ln N}$$
$$\approx \frac{n}{\ln n}.$$

Since there are $N/2 \approx n/2$ odd integers between N and $2N$, we will use the estimate

$$prob(\mathbf{a}) \approx 1 - \frac{2}{\ln n}.$$

Then we can compute as follows:

$$prob(\mathbf{a}|\mathbf{b}) = \frac{prob(\mathbf{b}|\mathbf{a})prob(\mathbf{a})}{prob(\mathbf{b})}$$

$$= \frac{prob(\mathbf{b}|\mathbf{a})prob(\mathbf{a})}{prob(\mathbf{b}|\mathbf{a})prob(\mathbf{a}) + prob(\mathbf{b}|\overline{\mathbf{a}})prob(\overline{\mathbf{a}})}$$

$$\approx \frac{prob(\mathbf{b}|\mathbf{a})\left(1 - \frac{2}{\ln n}\right)}{prob(\mathbf{b}|\mathbf{a})\left(1 - \frac{2}{\ln n}\right) + \frac{2}{\ln n}}$$

$$= \frac{prob(\mathbf{b}|\mathbf{a})(\ln n - 2)}{prob(\mathbf{b}|\mathbf{a})(\ln n - 2) + 2}$$

$$\leq \frac{2^{-m}(\ln n - 2)}{2^{-m}(\ln n - 2) + 2}$$

$$= \frac{\ln n - 2}{\ln n - 2 + 2^{m+1}}.$$

Note that in this computation, $\overline{\mathbf{a}}$ denotes the event

FIGURE 4.8
Error probabilities for the Solovay-Strassen test

m	2^{-m}	$\dfrac{175}{175 + 2^{m+1}}$
1	.500	.978
2	.250	.956
5	$.312 \times 10^{-1}$	.732
10	$.977 \times 10^{-3}$	$.787 \times 10^{-1}$
20	$.954 \times 10^{-6}$	$.834 \times 10^{-4}$
30	$.931 \times 10^{-9}$	$.815 \times 10^{-7}$
50	$.888 \times 10^{-15}$	$.777 \times 10^{-13}$
100	$.789 \times 10^{-30}$	$.690 \times 10^{-28}$

"a random odd integer n is prime."

It is interesting to compare the two quantities $(\ln n - 2)/(\ln n - 2 + 2^{m+1})$ and 2^{-m} as a function of m. Suppose that $n \approx 2^{256} \approx e^{177}$, since these are the sizes of primes that we seek for use in **RSA**. Then the first function is roughly $175/(175 + 2^{m+1})$. We tabulate the two functions for some values of m in Figure 4.8.

Although $175/(175 + 2^{m+1})$ approaches zero exponentially quickly, it does not do so as quickly as 2^{-m}. In practice, however, one would take m to be something like 50 or 100, which will reduce the probability of error to a very small quantity.

We conclude this section with another Monte Carlo algorithm for **Composites** which is known as the Miller-Rabin algorithm (it is also known as the "strong pseudo-prime test"). This algorithm is presented in Figure 4.9. It is clearly a polynomial-time algorithm: an elementary analysis shows that its complexity is $O((\log n)^3)$, as in the case of the Solovay-Strassen test. In fact, the Miller-Rabin algorithm performs better in practice than the Solovay-Strassen algorithm.

We show now that this algorithm cannot answer "n is composite" if n is prime, i.e., the algorithm is yes-biased.

THEOREM 4.10
The Miller-Rabin algorithm for **Composites** *is a yes-biased Monte Carlo algorithm.*

PROOF We will prove this by assuming the algorithm answers "n is composite" for some prime integer n, and obtain a contradiction. Since the algorithm answers "n is composite," it must be the case that $a^m \not\equiv 1 \pmod{n}$. Now consider the sequence of values b tested in the algorithm. Since b is squared in each iteration of the **for** loop, we are testing the values $a^m, a^{2m}, \ldots, a^{2^{k-1}m}$. Since the algorithm

FIGURE 4.9
The Miller-Rabin primality test for an odd integer n

---

1. write $n - 1 = 2^k m$, where m is odd
2. choose a random integer a, $1 \leq a \leq n - 1$
3. compute $b = a^m \bmod n$
4. **if** $b \equiv 1 \pmod{n}$ **then**
    answer "n is prime" and **quit**
5. **for** $i = 0$ **to** $k - 1$ **do**
6.   **if** $b \equiv -1 \pmod{n}$ **then**
        answer "n is prime" and **quit**
    else
        $b = b^2 \bmod n$
7. answer "n is composite"

---

answers "n is composite," we conclude that

$$a^{2^i m} \not\equiv -1 \pmod{n}$$

for $0 \leq i \leq k - 1$.

Now, using the assumption that n is prime, Fermat's theorem (Corollary 4.6) tells us that

$$a^{2^k m} \equiv 1 \pmod{n}$$

since $n - 1 = 2^k m$. Then $a^{2^{k-1} m}$ is a square root of 1 modulo n. Since n is prime, there are only two square roots of 1 modulo n, namely, $\pm 1 \bmod n$. This can be seen as follows: x is a square root of 1 modulo n if and only if

$$n \mid (x - 1)(x + 1).$$

Since n is prime, either $n \mid (x - 1)$ (i.e., $x \equiv 1 \pmod{n}$) or $n \mid (x + 1)$ (i.e., $x \equiv -1 \pmod{n}$).

We have that

$$a^{2^{k-1} m} \not\equiv -1 \pmod{n},$$

so it follows that

$$a^{2^{k-1} m} \equiv 1 \pmod{n}.$$

Then $a^{2^{k-2} m}$ must be a square root of 1. By the same argument,

$$a^{2^{k-2} m} \equiv 1 \pmod{n}.$$

Repeating this argument, we eventually obtain

$$a^m \equiv 1 \ (\text{mod } n),$$

which is a contradiction, since the algorithm would have answered "n is prime" in this case. ∎

It remains to consider the error probability of the Miller-Rabin algorithm. Although we will not prove it here, the error probability can be shown to be, at most, $1/4$.

4.6   Attacks On RSA

In this section, we address the question: are there possible attacks on **RSA** other than factoring n? Let us first observe that it is sufficient for the cryptanalyst to compute $\phi(n)$. For, if n and $\phi(n)$ are known, and n is the product of two primes p, q, then n can be easily factored, by solving the two equations

$$n = pq$$
$$\phi(n) = (p-1)(q-1)$$

for the two "unknowns" p and q. If we substitute $q = n/p$ into the second equation, we obtain a quadratic equation in the unknown value p:

$$p^2 - (n - \phi(n) + 1)p + n = 0.$$

The two roots of this equation will be p and q, the factors of n. Hence, if a cryptanalyst can learn the value of $\phi(n)$, then he can factor n and break the system. In other words, computing $\phi(n)$ is no easier than factoring n.

Here is an example to illustrate.

Example 4.9
Suppose the cryptanalyst has learned that $n = 84773093$ and $\phi(n) = 84754668$. This information gives rise to the following quadratic equation:

$$p^2 - 18426p + 84773093 = 0.$$

This can be solved by the quadratic formula, yielding the two roots 9539 and 8887. These are the two factors of n. ☐

4.6.1 The Decryption Exponent

We will now prove the very interesting result that any algorithm which computes the decryption exponent a can be used as a subroutine (or *oracle*) in a probabilistic algorithm that factors n. So we can say that computing a is no easier than factoring n. However, this does not rule out the possibility of breaking the cryptosystem without computing a.

Notice that this result is of much more than theoretical interest. It tells us that if a is revealed, then the value n is also compromised. If this happens, it is not sufficient for Bob to choose a new encryption exponent; he must also choose a new modulus n.

The algorithm we are going to describe is a probabilistic algorithm of the Las Vegas type. Here is the definition:

DEFINITION 4.5 *Suppose $0 \leq \epsilon < 1$ is a real number. A Las Vegas algorithm is a probabilistic algorithm such that, for any problem instance I, the algorithm may fail to give an answer with some probability ϵ (i.e., it can terminate with the message "no answer"). However, if the algorithm does return an answer, then the answer must be correct.*

REMARK A Las Vegas algorithm may not give an answer, but any answer it gives is correct. In contrast, a Monte Carlo algorithm always gives an answer, but the answer may be incorrect. ∎

If we have a Las Vegas algorithm to solve a problem, then we simply run the algorithm over and over again until it finds an answer. The probability that the algorithm will return "no answer" m times in succession is ϵ^m. The average (i.e., expected) number of times the algorithm must be run in order to obtain an answer is in fact $1/(1 - \epsilon)$ (see the exercises).

Suppose that **A** is a hypothetical algorithm that computes the decryption exponent a from b and n. We will describe a Las Vegas algorithm that uses **A** as an oracle. This algorithm will factor n with probability at least $1/2$. Hence, if the algorithm is run m times, then n will be factored with probability at least $1 - 1/2^m$.

The algorithm is based on certain facts concerning square roots of 1 modulo n, where $n = pq$ is the product of two distinct odd primes. Recall that the congruence $x^2 \equiv 1 \pmod{p}$ has two solutions modulo p, namely $x = \pm 1 \bmod p$. Similarly, the congruence $x^2 \equiv 1 \pmod{q}$ has two solutions, namely $x = \pm 1 \bmod q$.

Now, since $x^2 \equiv 1 \pmod{n}$ if and only if $x^2 \equiv 1 \pmod{p}$ and $x^2 \equiv 1 \pmod{q}$, it follows that $x^2 \equiv 1 \pmod{n}$ if and only if $x = \pm 1 \bmod p$ and $x = \pm 1 \bmod q$. Hence, there are four square roots of 1 modulo n, and they can be found using the Chinese remainder theorem. Two of these solutions are $x = \pm 1 \bmod n$; these are called the *trivial* square roots of 1 modulo p. The other two square roots are called *non-trivial*, and they are negatives of each other modulo n.

Here is a small example to illustrate.

Example 4.10
Suppose $n = 403 = 13 \times 31$. The four square roots of 1 modulo 403 are $1, 92, 311$ and 402. The square root 92 is obtained by solving the system $x \equiv 1 \pmod{13}$, $x \equiv -1 \pmod{31}$ using the Chinese remainder theorem. Having found this non-trivial square root, the other non-trivial square root must be $403 - 92 = 311$. It is the solution to the system $x \equiv -1 \pmod{13}$, $x \equiv 1 \pmod{31}$.   ▯

Suppose x is a non-trivial square root of 1 modulo n. Then we have

$$n \mid (x-1)(x+1),$$

but n divides neither factor on the right side. It follows that $\gcd(x+1, n) = p$ or q (and similarly, $\gcd(x-1, n) = p$ or q). Of course, a greatest common divisor can be computed using the Euclidean algorithm, without knowing the factorization of n. Hence, knowledge of a non-trivial square root of 1 modulo n yields the factorization of n with only a polynomial amount of computation. This important fact is the basis of many results in cryptography.

In Example 4.10 above, $\gcd(93, 403) = 31$ and $\gcd(312, 403) = 13$.

In Figure 4.10, we present an algorithm which, using the hypothetical algorithm **A** as a subroutine, attempts to factor n by finding a non-trivial square root of 1 modulo n. (Recall that **A** computes the decryption exponent a corresponding to the encryption exponent b.) We first do an example to illustrate the application of this algorithm.

Example 4.11
Suppose $n = 89855713$, $b = 34986517$ and $a = 82330933$, and the random value $w = 5$. We have

$$ab - 1 = 2^3 \times 360059073378795.$$

In step 6, $v = 85877701$, and in step 10, $v = 1$. In step 12, we compute

$$\gcd(85877702, n) = 9103.$$

This is one factor of n; the other is $n/9103 = 9871$.   ▯

Let's now proceed to the analysis of the algorithm. First, observe that if we are lucky enough to choose w to be a multiple of p or q, then we can factor n

FIGURE 4.10
Factoring algorithm, given the decryption exponent a

1.	choose w at random such that $1 \leq w \leq n-1$
2.	compute $x = \gcd(w, n)$
3.	**if** $1 < x < n$ **then quit** (success: $x = p$ or $x = q$)
4.	compute $a = \mathbf{A}(b)$
5.	write $ab - 1 = 2^s r$, r odd
6.	compute $v = w^r \bmod n$
7.	**if** $v \equiv 1 \pmod{n}$ **then quit** (failure)
8.	**while** $v \not\equiv 1 \pmod{n}$ **do**
9.	$\quad v_0 = v$
10.	$\quad v = v^2 \bmod n$
11.	**if** $v_0 \equiv -1 \pmod{n}$ **then**
	$\quad$ **quit** (failure)
	else
	$\quad$ compute $x = \gcd(v_0 + 1, n)$ (success: $x = p$ or $x = q$)

immediately. This is detected in step 2. If w is relatively prime to n, then we compute $w^r, w^{2r}, w^{4r}, \ldots$, by successive squaring, until

$$w^{2^t r} \equiv 1 \pmod{n}$$

for some t. Since

$$ab - 1 = 2^s r \equiv 0 \pmod{\phi(n)},$$

we know that $w^{2^s r} \equiv 1 \pmod{n}$. Hence, the **while** loop terminates after at most s iterations. At the end of the **while** loop, we have found a value v_0 such that $v_0^2 \equiv 1 \pmod{n}$ but $v_0 \not\equiv 1 \pmod{n}$. If $v_0 \equiv -1 \pmod{n}$, then the algorithm fails; otherwise, v_0 is a non-trivial square root of 1 modulo n and we are able to factor n (step 12).

The main task facing us now is to prove that the algorithm succeeds with probability at least $1/2$. There are two ways in which the algorithm can fail to factor n:

1. $w^r \equiv 1 \pmod{n}$ (step 7)
2. $w^{2^t r} \equiv -1 \pmod{n}$ for some t, $0 \leq t \leq s-1$ (step 11)

We have $s + 1$ congruences to consider. If a random value w is a solution to at least one of these $s + 1$ congruences, then it is a "bad" choice, and the algorithm fails. So we proceed by counting the number of solutions to each of these congruences.

First, consider the congruence $w^r \equiv 1 \pmod{n}$. The way to analyze a congruence such as this is to consider solutions modulo p and modulo q separately, and then combine them using the Chinese remainder theorem. Observe that $x \equiv 1 \pmod{n}$ if and only if $x \equiv 1 \pmod{p}$ and $x \equiv 1 \pmod{q}$.

So, we first consider $w^r \equiv 1 \pmod{p}$. Since p is prime, $\mathbb{Z}_p^*$ is a cyclic group by Theorem 4.7. Let g be a primitive element modulo p. We can write $w = g^u$ for a unique integer u, $0 \le u \le p - 2$. Then we have

$$w^r \equiv 1 \pmod{p}$$

$$g^{ur} \equiv 1 \pmod{p}$$

$$(p - 1) \mid ur.$$

Let us write

$$p - 1 = 2^i p_1$$

where p_1 is odd, and

$$q - 1 = 2^j q_1$$

where q_1 is odd. Since

$$\phi(n) = (p - 1)(q - 1) \mid (ab - 1) = 2^s r,$$

we have that

$$2^{i+j} p_1 q_1 \mid 2^s r.$$

Hence

$$i + j \le s$$

and

$$p_1 q_1 \mid r.$$

Now, the condition $(p - 1) \mid ur$ becomes $2^i p_1 \mid ur$. Since $p_1 \mid r$ and r is odd, it is necessary and sufficient that $2^i \mid u$. Hence, $u = k2^i$, $0 \le k \le p_1 - 1$, and the number of solutions to the congruence $w^r \equiv 1 \pmod{p}$ is p_1.

By an identical argument, the congruence $w^r \equiv 1 \pmod{q}$ has exactly q_1 solutions. We can combine any solution modulo p with any solution modulo q to obtain a unique solution modulo n, using the Chinese remainder theorem. Consequently, the number of solutions to the congruence $w^r \equiv 1 \pmod{n}$ is $p_1 q_1$.

The next step is to consider a congruence $w^{2^t r} \equiv -1 \pmod{n}$ for a fixed value t (where $0 \le t \le s - 1$). Again, we first look at the congruence modulo p and then modulo q (note that $w^{2^t r} \equiv -1 \pmod{n}$ if and only if $w^{2^t r} \equiv -1 \pmod{p}$)

and $w^{2^t r} \equiv -1 \pmod q$. First, consider $w^{2^t r} \equiv -1 \pmod p$. Writing $w = g^u$, as above, we get

$$g^{u2^t r} \equiv -1 \pmod p.$$

Since $g^{(p-1)/2} \equiv -1 \pmod p$, we have that

$$u2^t r \equiv \frac{p-1}{2} \pmod{p-1}$$

$$(p-1) \mid \left(u2^t r - \frac{p-1}{2}\right)$$

$$2(p-1) \mid (u2^{t+1}r - (p-1)).$$

Since $p - 1 = 2^i p_1$, we get

$$2^{i+1} p_1 \mid (u2^{t+1}r - 2^i p_1).$$

Taking out a common factor of p_1, this becomes

$$2^{i+1} \mid \left(\frac{u2^{t+1}r}{p_1} - 2^i\right).$$

Now, if $t \geq i$, then there can be no solutions since $2^{i+1} \mid 2^{t+1}$ but $2^{i+1} \nmid 2^i$. On the other hand, if $t \leq i - 1$, then u is a solution if and only if u is an odd multiple of 2^{i-t-1} (note that r/p_1 is an odd integer). So, the number of solutions in this case is

$$\frac{p-1}{2^{i-t-1}} \times \frac{1}{2} = 2^t p_1.$$

By similar reasoning, the congruence $w^{2^t r} \equiv -1 \pmod q$ has no solutions if $t \geq j$, and $2^t q_1$ solutions if $t \leq j - 1$. Using the Chinese remainder theorem, we see that the number of solutions of $w^{2^t r} \equiv -1 \pmod n$ is

$$0 \quad \text{if } t \geq \min\{i, j\}$$
$$2^{2t} p_1 q_1 \quad \text{if } t \leq \min\{i, j\} - 1.$$

Now, t can range from 0 to $s - 1$. Without loss of generality, suppose $i \leq j$; then the number of solutions is 0 if $t \geq i$. The total number of "bad" choices for w is *at most*

$$p_1 q_1 + p_1 q_1 (1 + 2^2 + 2^4 + \ldots + 2^{2i-2}) = p_1 q_1 \left(1 + \frac{2^{2i} - 1}{3}\right)$$

$$= p_1 q_1 \left(\frac{2}{3} + \frac{2^{2i}}{3}\right).$$

Recall that $p - 1 = 2^i p_1$ and $q - 1 = 2^j q_1$. Now, $j \geq i \geq 1$, so $p_1 q_1 < n/4$. We also have that

$$2^{2i} p_1 q_1 \leq 2^{i+j} p_1 q_1 = (p-1)(q-1) < n.$$

Hence, we obtain

$$p_1 q_1 \left(\frac{2}{3} + \frac{2^{2i}}{3} \right) < \frac{n}{6} + \frac{n}{3}$$

$$= \frac{n}{2}.$$

Since at most $(n-1)/2$ choices for w are "bad," it follows that at least $(n-1)/2$ choices are "good" and hence the probability of success of the algorithm is at least $1/2$.

4.6.2 Partial Information Concerning Plaintext Bits

The other result we will discuss concerns partial information about the plaintext that might be "leaked" by an **RSA** encryption. Two examples of partial information that we consider are the following:

1. given $y = e_K(x)$, compute $parity(y)$, where $parity(y)$ denotes the low-order bit of x

2. given $y = e_K(x)$, compute $half(y)$, where $half(y) = 0$ if $0 \le x < n/2$ and $half(y) = 1$ if $n/2 < x \le n - 1$.

We will prove that, given $y = e_K(x)$, any algorithm that computes $parity(y)$ or $half(y)$ can be used as an oracle to construct an algorithm that computes the plaintext x. What this means is that, given a ciphertext, computing the low-order bit of the plaintext is polynomially equivalent to determining the whole plaintext! First, we prove that computing $parity(y)$ is polynomially equivalent to computing $half(y)$. This follows from the following two easily proved identities (see the exercises):

$$half(y) = parity(y \times e_K(2) \bmod n) \tag{4.1}$$

$$parity(y) = half(y \times e_K(2^{-1}) \bmod n) \tag{4.2}$$

and from the multiplicative rule $e_K(x_1)e_K(x_2) = e_K(x_1 x_2)$.

We will show how to compute $x = d_K(y)$, given a hypothetical algorithm (oracle) which computes $half(y)$. The algorithm is presented in Figure 4.11. In steps 2–4, we compute

$$y_i = half(y \times (e_K(2))^i) = half(e_K(x \times 2^i)),$$

for $0 \le i \le \log_2 n$. We observe that

$$half(e_K(x)) = 0 \Leftrightarrow x \in \left[0, \frac{n}{2} \right)$$

$$half(e_K(2x)) = 0 \Leftrightarrow x \in \left[0, \frac{n}{4} \right) \cup \left[\frac{n}{2}, \frac{3n}{4} \right)$$

$$half(e_K(4x)) = 0 \Leftrightarrow x \in \left[0, \frac{n}{8} \right) \cup \left[\frac{n}{4}, \frac{3n}{8} \right) \cup \left[\frac{n}{2}, \frac{5n}{8} \right) \cup \left[\frac{3n}{4}, \frac{7n}{8} \right),$$

FIGURE 4.11
Decrypting RSA ciphertext, given an oracle for computing $half(y)$

1.   denote $k = \lfloor \log_2 n \rfloor$
2.   **for** $i = 0$ **to** k **do**
3. $y_i = half(y)$
4. $y = (y \times e_K(2)) \bmod n$
5. $lo = 0$
6. $hi = n$
7.   **for** $i = 0$ **to** k **do**
8. $mid = (hi + lo)/2$
9.   **if** $y_i = 1$ **then**
$lo = mid$
else
$hi = mid$
10. $x = \lfloor hi \rfloor$

and so on. Hence, we can find x by a binary search technique, which is done in steps 7–11. Here is a small example to illustrate.

Example 4.12
Suppose $n = 1457$, $b = 779$, and we have a ciphertext $y = 722$. $e_K(2)$ is computed to be 946. Suppose, using our oracle for *half*, that we obtain the following values y_i in step 3 of the algorithm:

i	0	1	2	3	4	5	6	7	8	9	10
y_i	1	0	1	0	1	1	1	1	1	0	0

Then the binary search proceeds as shown in Figure 4.12. Hence, the plaintext is $x = \lfloor 999.55 \rfloor = 999$.   ◻

4.7   The Rabin Cryptosystem

In this section, we describe the **Rabin Cryptosystem**, which is computationally secure against a chosen-plaintext attack provided that the modulus $n = pq$ cannot

FIGURE 4.12
Binary search for RSA decryption

i	lo	mid	hi
0	0.00	728.50	1457.00
1	728.50	1092.75	1457.00
2	728.50	910.62	1092.75
3	910.62	1001.69	1092.75
4	910.62	956.16	1001.69
5	956.16	978.92	1001.69
6	978.92	990.30	1001.69
7	990.30	996.00	1001.69
8	996.00	998.84	1001.69
9	998.84	1000.26	1001.69
10	998.84	999.55	1000.26
	998.84	999.55	999.55

be factored. The system is described in Figure 4.13.

We will show that the encryption function e_K is not an injection, so decryption cannot be done in an unambiguous fashion. In fact, there are four possible plaintexts that could be the encryption of any given ciphertext. More precisely, let ω be one of the four square roots of 1 modulo n. Let $x \in \mathbb{Z}_n$. Then, we can verify the following equations:

$$e_K\left(\omega\left(x + \frac{B}{2}\right) - \frac{B}{2}\right) = \left(\omega\left(x + \frac{B}{2}\right) - \frac{B}{2}\right)\left(\omega\left(x + \frac{B}{2}\right) + \frac{B}{2}\right)$$

$$= \omega^2\left(x + \frac{B}{2}\right)^2 - \left(\frac{B}{2}\right)^2$$

$$= x^2 + Bx + \frac{B^2}{4} - \frac{B^2}{4}$$

$$= x^2 + Bx$$

$$= e_K(x).$$

(Note that all arithmetic is being done in $\mathbb{Z}_n$, and division by 2 and 4 is the same as multiplication by 2^{-1} and 4^{-1} modulo n, respectively.)

The four plaintexts that encrypt to $e_K(x)$ are x, $-x - B$, $\omega(x + B/2) - B/2$ and $-\omega(x + B/2) - B/2$, where ω is a non-trivial square root of 1 modulo n. In general, there will be no way for Bob to distinguish which of these four possible plaintexts is the "right" plaintext, unless the plaintext contains sufficient redundancy to eliminate three of these four possible values.

FIGURE 4.13
Rabin Cryptosystem

Let n be the product of two distinct primes p and q, $p, q \equiv 3 \pmod{4}$. Let $\mathcal{P} = \mathcal{C} = \mathbb{Z}_n$, and define

$$\mathcal{K} = \{(n, p, q, B) : 0 \leq B \leq n - 1\}.$$

For $K = (n, p, q, B)$, define

$$e_K(x) = x(x + B) \bmod n$$

and

$$d_K(y) = \sqrt{\frac{B^2}{4} + y} - \frac{B}{2} \bmod n.$$

The values n and B are public, while p and q are secret.

Let us look at the decryption problem from Bob's point of view. He is given a ciphertext y and wants to determine x such that

$$x^2 + Bx \equiv y \pmod{n}.$$

This is a quadratic equation in the unknown x. We can eliminate the linear term by making the substitution $x_1 = x + B/2$, or equivalently, $x = x_1 - B/2$. Then the equation becomes

$$x_1^2 - Bx_1 + \frac{B^2}{4} + Bx_1 - \frac{B^2}{2} - y \equiv 0 \pmod{n},$$

or

$$x_1^2 \equiv \frac{B^2}{4} + y \pmod{n}.$$

If we define $C = B^2/4 + y$, then we can rewrite the congruence as

$$x_1^2 \equiv C \pmod{n}.$$

So, decryption reduces to extracting square roots modulo n. This is equivalent to solving the two congruences

$$x_1^2 \equiv C \pmod{p}$$

and

$$x_1^2 \equiv C \pmod{q}.$$

(There are two square roots of C modulo p and two square roots modulo q. Using the Chinese remainder theorem, these can be combined to yield four solutions modulo n.) We can use Euler's criterion to determine if C is a quadratic residue modulo p (and modulo q). In fact, C will be a quadratic residue modulo p and modulo q if encryption was performed correctly. But Euler's criterion does not help us find the square roots of C; it yields only an answer "yes" or "no."

When $p \equiv 3 \pmod 4$, there is a simple formula to compute square roots of quadratic residues modulo p. Suppose C is a quadratic residue and $p \equiv 3 \pmod 4$. Then we have that

$$(\pm C^{(p+1)/4})^2 \equiv C^{(p+1)/2} \pmod p$$
$$\equiv C^{(p-1)/2}C \pmod p$$
$$\equiv C \pmod p.$$

Here we again make use of Euler's criterion, which says that if C is a quadratic residue modulo p, then $C^{(p-1)/2} \equiv 1 \pmod p$. Hence, the two square roots of C modulo p are $\pm C^{(p+1)/4} \bmod p$. In a similar fashion, the two square roots of C modulo q are $\pm C^{(q+1)/4} \bmod q$. It is then straightforward to obtain the four square roots x_1 of C modulo n using the Chinese remainder theorem.

REMARK It is interesting that for $p \equiv 1 \pmod 4$ there is no known polynomial-time deterministic algorithm to compute square roots of quadratic residues modulo p. There is a polynomial-time Las Vegas algorithm, however. ∎

Once we have determined the four possible values for x_1, we compute x from the equation $x = x_1 - B/2$ to get the four possible plaintexts. This yields the decryption formula

$$d_K(y) = \sqrt{\frac{B^2}{4} + y} - \frac{B}{2}.$$

Example 4.13
Let's illustrate the encryption and decryption procedures for the **Rabin Cryptosystem** with a toy example. Suppose $n = 77 = 7 \times 11$ and $B = 9$. Then the encryption function is

$$e_K(x) = x^2 + 9x \bmod 77$$

and the decryption function is

$$d_K(y) = \sqrt{1 + y} - 43 \bmod 77.$$

Suppose Bob wants to decrypt the ciphertext $y = 22$. It is first necessary to find the square roots of 23 modulo 7 and modulo 11. Since 7 and 11 are both congruent to 3 modulo 4, we use our formula:

$$23^{(7+1)/4} \equiv 2^2 \equiv 4 \bmod 7$$

FIGURE 4.14
Factoring a Rabin modulus, given a decryption oracle

> 1. choose a random r, $1 \leq r \leq n - 1$
> 2. compute $y = r^2 - B^2/4 \bmod n$
> 3. call $\mathbf{A}(y)$, obtaining a decryption x
> 4. compute $x_1 = x + B/2$
> 5. **if** $x_1 \equiv \pm r \pmod{n}$ **then**
>   > **quit** (failure)
>   **else**
>   > $\gcd(x_1 + r, n) = p$ or q (success)

and
$$23^{(11+1)/4} \equiv 1^3 \equiv 1 \bmod 11.$$

Using the Chinese remainder theorem, we compute the four square roots of 23 modulo 77 to be $\pm 10, \pm 32 \bmod 77$. Finally, the four possible plaintexts are:

$$10 - 43 \bmod 77 = 44$$
$$67 - 43 \bmod 77 = 24$$
$$32 - 43 \bmod 77 = 66$$
$$45 - 43 \bmod 77 = 2.$$

It can be verified that each of these plaintexts encrypts to the ciphertext 22. $\quad\Box$

We now discuss the security of the **Rabin Cryptosystem**. We will prove that any hypothetical decryption algorithm $\mathbf{A}$ can be used as an oracle in a Las Vegas algorithm that factors the modulus n with probability at least $1/2$. This algorithm is depicted in Figure 4.14.

There are several points of explanation needed. First, observe that

$$y = e_K\left(r - \frac{B}{2}\right),$$

so a value x will be returned in step 3. Next, we look at step 4 and note that $x_1^2 \equiv r^2 \pmod{n}$. It follows that $x_1 \equiv \pm r \pmod{n}$ or $x_1 \equiv \pm \omega r \pmod{n}$, where ω is one of the non-trivial square roots of 1 modulo n. In the second case, we have

$$n \mid (x_1 - r)(x_1 + r),$$

but n does not divide either factor on the right side. Hence, computation of $\gcd(x_1 + r, n)$ (or $\gcd(x_1 - r, n)$) must yield either p or q, and the factorization of n is accomplished.

Let's compute the probability of success of this algorithm, over all $n-1$ choices for the random value r. For two non-zero residues r_1 and r_2, define

$$r_1 \sim r_2 \Leftrightarrow r_1^2 \equiv r_2^2 \pmod{n}.$$

It is easy to see that $r \sim r$ for all r; $r_1 \sim r_2$ implies $r_2 \sim r_1$; and $r_1 \sim r_2$ and $r_2 \sim r_3$ together imply $r_1 \sim r_3$. This says that the relation $\sim$ is an *equivalence relation*. The equivalence classes of $\mathbb{Z}_n \setminus \{0\}$ all have cardinality four: the equivalence class containing r is the set

$$[r] = \{\pm r, \pm \omega r \bmod n\},$$

where ω is a non-trivial square root of 1 modulo n.

In the algorithm presented in Figure 4.14, any two values r in the same equivalence class will yield the same value y. Now consider the value x returned by the oracle **A** when given y. We have

$$[y] = \{\pm y, \pm \omega y\}.$$

If $r = \pm y$, then the algorithm fails; while it succeeds if $r = \pm \omega y$. Since r is chosen at random, it is equally likely to be any of these four possible values. We conclude that the probability of success of the algorithm is $1/2$.

It is interesting that the **Rabin Cryptosystem** is provably secure against a chosen plaintext attack. However, the system is completely insecure against a chosen ciphertext attack. In fact the algorithm in Figure 4.14, that we used to prove security against a chosen plaintext attack, also can be used to break the **Rabin Cryptosystem** in a chosen ciphertext attack! In the chosen ciphertext attack, the oracle **A** is replaced by Bob's decryption algorithm.

4.8   Factoring Algorithms

There is a huge amount of literature on factoring algorithms, and a careful treatment would require more pages than we have in this book. We will just try to give a brief overview here, including an informal discussion of the best current factoring algorithms and their use in practice. The three algorithms that are most effective on very large numbers are the quadratic sieve, the elliptic curve algorithm and the number field sieve. Other well-known algorithms that were precursors include Pollard's rho-method and $p-1$ algorithm, Williams' $p+1$ algorithm, the continued fraction algorithm, and of course, trial division.

Throughout this section, we suppose that the integer n that we wish to factor is odd. *Trial division* consists of dividing n by every odd integer up to $\lfloor \sqrt{n} \rfloor$. If $n < 10^{12}$, say, this is a perfectly reasonable factorization method, but for larger n we generally need to use more sophisticated techniques.

FIGURE 4.15
The $p - 1$ **factoring algorithm**

Input: n and B
1. $a = 2$
2. **for** $j = 2$ **to** B **do**
$\qquad a = a^j \bmod n$
3. $d = \gcd(a - 1, n)$
4. **if** $1 < d < n$ **then**
$\qquad d$ is a factor of n (success)
else
$\qquad$ no factor of n is found (failure)

4.8.1 The $p - 1$ Method

As an example of a simple algorithm that can sometimes be applied to larger integers, we describe Pollard's $p - 1$ algorithm, which dates from 1974. This algorithm, presented in Figure 4.15, has two inputs: the (odd) integer n to be factored, and a "bound" B. Here is what is taking place in the $p - 1$ algorithm: Suppose p is a prime divisor of n, and $q \leq B$ for every prime power $q \mid (p - 1)$. Then it must be the case that

$$(p - 1) \mid B!$$

At the end of the **for** loop (step 2),

$$a \equiv 2^{B!} \pmod{n},$$

so

$$a \equiv 2^{B!} \pmod{p}$$

since $p \mid n$. Now,

$$2^{p-1} \equiv 1 \pmod{p}$$

by Fermat's theorem. Since $(p - 1) \mid B!$, we have that

$$a \equiv 1 \pmod{p}$$

(in step 3). Thus, in step 4,

$$p \mid (a - 1)$$

and

$$p \mid n,$$

so

$$p \mid d = \gcd(a - 1, n).$$

The integer d will be a non-trivial divisor of n (unless $a = 1$ in step 3). Having found a non-trivial factor d, we would then proceed to attempt to factor d and n/d if they are composite.

Here is an example to illustrate.

Example 4.14

Suppose $n = 15770708441$. If we apply the $p - 1$ algorithm with $B = 180$, then we find that $a = 11620221425$ in step 3, and d is computed to be 135979. In fact, the complete factorization of n into primes is

$$15770708441 = 135979 \times 115979.$$

in this case, the factorization succeeds because 135978 has only "small" prime factors:

$$135978 = 2 \times 3 \times 131 \times 173.$$

Hence, by taking $B \geq 173$, it will be the case that $135978 \mid B!$, as desired. ⬚

In the algorithm, there are $B - 1$ modular exponentiations, each requring at most $2 \log_2 B$ modular multiplications using square-and-multiply. The gcd computation can be done in time $O((\log n)^3)$ using the Euclidean algorithm. Hence, the complexity of the algorithm is $O(B \log B (\log n)^2 + (\log n)^3)$. If the integer B is $O((\log n)^i)$ for some fixed integer i, then the algorithm is indeed a polynomial-time algorithm; however, for such a choice of B the probability of success will be very small. On the other hand, if we increase the size of B drastically, say to $\sqrt{n}$, then the algorithm will be successful, but it will be no faster than trial division.

Thus, the drawback of this method is that it requires n to have a prime factor p such that $p - 1$ has only "small" prime factors. It would be very easy to construct an **RSA** modulus $n = pq$ which would resist factorization by this method. One would start by finding a large prime p_1 such that $p = 2p_1 + 1$ is also prime, and a large prime q_1 such that $q = 2q_1 + 1$ is also prime (using one of the Monte Carlo primality testing algorithms discussed in Section 4.5). Then the **RSA** modulus $n = pq$ will be resistant to factorization using the $p - 1$ method.

The more powerful elliptic curve algorithm, developed by Lenstra in the mid-1980's, is in fact a generalization of the $p - 1$ method. We will not discuss the theory at all here, but we do mention that the success of the elliptic curve method depends on the more likely situation that an integer "close to" p has only "small" prime factors. Whereas the $p - 1$ method depends on a relation that holds in the group $\mathbb{Z}_p$, the elliptic curve method involves groups defined on elliptic curves modulo p.

4.8.2 Dixon's Algorithm and the Quadratic Sieve

Dixon's algorithm is based on a very simple idea that we already saw in connection with the **Rabin Cryptosystem**. Namely, if we can find $x \not\equiv \pm y \pmod{n}$ such that $x^2 \equiv y^2 \pmod{n}$, then $\gcd(x - y, n)$ is a non-trivial factor of n.

The method uses a *factor base*, which is a set $\mathcal{B}$ of "small" primes. We first obtain several integers x such that all the prime factors of $x^2 \bmod n$ occur in the factor base $\mathcal{B}$. (How this is done will be discussed a bit later.) The idea is to then take the product of several of these x's in such a way that every prime in the factor base is used an even number of times. This then gives us a congruence of the desired type $x^2 \equiv y^2 \pmod{n}$, which (we hope) will lead to a factorization of n.

We illustrate with a carefully contrived example.

Example 4.15
Suppose $n = 15770708441$ (this was the same n that we used in Example 4.14). Let $\mathcal{B} = \{2, 3, 5, 7, 11, 13\}$. Consider the three congruences:

$$8340934156^2 \equiv 3 \times 7 \pmod{n}$$
$$12044942944^2 \equiv 2 \times 7 \times 13 \pmod{n}$$
$$2773700011^2 \equiv 2 \times 3 \times 13 \pmod{n}.$$

If we take the product of these three congruences, then we have

$$(8340934156 \times 12044942944 \times 2773700011)^2 \equiv (2 \times 3 \times 7 \times 13)^2 \pmod{n}.$$

Reducing the expressions inside the parentheses modulo n, we have

$$9503435785^2 \equiv 546^2 \pmod{n}.$$

Then we compute

$$\gcd(9503435785 - 546, 15770708441) = 115759,$$

finding the factor 115759 of n. ❑

Suppose $\mathcal{B} = \{p_1, \ldots, p_B\}$ is the factor base. Let C be slightly larger than B (say $C = B + 10$), and suppose we have obtained C congruences:

$$x_j{}^2 \equiv p_1{}^{\alpha_{1j}} \times p_2{}^{\alpha_{2j}} \ldots \times p_B{}^{\alpha_{Bj}} \pmod{n},$$

for $1 \leq j \leq C$. For each j, consider the vector

$$a_j = (\alpha_{1j} \bmod 2, \ldots, \alpha_{Bj} \bmod 2) \in (\mathbb{Z}_2)^B.$$

If we can find a subset of the a_j's that sum modulo 2 to the vector $(0, \ldots, 0)$, then the product of the corresponding x_j's will use each factor in B an even number of times.

We illustrate by returning to Example 4.15, where there exists a dependence even though $C < B$ in this case.

Example 4.15   (Cont.)
The three vectors a_1, a_2, a_3 are as follows:

$$a_1 = (0, 1, 0, 1, 0, 0)$$
$$a_2 = (1, 0, 0, 1, 0, 1)$$
$$a_3 = (1, 1, 0, 0, 0, 1).$$

It is easy to see that

$$a_1 + a_2 + a_3 = (0, 0, 0, 0, 0, 0) \bmod 2.$$

This gives rise to the congruence we saw earlier that successfully factored n.   ⬚

Observe that finding a subset of the C vectors $a_1, \ldots, a_C$ that sums modulo 2 to the all-zero vector is nothing more than finding a linear dependence (over $\mathbb{Z}_2$) of these vectors. Provided $C > B$, such a linear dependence must exist, and it can be found easily using the standard method of Gaussian elimination. The reason why we take $C > B + 1$ is that there is no guarantee that any given congruence will yield the factorization of n. Approximately 50% of the time it will turn out that $x \equiv \pm y \pmod{n}$. But if $C > B + 1$, then we can obtain several such congruences (arising from different linear dependencies among the a_j's). Hopefully, at least one of the resulting congruences will yield the factorization.

It remains to discuss how we obtain integers x_j such that the values $x_j^2 \bmod n$ factor completely over the factor base B. There are several methods of doing this. One common approach is the Quadratic Sieve due to Pomerance, which uses integers of the form $x_j = j + \lfloor \sqrt{n} \rfloor$, $j = 1, 2, \ldots$. The name "quadratic sieve" comes from a sieving procedure (which we will not describe here) that is used to determine those x_j's that factor over B.

There is, of course, a trade-off here: if $B = |B|$ is large, then it is more likely that an integer x_j factors over B. But the larger B is, the more congruences we need to accumulate before we are able to find a dependence relation. The optimal choice for B is approximately

$$\sqrt{e^{\sqrt{\ln n \ln \ln n}}},$$

and this leads to an expected running time of

$$O\left(e^{(1+o(1))\sqrt{\ln n \ln \ln n}}\right).$$

The number field sieve is a more recent factoring algorithm from the late 1980's. It also factors n by constructing a congruence $x^2 \equiv y^2 \pmod{n}$, but it does so by means of computations in rings of algebraic integers.

4.8.3 Factoring Algorithms in Practice

The asymptotic running times of the quadratic sieve, elliptic curve and number field sieve are as follows:

quadratic sieve	$O\left(e^{(1+o(1))\sqrt{\ln n \ln \ln n}}\right)$
elliptic curve	$O\left(e^{(1+o(1))\sqrt{2\ln p \ln \ln p}}\right)$
number field sieve	$O\left(e^{(1.92+o(1))(\ln n)^{1/3}(\ln \ln n)^{2/3}}\right)$

The notation $o(1)$ denotes a function of n that approaches 0 as $n \to \infty$, and p denotes the smallest prime factor of n.

In the worst case, $p \approx \sqrt{n}$ and the asymptotic running times of the quadratic sieve and elliptic curve algorithms are essentially the same. But in such a situation, quadratic sieve generally outperforms elliptic curve. The elliptic curve method is more useful if the prime factors of n are of differing size. One very large number that was factored using the elliptic curve method was the Fermat number $2^{2^{11}} - 1$ in 1988 by Brent.

For factoring **RSA** moduli (where $n = pq$, p, q are prime, and p and q are roughly the same size), the quadratic sieve is currently the most successful algorithm. Some notable milestones have included the following factorizations. In 1983, the quadratic sieve successfully factored a 69-digit number that was a (composite) factor of $2^{251} - 1$ (this computation was done by Davis, Holdridge, and Simmons). Progress continued throughout the 1980's, and by 1989, numbers having up to 106 digits were factored by this method by Lenstra and Manasse, by distributing the computations to hundreds of widely separated workstations (they called this approach "factoring by electronic mail").

More recently, in April 1994, a 129-digit number known as **RSA**-129 was factored by Atkins, Graff, Lenstra, and Leyland using the quadratic sieve. (The numbers **RSA**-100, **RSA**-110, ..., **RSA**-500 are a list of **RSA** moduli publicized on the Internet as "challenge" numbers for factoring algorithms. Each number **RSA**-d is a d-digit number that is the product of two primes of approximately the same length.) The factorization of **RSA**-129 required 5000 MIPS-years of computing time donated by over 600 researchers around the world.

The number field sieve is the most recent of the three algorithms. It seems to have great potential since its asymptotic running time is faster than either quadratic sieve or the elliptic curve. It is still in developmental stages, but people have speculated that number field sieve might prove to be faster for numbers having more than about 125–130 digits. In 1990, the number field sieve was used by Lenstra, Lenstra, Manasse, and Pollard to factor $2^{2^9} - 1$ into three primes having 7, 49 and 99 digits.

4.9 Notes and References

The idea of public-key cryptography was introduced by Diffie and Hellman in 1976. Although [DH76A] is the most cited reference, the conference paper [DH76] actually appeared a bit earlier. The **RSA Cryptosystem** was discovered by Rivest, Shamir and Adleman [RSA78]. The **Rabin Cryptosystem** was described in Rabin [RA79]; a similar provably secure system in which decryption is unambiguous was found by Williams [WI80]. For a general survey article on public-key cryptography, we recommend Diffie [DI92].

The Solovay-Strassen test was first described in [SS77]. The Miller-Rabin test was given in [MI76] and [RA80]. Our discussion of error probabilities is motivated by observations of Brassard and Bratley [BB88A, §8.6] (see also [BBCGP88]). The best current bounds on the error probability of the Miller-Rabin algorithm can be found in [DLP93].

The material in Section 4.6 is based on the treatment by Salomaa [SA90, pp. 143–154]. The factorization of n given the decryption exponent was proved in [DE84]; the results on partial information revealed by **RSA** is from [GMT82].

As mentioned earlier, there are many sources of information on factoring algorithms. Pomerance [PO90] is a good survey on factoring, and Lenstra and Lenstra [LL90] is a good article on number-theoretic algorithms in general. Bressoud [BR89] is an elementary textbook devoted to factoring and primality testing. Cryptography textbooks that emphasize number theory include Koblitz [KO94] and Kranakis [KR86]. Lenstra and Lenstra [LL93] is a monograph on the number field sieve.

Exercises 4.7–4.9 give some examples of protocol failures. For a nice article on this subject, see Moore [MO92].

Exercises

4.1 Use the Extended Euclidean algorithm to compute the following multiplicative inverses:

(a) $17^{-1} \bmod 101$

(b) $357^{-1} \bmod 1234$

(c) $3125^{-1} \bmod 9987$.

4.2 Solve the following system of congruences:

$$x \equiv 12 \pmod{25}$$

$$x \equiv 9 \pmod{26}$$

$$x \equiv 23 \pmod{27}.$$

4.3 Solve the following system of congruences:

$$13x \equiv 4 \pmod{99}$$

$$15x \equiv 56 \pmod{101}.$$

HINT First use the Extended Euclidean algorithm, and then apply the Chinese remainder theorem.

4.4 Here we investigate some properties of primitive roots.

(a) The integer 97 is prime. Prove that $x \neq 0$ is a primitive root modulo 97 if and only if $x^{32} \not\equiv 1 \pmod{97}$ and $x^{48} \not\equiv 1 \pmod{97}$.

(b) Use this method to find the smallest primitive root modulo 97.

(c) Suppose p is prime, and $p - 1$ has prime power factorization

$$p - 1 = \prod_{i=1}^{n} p_i^{e_i},$$

where the p_i's are distinct primes. Prove that $x \neq 0$ is a primitive root modulo p if and only if $x^{(p-1)/p_i} \not\equiv 1 \pmod{p}$ for $1 \leq i \leq n$.

4.5 Suppose that $n = pq$, where p and q are distinct odd primes and $ab \equiv 1 \pmod{(p-1)(q-1)}$. The **RSA** encryption operation is $e(x) = x^b \bmod n$ and the decryption operation is $d(y) = y^a \bmod n$. We proved that $d(e(x)) = x$ if $x \in \mathbb{Z}_n^*$. Prove that the same statement is true for any $x \in \mathbb{Z}_n$.

HINT Use the fact that $x_1 \equiv x_2 \pmod{pq}$ if and only if $x_1 \equiv x_2 \pmod{p}$ and $x_1 \equiv x_2 \pmod{q}$. This follows from the Chinese remainder theorem.

4.6 Two samples of **RSA** ciphertext are presented in Tables 4.1 and 4.2. Your task is to decrypt them. The public parameters of the system are $n = 18923$ and $b = 1261$ (for Table 4.1) and $n = 31313$ and $b = 4913$ (for Table 4.2). This can be accomplished as follows. First, factor n (which is easy because it is so small). Then compute the exponent a from $\phi(n)$, and, finally, decrypt the ciphertext. Use the square-and-multiply algorithm to exponentiate modulo n.

In order to translate the plaintext back into ordinary English text, you need to know how alphabetic characters are "encoded" as elements in $\mathbb{Z}_n$. Each element of

TABLE 4.1
RSA Ciphertext

12423	11524	7243	7459	14303	6127	10964	16399
9792	13629	14407	18817	18830	13556	3159	16647
5300	13951	81	8986	8007	13167	10022	17213
2264	961	17459	4101	2999	14569	17183	15827
12693	9553	18194	3830	2664	13998	12501	18873
12161	13071	16900	7233	8270	17086	9792	14266
13236	5300	13951	8850	12129	6091	18110	3332
15061	12347	7817	7946	11675	13924	13892	18031
2620	6276	8500	201	8850	11178	16477	10161
3533	13842	7537	12259	18110	44	2364	15570
3460	9886	8687	4481	11231	7547	11383	17910
12867	13203	5102	4742	5053	15407	2976	9330
12192	56	2471	15334	841	13995	17592	13297
2430	9741	11675	424	6686	738	13874	8168
7913	6246	14301	1144	9056	15967	7328	13203
796	195	9872	16979	15404	14130	9105	2001
9792	14251	1498	11296	1105	4502	16979	1105
56	4118	11302	5988	3363	15827	6928	4191
4277	10617	874	13211	11821	3090	18110	44
2364	15570	3460	9886	9988	3798	1158	9872
16979	15404	6127	9872	3652	14838	7437	2540
1367	2512	14407	5053	1521	297	10935	17137
2186	9433	13293	7555	13618	13000	6490	5310
18676	4782	11374	446	4165	11634	3846	14611
2364	6789	11634	4493	4063	4576	17955	7965
11748	14616	11453	17666	925	56	4118	18031
9522	14838	7437	3880	11476	8305	5102	2999
18628	14326	9175	9061	650	18110	8720	15404
2951	722	15334	841	15610	2443	11056	2186

$\mathbb{Z}_n$ represents three alphabetic characters as in the following examples:

$$DOG \quad \rightarrow \quad 3 \times 26^2 + 14 \times 26 + 6 \quad = \quad 2398$$
$$CAT \quad \rightarrow \quad 2 \times 26^2 + 0 \times 26 + 19 \quad = \quad 1371$$
$$ZZZ \quad \rightarrow \quad 25 \times 26^2 + 25 \times 26 + 25 \quad = \quad 17575.$$

You will have to invert this process as the final step in your program.

The first plaintext was taken from "The Diary of Samuel Marchbanks," by Robert-son Davies, 1947, and the second was taken from "Lake Wobegon Days," by Garri-son Keillor, 1985.

4.7 This exercise exhibits what is called a *protocol failure*. It provides an example where ciphertext can be decrypted by an opponent, without determining the key, if a cryptosystem is used in a careless way. (Since the opponent does not determine the key, it is not accurate to call it cryptanalysis.) The moral is that it is not sufficient to use a "secure" cryptosystem in order to guarantee "secure" communication.

Suppose Bob has an **RSA Cryptosystem** with a large modulus n for which the factorization cannot be found in a reasonable amount of time. Suppose Alice sends

TABLE 4.2
RSA Ciphertext

6340	8309	14010	8936	27358	25023	16481	25809
23614	7135	24996	30590	27570	26486	30388	9395
27584	14999	4517	12146	29421	26439	1606	17881
25774	7647	23901	7372	25774	18436	12056	13547
7908	8635	2149	1908	22076	7372	8686	1304
4082	11803	5314	107	7359	22470	7372	22827
15698	30317	4685	14696	30388	8671	29956	15705
1417	26905	25809	28347	26277	7897	20240	21519
12437	1108	27106	18743	24144	10685	25234	30155
23005	8267	9917	7994	9694	2149	10042	27705
15930	29748	8635	23645	11738	24591	20240	27212
27486	9741	2149	29329	2149	5501	14015	30155
18154	22319	27705	20321	23254	13624	3249	5443
2149	16975	16087	14600	27705	19386	7325	26277
19554	23614	7553	4734	8091	23973	14015	107
3183	17347	25234	4595	21498	6360	19837	8463
6000	31280	29413	2066	369	23204	8425	7792
25973	4477	30989					

a message to Bob by representing each alphabetic character as an integer between 0 and 25 (i.e., $A \leftrightarrow 0$, $B \leftrightarrow 1$, etc.), and then encrypting each residue modulo 26 as a separate plaintext character.

   (a) Describe how Oscar can easily decrypt a message which is encrypted in this way.

   (b) Illustrate this attack by decrypting the following ciphertext (which was encrypted using an **RSA Cryptosystem** with $n = 18721$ and $b = 25$) without factoring the modulus:

$$365, 0, 4845, 14930, 2608, 2608, 0.$$

4.8 This exercise illustrates another example of a protocol failure (due to Simmons) involving **RSA**; it is called the *common modulus* protocol failure. Suppose Bob has an **RSA Cryptosystem** with modulus n and decryption exponent b_1, and Charlie has an **RSA Cryptosystem** with (the same) modulus n and decryption exponent b_2. Suppose also that $\gcd(b_1, b_2) = 1$. Now, consider the situation that arises if Alice encrypts the same plaintext x to send to both Bob and Charlie. Thus, she computes $y_1 = x^{b_1} \bmod n$ and $y_2 = x^{b_2} \bmod n$, and then she sends y_1 to Bob and y_2 to Charlie. Suppose Oscar intercepts y_1 and y_2, and performs the computations indicated in Figure 4.16.

   (a) Prove that the value x_1 computed in step 3 of Figure 4.16 is in fact Alice's plaintext, x. Thus, Oscar can decrypt the message Alice sent, even though the cryptosystem may be "secure."

   (b) Illustrate the attack by computing x by this method if $n = 18721$, $b_1 = 43$, $b_2 = 7717$, $y_1 = 12677$ and $y_2 = 14702$.

4.9 We give yet another protocol failure involving **RSA**. Suppose that three users in a network, say Bob, Bart and Bert, all have public encryption exponents $b = 3$.

FIGURE 4.16
RSA common modulus protocol failure

> Input: n, b_1, b_2, y_1, y_2
> 1. compute $c_1 = b_1^{-1} \bmod b_2$
> 2. compute $c_2 = (c_1 b_1 - 1)/b_2$
> 3. compute $x_1 = y_1^{c_1}(y_2^{c_2})^{-1} \bmod n$

Let their moduli be denoted by n_1, n_2, n_3. Now suppose Alice encrypts the same plaintext x to send to Bob, Bart and Bert. That is, Alice computes $y_i = x^3 \bmod n_i$, $1 \le i \le 3$. Describe how Oscar can compute x, given y_1, y_2 and y_3, without factoring any of the moduli.

4.10 A plaintext x is said to be *fixed* if $e_K(x) = x$. Show that, for the **RSA Cryptosystem**, the number of fixed plaintexts $x \in \mathbb{Z}_n^*$ is equal to $\gcd(b-1, p-1) \times \gcd(b-1, q-1)$.

> **HINT** Consider the system of two congruences $e_K(x) \equiv x \pmod{p}, e_K(x) \equiv x \pmod{q}$.

4.11 Suppose **A** is a deterministic algorithm which is given as input an **RSA** modulus n, an encryption exponent b, and a ciphertext y. **A** will either decrypt y or return no answer. Supposing that there are $\epsilon(n-1)$ ciphertexts which **A** is able to decrypt, show how to use **A** as an oracle in a Las Vegas decryption algorithm having success probability ϵ.

> **HINT** Use the multiplicative property of **RSA** that $e_K(x_1)e_K(x_2) = e_K(x_1 x_2)$, where all arithmetic operations are modulo n.

4.12 Write a program to evaluate Jacobi symbols using the four properties presented in Section 4.5. The program should not do any factoring, other than dividing out powers of two. Test your program by computing the following Jacobi symbols:
$$\left(\frac{610}{987}\right), \left(\frac{20964}{1987}\right), \left(\frac{1234567}{11111111}\right).$$

4.13 For $n = 837$, 851 and 1189, find the number of bases b such that n is an Euler pseudo-prime to the base b.

4.14 The purpose of this question is to prove that the error probability of the Solovay-Strassen primality test is at most $1/2$. Let $\mathbb{Z}_n^*$ denote the group of units modulo n. Define
$$G(n) = \left\{ a : a \in \mathbb{Z}_n^*, \left(\frac{a}{n}\right) \equiv a^{(n-1)/2} \pmod{n} \right\}.$$

(a) Prove that $G(n)$ is a subgroup of $\mathbb{Z}_n^*$. Hence, by Lagrange's theorem, if $G(n) \ne \mathbb{Z}_n^*$, then
$$|G(n)| \le \frac{|\mathbb{Z}_n^*|}{2} \le \frac{n-1}{2}.$$

(b) Suppose $n = p^k q$, where p and q are odd, p is prime, $k \ge 2$, and $\gcd(p, q) = 1$. Let $a = 1 + p^{k-1}q$. Prove that
$$\left(\frac{a}{n}\right) \not\equiv a^{(n-1)/2} \pmod{n}.$$

HINT Use the binomial theorem to compute $a^{(n-1)/2}$.

(c) Suppose $n = p_1 \ldots p_s$, where the p_i's are distinct odd primes. Suppose $a \equiv u \pmod{p_1}$ and $a \equiv 1 \pmod{p_2 p_3 \ldots p_s}$, where u is a quadratic non-residue modulo p_1 (note that such an a exists by the Chinese remainder theorem). Prove that
$$\left(\frac{a}{n}\right) \equiv -1 \pmod{n},$$
but
$$a^{(n-1)/2} \equiv 1 \pmod{p_2 p_3 \ldots p_s},$$
so
$$a^{(n-1)/2} \not\equiv -1 \pmod{n}.$$

(d) If n is odd and composite, prove that $|G(n)| \leq (n-1)/2$.

(e) Summarize the above: prove that the error probability of the Solovay-Strassen primality test is at most $1/2$.

4.15 Suppose we have a Las Vegas algorithm with failure probability ϵ.

(a) Prove that the probability of first achieving success on the nth trial is $p_n = \epsilon^{n-1}(1-\epsilon)$.

(b) The average (expected) number of trials to achieve success is
$$\sum_{n=1}^{\infty} (n \times p_n).$$
Show that this average is equal to $1/(1-\epsilon)$.

(c) Let δ be a positive real number less than 1. Show that the number of iterations required in order to reduce the proability of failure to at most δ is
$$\left\lfloor \frac{\log_2 \delta}{\log_2 \epsilon} \right\rfloor.$$

4.16 Suppose Bob has carelessly revealed his decryption exponent to be $a = 14039$ in an **RSA Cryptosystem** with public key $n = 36581$ and $b = 4679$. Implement the probablistic algorithm to factor n given this information. Test your algorithm with the "random" choices $w = 9983$ and $w = 13461$. Show all computations.

4.17 Prove Equations 4.1 and 4.2 relating the functions *half* and *parity*.

4.18 Suppose $p = 199$, $q = 211$ and $B = 1357$ in the **Rabin Cryptosystem**. Perform the following computations.

(a) Determine the four square roots of 1 modulo n, where $n = pq$.

(b) Compute the encryption $y = e_K(32767)$.

(c) Determine the four possible decryptions of this given ciphertext y.

4.19 Factor 262063 and 9420457 using the $p-1$ method. How big does B have to be in each case to be successful?

5

Other Public-key Cryptosystems

In this chapter, we look at several other public-key cryptosystems. The **ElGamal Cryptosystem** is based on the **Discrete Logarithm** problem, which we will have occasion to use in numerous cryptographic protocols throughout the rest of the text. Thus we devote a considerable amount of time to discussion of this important problem. In later sections, we give relatively brief treatments of some other well-known public-key cryptosystems. These include ElGamal-type systems based on finite fields and elliptic curves, the (broken) **Merkle-Hellman Knapsack Cryptosystem** and the **McEliece Cryptosystem**.

5.1 The ElGamal Cryptosystem and Discrete Logs

The **ElGamal Cryptosystem** is based on the **Discrete Logarithm** problem. We begin by describing this problem in the setting of a finite field $\mathbb{Z}_p$, where p is prime, in Figure 5.1. (Recall that the multiplicative group $\mathbb{Z}_p{}^*$ is cyclic, and a generator of $\mathbb{Z}_p{}^*$ is called a primitive element.)

The **Discrete Logarithm** problem in $\mathbb{Z}_p$ has been the object of much study. The problem is generally regarded as being difficult if p is carefully chosen. In particular, there is no known polynomial-time algorithm for the **Discrete Logarithm** problem. To thwart known attacks, p should have at least 150 digits, and $p-1$ should have at least one "large" prime factor. The utility of the **Discrete Logarithm** problem in a cryptographic setting is that finding discrete logs is (probably) difficult, but the inverse operation of exponentiation can be computed efficiently by using the square-and-multiply method described earlier. Stated another way, exponentiation modulo p is a one-way function for suitable primes p.

ElGamal has developed a public-key cryptosystem based on the **Discrete Logarithm** problem. This system is presented in Figure 5.2.

The **ElGamal Cryptosystem** is non-deterministic, since the ciphertext depends on both the plaintext x and on the random value k chosen by Alice. So there will

FIGURE 5.1
The discrete logarithm problem in $\mathbb{Z}_p$

> **Problem Instance** $I = (p, \alpha, \beta)$, where p is prime, $\alpha \in \mathbb{Z}_p$ is a primitive element, and $\beta \in \mathbb{Z}_p{}^*$.
>
> **Objective** Find the unique integer a, $0 \le a \le p - 2$, such that
>
> $$\alpha^a \equiv \beta \pmod{p}.$$
>
> We will denote this integer a by $\log_\alpha \beta$.

FIGURE 5.2
ElGamal Public-key Cryptosystem in $\mathbb{Z}_p{}^*$

> Let p be a prime such that the discrete log problem in $\mathbb{Z}_p$ is intractable, and let $\alpha \in \mathbb{Z}_p{}^*$ be a primitive element. Let $\mathcal{P} = \mathbb{Z}_p{}^*$, $\mathcal{C} = \mathbb{Z}_p{}^* \times \mathbb{Z}_p{}^*$, and define
> $$\mathcal{K} = \{(p, \alpha, a, \beta) : \beta \equiv \alpha^a \pmod{p}\}.$$
> The values p, α and β are public, and a is secret.
>
> For $K = (p, \alpha, a, \beta)$, and for a (secret) random number $k \in \mathbb{Z}_{p-1}$, define
> $$e_K(x, k) = (y_1, y_2),$$
>
> where
> $$y_1 = \alpha^k \bmod p$$
> and
> $$y_2 = x\beta^k \bmod p.$$
>
> For $y_1, y_2 \in \mathbb{Z}_p{}^*$, define
> $$d_K(y_1, y_2) = y_2(y_1{}^a)^{-1} \bmod p.$$

be many ciphertexts that are encryptions of the same plaintext.

Informally, this is how the **ElGamal Cryptosystem** works. The plaintext x is "masked" by multiplying it by β^k, yielding y_2. The value α^k is also transmitted as part of the ciphertext. Bob, who knows the secret exponent a, can compute β^k from α^k. Then he can "remove the mask" by dividing y_2 by β^k to obtain x.

A small example will illustrate.

Example 5.1

Suppose $p = 2579$, $\alpha = 2$, $a = 765$, and hence

$$\beta = 2^{765} \bmod 2579 = 949.$$

Now, suppose that Alice wishes to send the message $x = 1299$ to Bob. Say $k = 853$ is the random integer she chooses. Then she computes

$$y_1 = 2^{853} \bmod 2579$$

$$= 435$$

and

$$y_2 = 1299 \times 949^{853} \bmod 2579$$

$$= 2396.$$

When Bob receives the ciphertext $y = (435, 2396)$, he computes

$$x = 2396 \times (435^{765})^{-1} \bmod 2579$$

$$= 1299,$$

which was the plaintext that Alice encrypted.   □

5.1.1 Algorithms for the Discrete Log Problem

Throughout this section, we assume that p is prime and α is a primitive element modulo p. We take p and α to be fixed. Hence the **Discrete Logarithm** problem can be phrased in the following form: Given $\beta \in \mathbb{Z}_p{}^*$, find the unique exponent a, $0 \le a \le p - 2$, such that $\alpha^a \equiv \beta \pmod{p}$.

Clearly, the **Discrete Logarithm** problem can be solved by exhaustive search in $O(p)$ time and $O(1)$ space (neglecting logarithmic factors). By precomputing all possible values α^a, and sorting the ordered pairs $(a, \alpha^a \bmod p)$ with respect to their second coordinates, we can solve the discrete log problem in $O(1)$ time with $O(p)$ precomputation and $O(p)$ memory (again, neglecting logarithmic factors). The first non-trivial algorithm we describe is a time-memory trade-off due to Shanks.

FIGURE 5.3
Shanks' algorithm for the discrete logarithm problem

1.	Compute $\alpha^{mj} \bmod p$, $0 \leq j \leq m - 1$
2.	Sort the m ordered pairs $(j, \alpha^{mj} \bmod p)$ with respect to their second coordinates, obtaining a list L_1
3.	Compute $\beta\alpha^{-i} \bmod p$, $0 \leq i \leq m - 1$
4.	Sort the m ordered pairs $(i, \beta\alpha^{-i} \bmod p)$ with respect to their second coordinates, obtaining a list L_2
5.	Find a pair $(j, y) \in L_1$ and a pair $(i, y) \in L_2$ (i.e., a pair having identical second coordinates)
6.	define $\log_\alpha \beta = mj + i \bmod (p - 1)$.

Shanks' Algorithm

Denote $m = \lceil \sqrt{p - 1} \rceil$. Shanks' algorithm is presented in Figure 5.3. Some comments are in order. First, steps 1 and 2 can be precomputed, if desired (this will not affect the asymptotic running time, however). Next, observe that if $(j, y) \in L_1$ and $(i, y) \in L_2$, then

$$\alpha^{mj} = y = \beta\alpha^{-i},$$

so

$$\alpha^{mj+i} = \beta,$$

as desired. Conversely, for any β, we can write

$$\log_\alpha \beta = mj + i,$$

where $0 \leq j, i \leq m - 1$. Hence, the search in step 5 will be successful.

It is not difficult to implement the algorithm to run in $O(m)$ time with $O(m)$ memory (neglecting logarithmic factors). Note that step 5 can be done with one (simultaneous) pass through each of the two lists L_1 and L_2.

Here is a small example to illustrate.

Example 5.2
Suppose $p = 809$, and we wish to find $\log_3 525$. So we have $\alpha = 3$, $\beta = 525$ and $m = \lceil \sqrt{808} \rceil = 29$. Then

$$\alpha^{29} \bmod 809 = 99.$$

First, we compute the ordered pairs $(j, 99^j \bmod 809)$ for $0 \le j \le 28$. We obtain the list

$$
\begin{array}{lllll}
(0,1) & (1,99) & (2,93) & (3,308) & (4,559) \\
(5,329) & (6,211) & (7,664) & (8,207) & (9,268) \\
(10,644) & (11,654) & (12,26) & (13,147) & (14,800) \\
(15,727) & (16,781) & (17,464) & (18,632) & (19,275) \\
(20,528) & (21,496) & (22,564) & (23,15) & (24,676) \\
(25,586) & (26,575) & (27,295) & (28,81) &
\end{array}
$$

which is then sorted to produce L_1.

The second list contains the ordered pairs $(i, 525 \times (3^i)^{-1} \bmod 809), 0 \le j \le 28$. It is as follows:

$$
\begin{array}{lllll}
(0,525) & (1,175) & (2,328) & (3,379) & (4,396) \\
(5,132) & (6,44) & (7,554) & (8,724) & (9,511) \\
(10,440) & (11,686) & (12,768) & (13,256) & (14,355) \\
(15,388) & (16,399) & (17,133) & (18,314) & (19,644) \\
(20,754) & (21,521) & (22,713) & (23,777) & (24,259) \\
(25,356) & (26,658) & (27,489) & (28,163) &
\end{array}
$$

After sorting this list, we get L_2.

Now, if we proceed simultaneously through the two sorted lists, we find that $(10,644)$ is in L_1 and $(19,644)$ is in L_2. Hence, we can compute

$$
\log_3 525 = 29 \times 10 + 19
$$

$$
= 309.
$$

As a check, it can be verified that indeed $3^{309} \equiv 525 \pmod{809}$. $\quad\square$

The Pohlig-Hellman Algorithm

The next algorithm we study is the Pohlig-Hellman algorithm. Suppose

$$
p - 1 = \prod_{i=1}^{k} p_i^{c_i},
$$

where the p_i's are distinct primes. The value $a = \log_\alpha \beta$ is determined (uniquely) modulo $p - 1$. We first observe that if we can compute $a \bmod p_i^{c_i}$ for each $i, 1 \le i \le k$, then we can compute $a \bmod (p - 1)$ by the Chinese remainder theorem. So, let's suppose that q is prime,

$$
p - 1 \equiv 0 \pmod{q^c}
$$

and

$$p - 1 \not\equiv 0 \pmod{q^{c+1}}.$$

We will show how to compute the value

$$x = a \bmod q^c,$$

where $0 \leq x \leq q^c - 1$. We can express x in radix q representation as

$$x = \sum_{i=0}^{c-1} a_i q^i,$$

where $0 \leq a_i \leq q - 1$ for $0 \leq i \leq c - 1$. Also, observe that we can express a as

$$a = x + q^c s$$

for some integer s.

The first step of the algorithm is to compute a_0. The main observation is that

$$\beta^{(p-1)/q} \equiv \alpha^{(p-1)a_0/q} \pmod{p}.$$

To see this, note that

$$\beta^{(p-1)/q} \equiv \alpha^{(p-1)(x+q^c s)/q} \pmod{p},$$

so it suffices to show that

$$\alpha^{(p-1)(x+q^c s)/q} \equiv \alpha^{(p-1)a_0/q} \pmod{p}.$$

This will be true if and only if

$$\frac{(p-1)(x+q^c s)}{q} \equiv \frac{(p-1)a_0}{q} \pmod{p-1}.$$

However, we have

$$\frac{(p-1)(x+q^c s)}{q} - \frac{(p-1)a_0}{q} = \frac{p-1}{q}(x + q^c s - a_0)$$

$$= \frac{p-1}{q}\left(\sum_{i=0}^{c-1} a_i q^i + q^c s - a_0\right)$$

$$= \frac{p-1}{q}\left(\sum_{i=1}^{c-1} a_i q^i + q^c s\right)$$

$$= (p-1)\left(\sum_{i=1}^{c-1} a_i q^{i-1} + q^{c-1} s\right)$$

$$\equiv 0 \pmod{p-1},$$

which was what we wanted to prove.

Hence, we begin by computing $\beta^{(p-1)/q} \bmod p$. If

$$\beta^{(p-1)/q} \equiv 1 \pmod{p},$$

then $a_0 = 0$. Otherwise, we successively compute

$$\gamma = \alpha^{(p-1)/q} \bmod p, \gamma^2 \bmod p, \ldots,$$

until

$$\gamma^i \equiv \beta^{(p-1)/q} \pmod{p}$$

for some i. When this happens, we have $a_0 = i$.

Now, if $c = 1$, we're done. Otherwise $c > 1$, and we proceed to determine a_1. To do this, we define

$$\beta_1 = \beta\alpha^{-a_0}$$

and denote

$$x_1 = \log_\alpha \beta_1 \bmod q^c.$$

It is not hard to see that

$$x_1 = \sum_{i=1}^{c-1} a_i q^i.$$

Hence, it follows that

$$\beta_1^{(p-1)/q^2} \equiv \alpha^{(p-1)a_1/q} \pmod{p}.$$

So, we will compute $\beta_1^{(p-1)/q^2} \bmod p$, and then find i such that

$$\gamma^i \equiv \beta_1^{(p-1)/q^2} \pmod{p}.$$

Then we have $a_1 = i$.

If $c = 2$, we are now finished; otherwise, we repeat this process $c - 2$ more times, obtaining $a_2, \ldots, a_{c-1}$.

A pseudo-code description of the Pohlig-Hellman algorithm is given in Figure 5.4. In this algorithm, α is a primitive element modulo p, q is prime,

$$p - 1 \equiv 0 \pmod{q^c}$$

and

$$p - 1 \not\equiv 0 \pmod{q^{c+1}}.$$

The algorithm calculates $a_0, \ldots, a_{c-1}$, where

$$\log_\alpha \beta \bmod q^c = \sum_{i=0}^{c-1} a_i q^i.$$

We illustrate the Pohlig-Hellman algorithm with a small example.

FIGURE 5.4
Pohlig-Hellman algorithm to compute $\log_\alpha \beta \bmod q^c$

1.   compute $\gamma_i = \alpha^{(p-1)i/q} \bmod p$ for $0 \le i \le q-1$
2.   set $j = 0$ and $\beta_j = \beta$
3.   **while** $j \le c-1$ **do**
4.   compute $\delta = \beta_j^{\,(p-1)/q^{j+1}} \bmod p$
5.   find i such that $\delta = \gamma_i$
6. $a_j = i$
7. $\beta_{j+1} = \beta_j \alpha^{-a_j q^j} \bmod p$
8. $j = j + 1$

Example 5.3
Suppose $p = 29$; then
$$n = p - 1 = 28 = 2^2 7^1.$$

Suppose $\alpha = 2$ and $\beta = 18$, so we want to determine $a = \log_2 18$. We proceed by first computing $a \bmod 4$ and then computing $a \bmod 7$.

We start by setting $q = 2$ and $c = 2$. First,

$$\gamma_0 = 1$$

and

$$\gamma_1 = \alpha^{28/2} \bmod 29$$
$$= 2^{14} \bmod 29$$
$$= 28.$$

Next,

$$\delta = \beta^{28/2} \bmod 29$$
$$= 18^{14} \bmod 29$$
$$= 28.$$

Hence, $a_0 = 1$. Next, we compute

$$\beta_1 = \beta_0 \alpha^{-1} \bmod 29$$
$$= 9.$$

and

$$\beta_1{}^{28/4} \bmod 29 = 9^7 \bmod 29$$
$$= 28.$$

Since

$$\gamma_1 \equiv 28 \bmod 29,$$

we have $a_1 = 1$. Hence, $a \equiv 3 \pmod 4$.

Next, we set $q = 7$ and $c = 1$. We have

$$\beta^{28/7} \bmod 29 = 18^4 \bmod 29$$
$$= 25$$

and

$$\gamma_1 = \alpha^{28/7} \bmod 29$$
$$= 2^4 \bmod 29$$
$$= 16.$$

Then we would compute

$$\gamma_2 = 24$$
$$\gamma_3 = 7$$
$$\gamma_4 = 25.$$

Hence, $a_0 = 4$ and $a \equiv 4 \pmod 7$.

Finally, solving the system

$$a \equiv 3 \pmod 4$$
$$a \equiv 4 \pmod 7$$

using the Chinese remainder theorem, we get $a \equiv 11 \pmod{28}$. That is, we have computed $\log_2 18$ in $\mathbb{Z}_{29}$ to be 11.   ▯

The Index Calculus Method

The index calculus method for computing discrete logs bears considerable resemblance to many of the best factoring algorithms. We give a very brief overview in this section. The method uses a *factor base*, which, as before, is a set B of "small" primes. Suppose $B = \{p_1, p_2, \ldots, p_B\}$. The first step (a preprocessing step) is to find the logarithms of the B primes in the factor base. The second step is to compute a discrete log of a desired element β, using the knowledge of the discrete logs of the elements in the factor base.

In the precomputation, we construct $C = B + 10$ congruences modulo p, as follows:

$$\alpha^{x_j} \equiv p_1{}^{a_{1j}} p_2{}^{a_{2j}} \ldots p_B{}^{a_{Bj}} \pmod{p},$$

$1 \leq j \leq C$. Notice these congruences can be written equivalently as

$$x_j \equiv a_{1j} \log_\alpha p_1 + \ldots + a_{Bj} \log_\alpha p_B \pmod{p-1},$$

$1 \leq j \leq C$. Given C congruences in the B "unknowns" $\log_\alpha p_i$ $(1 \leq i \leq B)$, we hope that there is a unique solution modulo $p - 1$. If this is the case, then we can compute the logarithms of the elements in the factor base.

How do we generate congruences of the desired form? One elementary way is to take a random value x, compute $\alpha^x \bmod p$, and then determine if $\alpha^x \bmod p$ has all its factors in B (using trial division, for example).

Now, given that we have already successfully carried out the precomputation step, we compute a desired logarithm $\log_\alpha \beta$ by means of a Las Vegas type probabilistic algorithm. Choose a random integer s $(1 \leq s \leq p - 2)$ and compute

$$\gamma = \beta \alpha^s \bmod p.$$

Now attempt to factor γ over the factor base B. If this can be done, then we obtain a congruence of the form

$$\beta \alpha^s \equiv p_1{}^{c_1} p_2{}^{c_2} \ldots p_B{}^{c_B} \pmod{p}.$$

This can be written equivalently as

$$\log_\alpha \beta + s \equiv c_1 \log_\alpha p_1 + \ldots + c_B \log_\alpha p_B \pmod{p-1}.$$

Since everything is now known except $\log_\alpha \beta$, we can easily solve for $\log_\alpha \beta$.

Here is a small, very artificial, example to illustrate the two steps in the algorithm.

Example 5.4
Suppose $p = 10007$ and $\alpha = 5$ is the primitive element used as the base of logarithms modulo p. Suppose we take $B = \{2, 3, 5, 7\}$ as the factor base. Of course $\log_5 5 = 1$, so there are three logs of factor base elements to be determined.

Some examples of "lucky" exponents that might be chosen are 4063, 5136 and 9865.

With $x = 4063$, we compute

$$5^{4063} \bmod 10007 = 42 = 2 \times 3 \times 7.$$

This yields the congruence

$$\log_5 2 + \log_5 3 + \log_5 7 \equiv 4063 \pmod{10006}.$$

Similarly, since

$$5^{5136} \bmod 10007 = 54 = 2 \times 3^3$$

and

$$5^{9865} \bmod 10007 = 189 = 3^3 \times 7,$$

we obtain two more congruences:

$$\log_5 2 + 3\log_5 3 \equiv 5136 \pmod{10006}$$

and

$$3\log_5 3 + \log_5 7 \equiv 9865 \pmod{10006}.$$

We now have three congruences in three unknowns, and there happens to be a unique solution modulo 10006, namely $\log_5 2 = 6578$, $\log_5 3 = 6190$ and $\log_5 7 = 1301$.

Now, let's suppose that we wish to find $\log_5 9451$. Suppose we choose the "random" exponent $s = 7736$, and compute

$$9451 \times 5^{7736} \bmod 10007 = 8400.$$

Since $8400 = 2^4 3^1 5^2 7^1$ factors over $\mathcal{B}$, we obtain

$$\log_5 9451 = 4\log_5 2 + \log_5 3 + 2\log_5 5 + \log_5 7 - s \bmod 10006$$

$$= 4 \times 6578 + 6190 + 2 \times 1 + 1301 - 7736 \bmod 10006$$

$$= 6057.$$

To verify, we can check that $5^{6057} \equiv 9451 \pmod{10007}$.   ⬚

Heuristic analyses of various versions of the algorithm have been done. Under reasonable assumptions, the asymptotic running time of the precomputation phase is $O\left(e^{(1+o(1))\sqrt{\ln p \ln \ln p}}\right)$, and the time to find an individual discrete log is $O\left(e^{(1/2+o(1))\sqrt{\ln p \ln \ln p}}\right)$.

5.1.2   Bit Security of Discrete Logs

We now look at the question of partial information about discrete logs. In particular, we consider whether individual bits of a discrete logarithm are easy or hard to compute. To be precise, consider the problem presented in Figure 5.5, which we call the ith **Bit** problem.

FIGURE 5.5
ith bit of discrete logarithm

> **Problem Instance** $I = (p, \alpha, \beta, i)$, where p is prime, $\alpha \in \mathbb{Z}_p^*$ is a primitive element, $\beta \in \mathbb{Z}_p^*$, and i is an integer such that $1 \leq i \leq \lceil \log_2(p-1) \rceil$.
>
> **Objective**   Compute $L_i(\beta)$, which (for the specified α and p) denotes the ith least significant bit of $\log_\alpha \beta$.

We will first show that computing the least significant bit of a discrete logarithm is easy. In other words, if $i = 1$, the ith **Bit** problem can be solved efficiently. This follows from Euler's criterion concerning quadratic residues modulo p, where p is prime.

Consider the mapping $f : \mathbb{Z}_p^* \to \mathbb{Z}_p^*$ defined by

$$f(x) = x^2 \bmod p.$$

Denote by $\mathrm{QR}(p)$ the set of quadratic residues modulo p; then

$$\mathrm{QR}(p) = \{x^2 \bmod p : x \in \mathbb{Z}_p^*\}.$$

First, observe that $f(x) = f(p - x)$. Next note that

$$w^2 \equiv x^2 \pmod{p}$$

if and only if

$$p \mid (w - x)(w + x),$$

which happens if and only if

$$w \equiv \pm x \pmod{p}.$$

It follows that

$$|f^{-1}(y)| = 2$$

for every $y \in \mathrm{QR}(p)$, and hence

$$|\mathrm{QR}(p)| = \frac{p-1}{2}.$$

That is, exactly half the residues in $\mathbb{Z}_p^*$ are quadratic residues and half are not.

Now, suppose α is a primitive element of $\mathbb{Z}_p$. Then $\alpha^a \in \mathrm{QR}(p)$ if a is even. Since the $(p-1)/2$ elements $\alpha^0 \bmod p, \alpha^2 \bmod p, \ldots, \alpha^{p-3} \bmod p$ are all distinct, it follows that

$$\mathrm{QR}(p) = \{\alpha^{2i} \bmod p : 0 \leq i \leq (p-3)/2\}.$$

Hence, β is a quadratic residue if and only if $\log_\alpha \beta$ is even, that is, if and only if $L_1(\beta) = 0$. But we already know, by Euler's criterion, that β is a quadratic residue if and only if

$$\beta^{(p-1)/2} \equiv 1 \pmod{p}.$$

So we have the following efficient formula to calculate $L_1(\beta)$:

$$L_1(\beta) = \begin{cases} 0 & \text{if } \beta^{(p-1)/2} \equiv 1 \pmod{p} \\ 1 & \text{otherwise.} \end{cases}$$

Let's now consider the computation of $L_i(\beta)$ for values of i exceeding 1. Suppose

$$p - 1 = 2^s t$$

where t is odd. Then it can be shown that it is easy to compute $L_i(\beta)$ if $i \leq s$. On the other hand, computing $L_{s+1}(\beta)$ is (probably) difficult, in the sense that any hypothetical algorithm (or oracle) to compute $L_{s+1}(\beta)$ could be used to find discrete logarithms in $\mathbb{Z}_p$.

We shall prove this result in the case $s = 1$. More precisely, if $p \equiv 3 \pmod 4$ is prime, then we show how any oracle for computing $L_2(\beta)$ can be used to solve the **Discrete Log** problem in $\mathbb{Z}_p$.

Recall that, if β is a quadratic residue in $\mathbb{Z}_p$ and $p \equiv 3 \pmod 4$, then the two square roots of β modulo p are $\pm \beta^{(p+1)/4} \bmod p$. It is also important that, for any $\beta \neq 0$,

$$L_1(\beta) \neq L_1(p - \beta)$$

if $p \equiv 3 \pmod 4$. We see this as follows. Suppose

$$\alpha^a \equiv \beta \pmod{p};$$

then

$$\alpha^{a+(p-1)/2} \equiv -\beta \pmod{p}.$$

Since $p \equiv 3 \pmod 4$, the integer $(p-1)/2$ is odd, and the result follows.

Now, suppose that $\beta = \alpha^a$ for some (unknown) even exponent a. Then either

$$\beta^{(p+1)/4} \equiv \alpha^{a/2} \pmod{p}$$

or

$$-\beta^{(p+1)/4} \equiv \alpha^{a/2} \pmod{p}.$$

We can determine which of these two possibilities is correct if we know the value $L_2(\beta)$, since

$$L_2(\beta) = L_1(\alpha^{a/2}).$$

This fact is exploited in our algorithm, which we present in Figure 5.6.

At the end of the algorithm, the x_i's comprise the bits in the binary representation of $\log_\alpha \beta$; that is,

$$\log_\alpha \beta = \sum_{i \geq 0} x_i 2^i.$$

We will work out a small example to illustrate the algorithm.

FIGURE 5.6
Computing discrete logs in $\mathbb{Z}_p$ for $p \equiv 3 \pmod 4$, given an oracle for $L_2(\beta)$

1.	$x_0 = L_1(\beta)$
2.	$\beta = \beta/\alpha^{x_0} \bmod p$
3.	$i = 1$
4.	**while** $\beta \neq 1$ **do**
5.	$\quad x_i = L_2(\beta)$
6.	$\quad \gamma = \beta^{(p+1)/4} \bmod p$
7.	$\quad$ **if** $L_1(\gamma) = x_i$ **then**
8.	$\quad\quad \beta = \gamma$
9.	$\quad$ **else**
10.	$\quad\quad \beta = p - \gamma$
11.	$\quad \beta = \beta/\alpha^{x_i} \bmod p$
12.	$\quad i = i + 1$

Example 5.5
Suppose $p = 19$, $\alpha = 2$ and $\beta = 6$. Since the example is so small, we can tabulate the values of $L_1(\gamma)$ and $L_2(\gamma)$ for all $\gamma \in \mathbb{Z}_{19}^*$. (In general, L_1 can be computed efficiently using Euler's criterion and L_2 is an oracle.) These values are given in Table 5.1. The algorithm now proceeds as shown in Figure 5.7.

Hence, $\log_2 6 = 1110_2 = 14$, as can easily be verified. ▯

It is possible to give formal proof of the algorithm's correctness using mathematical induction. Denote

$$x = \log_\alpha \beta = \sum_{i \geq 0} x_i 2^i.$$

For $i \geq 0$, define

$$Y_i = \left\lfloor \frac{x}{2^{i+1}} \right\rfloor.$$

Also, define β_0 to be the value of β in step 2 of the algorithm; and, for $i \geq 1$, define β_i to be the value of β in step 11 during the ith iteration of the **while** loop. It can be proved by induction that

$$\beta_i \equiv \alpha^{2Y_i} \pmod p$$

TABLE 5.1
Values of L_1 and L_2 for $p = 19$, $\alpha = 2$

γ	$L_1(\gamma)$	$L_2(\gamma)$	γ	$L_1(\gamma)$	$L_2(\gamma)$	γ	$L_1(\gamma)$	$L_2(\gamma)$
1	0	0	7	0	1	13	1	0
2	1	0	8	1	1	14	1	1
3	1	0	9	0	0	15	1	1
4	0	1	10	1	0	16	0	0
5	0	0	11	0	0	17	0	1
6	0	1	12	1	1	18	1	0

FIGURE 5.7
Computation of $\log_2 6$ in $\mathbb{Z}_{19}$

1.	$x_0 = 0$
2.	$\beta = 6$
3.	$i = 1$
5.	$x_1 = L_2(6) = 1$
6.	$\gamma = 5$
7.	$L_1(5) = 0 \neq x_1$
10.	$\beta = 14$
11.	$\beta = 7$
12.	$i = 2$
5.	$x_2 = L_2(7) = 1$
6.	$\gamma = 11$
7.	$L_1(11) = 0 \neq x_2$
10.	$\beta = 8$
11.	$\beta = 4$
12.	$i = 3$
5.	$x_3 = L_2(4) = 1$
6.	$\gamma = 17$
7.	$L_1(17) = 0 \neq x_3$
10.	$\beta = 2$
11.	$\beta = 1$
12.	$i = 4$
4.	DONE

FIGURE 5.8
The discrete logarithm problem in $(G, \circ)$

Problem Instance $I = (G, \alpha, \beta)$, where G is a finite group with group operation $\circ$, $\alpha \in G$ and $\beta \in H$, where $H = \{\alpha^i : i \geq 0\}$ is the subgroup generated by α.

Objective   Find the unique integer a such that $0 \leq a \leq |H| - 1$ and $\alpha^a = \beta$, where the notation α^a means

$$\underbrace{\alpha \circ \ldots \circ \alpha}_{a \text{ times}}.$$

We will denote this integer a by $\log_\alpha \beta$.

---

for all $i \geq 0$. Now, with the observation that

$$2Y_i = Y_{i-1} - x_i,$$

it follows that

$$x_{i+1} = L_2(\beta_i),$$

$i \geq 0$. Since

$$x_0 = L_1(\beta),$$

the algorithm is correct. The details are left to the reader.

5.2   Finite Field and Elliptic Curve Systems

We have spent a considerable amount of time looking at the **Discrete Logarithm** problem and the factoring. We will see these two problems again and again, underlying various types of cryptosystems and cryptographic protocols. So far, we have considered the **Discrete Logarithm** problem in the finite field $\mathbb{Z}_p$, but it is also useful to consider the problem in other settings. This is the theme of this section.

The **ElGamal Cryptosystem** can be implemented in any group where the **Discrete Log** problem is intractible. We used the multiplicative group $\mathbb{Z}_p^*$, but other groups are also suitable candidates. First, we phrase the **Discrete Logarithm** problem in a general (finite) group G, where we will denote the group operation by $\circ$. This generalized version of the problem is presented in Figure 5.8.

It is easy to define an **ElGamal Cryptosystem** in the subgroup H in a similar fashion as it was originally described in $\mathbb{Z}_p^*$. This is done in Figure 5.9. Note

FIGURE 5.9
Generalized ElGamal Public-key Cryptosystem

Let G be a finite group with group operation $\circ$, and let $\alpha \in G$ be an element such that the discrete log problem in H is intractible, where $H = \{\alpha^i : i \geq 0\}$ is the subgroup generated by α. Let $\mathcal{P} = G$, $\mathcal{C} = G \times G$, and define

$$\mathcal{K} = \{(G, \alpha, a, \beta) : \beta = \alpha^a\}.$$

The values α and β are public, and a is secret.

For $K = (G, \alpha, a, \beta)$, and for a (secret) random number $k \in \mathbb{Z}_{|H|}$, define

$$e_K(x, k) = (y_1, y_2),$$

where

$$y_1 = \alpha^k$$

and

$$y_2 = x \circ \beta^k.$$

For a ciphertext $y = (y_1, y_2)$, define

$$d_K(y) = y_2 \circ (y_1{}^a)^{-1}.$$

that encryption requires the use of a random integer k such that $0 \leq k \leq |H| - 1$. However, if Alice does not know the order of the subgroup H, she can generate an integer k such that $0 \leq k \leq |G| - 1$, and encryption and decryption will work without any changes. Also note that the group G need not be an abelian group (of course H is abelian since it is cyclic).

Let's now turn to the "generalized" **Discrete Log** problem. The subgroup H generated by any $\alpha \in G$ is of course a cyclic group of order $|H|$. So any version of the problem is equivalent, in some sense, to the **Discrete Log** problem in a cyclic group. However, the difficulty of the **Discrete Log** problem seems to depend in an essential way on the representation of the group that is used.

As an example to illustrate a representation where the problem is easy to solve, consider the additive cyclic group $\mathbb{Z}_n$, and suppose $\gcd(\alpha, n) = 1$, so α is a generator of $\mathbb{Z}_n$. Since the group operation is addition modulo n, an "exponentiation" operation, α^a, corresponds to multiplication by a modulo n. Hence, in this setting, the **Discrete Log** problem is to find the integer a such that

$$\alpha a \equiv \beta \pmod{n}.$$

Since $\gcd(\alpha, n) = 1$, α has a multiplicative inverse modulo n, and we can compute $\alpha^{-1} \bmod n$ easily using the Euclidean algorithm. Then we can solve for a, obtaining

$$\log_\alpha \beta = \beta \alpha^{-1} \bmod n.$$

We previously discussed the **Discrete Log** problem in the multiplicative group $\mathbb{Z}_p^*$, where p is prime. This group is a cyclic group of order $p - 1$, and hence it is isomorphic to the additive group $\mathbb{Z}_{p-1}$. By the discussion above, we know how to compute discrete logs efficiently in this additive group. This suggests that we could solve the **Discrete Log** problem in $\mathbb{Z}_p^*$ by "reducing" the problem to the the easily solved formulation in $\mathbb{Z}_{p-1}$.

Let us think about how this could be done. The statement that $(\mathbb{Z}_p^*, \times)$ is isomorphic to $(\mathbb{Z}_{p-1}, +)$ means that there is a bijection

$$\phi : \mathbb{Z}_p^* \rightarrow \mathbb{Z}_{p-1}$$

such that

$$\phi(xy \bmod p) = (\phi(x) + \phi(y)) \bmod (p - 1).$$

It follows easily that

$$\phi(\alpha^a \bmod p) = a\phi(\alpha) \bmod (p - 1),$$

so we have that

$$\beta \equiv \alpha^a \ (\bmod \ p) \Leftrightarrow a\phi(\alpha) \equiv \phi(\beta) \ (\bmod \ p - 1).$$

Hence, solving for a as described above, we have that

$$\log_\alpha \beta = \phi(\beta)(\phi(\alpha))^{-1} \bmod (p - 1).$$

Consequently, if we have an efficient method of computing the isomorphism ϕ, then we would have an efficient algorithm to compute discrete logs in $\mathbb{Z}_p^*$. The catch is that there is no known general method to efficiently compute the isomorphism ϕ for an arbitrary prime p. Even though we know the two groups in question are isomorphic, we do not know an efficient algorithm to explicitly describe the isomorphism.

This method can be applied to the **Discrete Log** problem in any group G. If there is an efficient method of computing the isomorphism between H and $\mathbb{Z}_{|H|}$, then the discrete log problem in G described above can be solved efficiently. Conversely, it is not hard to see that an efficient method of computing discrete logs yields an efficient algorithm to compute the isomorphism between the two groups.

This discussion has shown that the **Discrete Log** problem may be easy or (apparently) difficult, depending on the representation of the (cyclic) group that is used. So it may be useful to look at other groups in the hope of finding other settings where the **Discrete Log** problem seems to be intractible.

Two such classes of groups are

1. the multiplicative group of the Galois field $GF(p^n)$
2. the group of an elliptic curve defined over a finite field.

We will discuss these two classes of groups in the next subsections.

5.2.1 Galois Fields

We have already discussed the fact that $\mathbb{Z}_p$ is a field if p is prime. However, there are other examples of finite fields not of this form. In fact, there is a finite field with q elements if $q = p^n$ where p is prime and $n \geq 1$ is an integer. We will now describe very briefly how to construct such a field. First, we need several definitions.

DEFINITION 5.1 *Suppose p is prime. Define $\mathbb{Z}_p[x]$ to be the set of all polynomials in the indeterminate x. By defining addition and multiplication of polynomials in the usual way (and reducing coefficients modulo p), we construct a ring.*

For $f(x), g(x) \in \mathbb{Z}_p[x]$, we say that $f(x)$ divides $g(x)$ (notation: $f(x) \mid g(x)$) if there exists $q(x) \in \mathbb{Z}_p[x]$ such that

$$g(x) = q(x)f(x).$$

For $f(x) \in \mathbb{Z}_p[x]$, define $\deg(f)$, the degree of f, to be the highest exponent in a term of f.

Suppose $f(x), g(x), h(x) \in \mathbb{Z}_p[x]$, and $\deg(f) = n \geq 1$. We define

$$g(x) \equiv h(x) \pmod{f(x)}$$

if

$$f(x) \mid (g(x) - h(x)).$$

Notice the resemblance of the definition of congruence of polynomials to that of congruence of integers.

We are now going to define a ring of polynomials "modulo $f(x)$" which we denote by $\mathbb{Z}_p[x]/(f(x))$. The construction of $\mathbb{Z}_p[x]/(f(x))$ from $\mathbb{Z}_p[x]$ is based on the idea of congruences modulo $f(x)$ and is analogous to the construction of $\mathbb{Z}_m$ from $\mathbb{Z}$.

Suppose $\deg(f) = n$. If we divide $g(x)$ by $f(x)$, we obtain a (unique) *quotient* $q(x)$ and *remainder* $r(x)$, where

$$g(x) = q(x)f(x) + r(x)$$

and

$$\deg(r) < n.$$

This can be done by usual long division of polynomials. Hence any polynomial in $\mathbb{Z}_p[x]$ is congruent modulo $f(x)$ to a unique polynomial of degree at most $n - 1$.

Now we define the elements of $\mathbb{Z}_p[x]/(f(x))$ to be the p^n polynomials in $\mathbb{Z}_p[x]$ of degree at most $n - 1$. Addition and multiplication in $\mathbb{Z}_p[x]/(f(x))$ is defined as in $\mathbb{Z}_p[x]$, followed by a reduction modulo $f(x)$. Equipped with these operations, $\mathbb{Z}_p[x]/(f(x))$ is a ring.

Recall that $\mathbb{Z}_m$ is a field if and only if m is prime, and multiplicative inverses can be found using the Euclidean algorithm. A similar situation holds for $\mathbb{Z}_p[x]/(f(x))$. The analog of primality for polynomials is irreducibility, which we define as follows:

DEFINITION 5.2 *A polynomial $f(x) \in \mathbb{Z}_p[x]$ is said to be irreducible if there do not exist polynomials $f_1(x), f_2(x) \in \mathbb{Z}_p[x]$ such that*

$$f(x) = f_1(x)f_2(x),$$

where $deg(f_1) > 0$ and $deg(f_2) > 0$.

A very important fact is that $\mathbb{Z}_p[x]/(f(x))$ is a field if and only if $f(x)$ is irreducible. Further, multiplicative inverses in $\mathbb{Z}_p[x]/(f(x))$ can be computed using a straightforward modification of the (extended) Euclidean algorithm.

Here is an example to illustrate the concepts described above.

Example 5.6
Let's attempt to construct a field having eight elements. This can be done by finding an irreducible polynomial of degree three in $\mathbb{Z}_2[x]$. It is sufficient to consider the polynomials having constant term equal to 1, since any polynomial with constant term 0 is divisible by x and hence is reducible. There are four such polynomials:

$$f_1(x) = x^3 + 1$$
$$f_2(x) = x^3 + x + 1$$
$$f_3(x) = x^3 + x^2 + 1$$
$$f_4(x) = x^3 + x^2 + x + 1.$$

Now, $f_1(x)$ is reducible, since

$$x^3 + 1 = (x + 1)(x^2 + x + 1)$$

(remember that all coefficients are to be reduced modulo 2). Also, f_4 is reducible since

$$x^3 + x^2 + x + 1 = (x + 1)(x^2 + 1).$$

However, $f_2(x)$ and $f_3(x)$ are both irreducible, and either one can be used to construct a field having eight elements.

Let us use $f_2(x)$, and thus construct the field $\mathbb{Z}_2[x]/(x^3 + x + 1)$. The eight field elements are the eight polynomials 0, 1, x, $x + 1$, x^2, $x^2 + 1$, $x^2 + x$ and $x^2 + x + 1$.

To compute a product of two field elements, we multiple the two polynomials together, and reduce modulo $x^3 + x + 1$ (i.e., divide by $x^3 + x + 1$ and find the remainder polynomial). Since we are dividing by a polynomial of degree three, the remainder will have degree at most two and hence is an element of the field.

For example, to compute $(x^2 + 1)(x^2 + x + 1)$ in $\mathbb{Z}_2[x]/(x^3 + x + 1)$, we first compute the product in $\mathbb{Z}_2[x]$, which is $x^4 + x^3 + x + 1$. Then we divide by $x^3 + x + 1$, obtaining the expression

$$x^4 + x^3 + x + 1 = (x + 1)(x^3 + x + 1) + x^2 + x.$$

Hence, in the field $\mathbb{Z}_2[x]/(x^3 + x + 1)$, we have that

$$(x^2 + 1)(x^2 + x + 1) = x^2 + x.$$

Below, we present a complete multiplication table for the non-zero field elements. To save space, we write a polynomial $a_2x^2 + a_1x + a_0$ as the ordered triple $a_2a_1a_0$.

	001	010	011	100	101	110	111
001	001	010	011	100	101	110	111
010	010	100	110	011	001	111	101
011	011	110	101	111	100	001	010
100	100	011	111	110	010	101	001
101	101	001	100	010	111	011	110
110	110	111	001	101	011	010	100
111	111	101	010	001	110	100	011

Computation of inverses can be done by using a straightforward adaptation of the extended Euclidean algorithm.

Finally, the multiplicative group of the non-zero polynomials in the field is a cyclic group of order seven. Since 7 is prime, it follows that any non-zero field element is a generator of this group, i.e., a primitive element of the field.

For example, if we compute the powers of x, we obtain

$$x^1 = x$$

$$x^2 = x^2$$

$$x^3 = x + 1$$

$$x^4 = x^2 + x$$

$$x^5 = x^2 + x + 1$$

$$x^6 = x^2 + 1$$

$$x^7 = 1,$$

which comprise all the non-zero field elements. ☐

It remains to discuss existence and uniqueness of fields of this type. It can be shown that there is at least one irreducible polynomial of any given degree $n \geq 1$ in $\mathbb{Z}_p[x]$. Hence, there is a finite field with p^n elements for all primes p and all integers $n \geq 1$. There are usually many irreducible polynomials of degree n in $\mathbb{Z}_p[x]$. But the finite fields constructed from any two irreducible polynomials of degree n can be shown to be isomorphic. Thus there is a unique finite field of any size p^n (p prime, $n \geq 1$), which is denoted by $\mathrm{GF}(p^n)$. In the case $n = 1$, the resulting field $\mathrm{GF}(p)$ is the same thing as $\mathbb{Z}_p$. Finally, it can be shown that there does not exist a finite field with r elements unless $r = p^n$ for some prime p and some integer $n \geq 1$.

We have already noted that the multiplicative group $\mathbb{Z}_p^*$ (p prime) is a cyclic group of order $p - 1$. In fact, the multiplicative group of any finite field is cyclic: $\mathrm{GF}(p^n) \backslash \{0\}$ is a cyclic group of order $p^n - 1$. This provides further examples of cyclic groups in which the discrete log problem can be studied.

In practice, the finite fields $\mathrm{GF}(2^n)$ have been most studied. Both the Shanks and Pohlig-Hellman discrete logarithm algorithms work for fields $\mathrm{GF}(2^n)$. The index calculus method can be modified to work in these fields. The precomputation time of the index calculus algorithm turns out to be

$$O\left(e^{(1.405+o(1))n^{1/3}(\ln n)^{2/3}}\right),$$

and the time to find an individual discrete log is

$$O\left(e^{(1.098+o(1))n^{1/3}(\ln n)^{2/3}}\right).$$

However, for large values of n (say $n > 800$), the discrete log problem in $\mathrm{GF}(2^n)$ is thought to be intractible provided $2^n - 1$ has at least one "large" prime factor (in order to thwart a Pohlig-Hellman attack).

5.2.2 Elliptic Curves

We begin by defining the concept of an elliptic curve.

DEFINITION 5.3 *Let $p > 3$ be prime. The elliptic curve $y^2 = x^3 + ax + b$ over $\mathbb{Z}_p$ is the set of solutions $(x, y) \in \mathbb{Z}_p \times \mathbb{Z}_p$ to the congruence*

$$y^2 \equiv x^3 + ax + b \pmod{p}, \tag{5.1}$$

where $a, b \in \mathbb{Z}_p$ are constants such that $4a^3 + 27b^2 \not\equiv 0 \pmod{p}$, together with a special point $\mathcal{O}$ called the point at infinity.[1]

---

[1] Equation 5.1 can be used to define an elliptic curve over any field $\mathrm{GF}(p^n)$, for $p > 3$ prime. An elliptic curve over $\mathrm{GF}(2^n)$ or $\mathrm{GF}(3^n)$ is defined by a slightly different equation.

An elliptic curve E can be made into an abelian group by defining a suitable operation on its points. The operation is written additively, and is defined as follows (where all arithmetic operations are performed in $\mathbb{Z}_p$): Suppose

$$P = (x_1, y_1)$$

and

$$Q = (x_2, y_2)$$

are points on E. If $x_2 = x_1$ and $y_2 = -y_1$, then $P + Q = \mathcal{O}$; otherwise $P + Q = (x_3, y_3)$, where

$$x_3 = \lambda^2 - x_1 - x_2$$

$$y_3 = \lambda(x_1 - x_3) - y_1,$$

and

$$\lambda = \begin{cases} \dfrac{y_2 - y_1}{x_2 - x_1}, & \text{if } P \neq Q \\[2mm] \dfrac{3x_1{}^2 + a}{2y_1}, & \text{if } P = Q. \end{cases}$$

Finally, define

$$P + \mathcal{O} = \mathcal{O} + P = P$$

for all $P \in E$. With this definition of addition, it can be shown that E is an abelian group with identity element $\mathcal{O}$ (most of the verifications are tedious but straightforward, but proving associativity is quite difficult).

Note that inverses are very easy to compute. The inverse of (x, y) (which we write as $-(x, y)$ since the group operation is additive) is $(x, -y)$, for all $(x, y) \in E$.

Let us look at a small example.

Example 5.7
Let E be the elliptic curve $y^2 = x^3 + x + 6$ over $\mathbb{Z}_{11}$. Let's first determine the points on E. This can be done by looking at each possible $x \in \mathbb{Z}_{11}$, computing $x^3 + x + 6 \bmod 11$, and then trying to solve Equation 5.1 for y. For a given x we can test to see if $z = x^3 + x + 6 \bmod 11$ is a quadratic residue by applying Euler's criterion. Recall that there is an explicit formula to compute square roots of quadratic residues modulo p for primes $p \equiv 3 \pmod 4$. Applying this formula, we have that the square roots of a quadratic residue z are

$$\pm z^{(11+1)/4} \bmod 11 = \pm z^3 \bmod 11.$$

The results of these computations are tabulated in Table 5.2.

Thus E has 13 points on it. Since any group of prime order is cyclic, it follows that E is isomorphic to $\mathbb{Z}_{13}$, and any point other than the point at infinity is a generator of E. Suppose we take the generator $\alpha = (2, 7)$. Then we can compute the

TABLE 5.2
Points on the elliptic curve $y^2 = x^3 + x + 6$ over $\mathbb{Z}_{11}$

x	$x^3 + x + 6 \bmod 11$	in QR(11)?	y
0	6	no	
1	8	no	
2	5	yes	4, 7
3	3	yes	5, 6
4	8	no	
5	4	yes	2, 9
6	8	no	
7	4	yes	2, 9
8	9	yes	3, 8
9	7	no	
10	4	yes	2, 9

"powers" of α (which we will write as multiples of α, since the group operation is additive). To compute $2\alpha = (2, 7) + (2, 7)$, we first compute

$$\lambda = (3 \times 2^2 + 1)(2 \times 7)^{-1} \bmod 11$$

$$= 2 \times 3^{-1} \bmod 11$$

$$= 2 \times 4 \bmod 11$$

$$= 8.$$

Then we have

$$x_3 = 8^2 - 2 - 2 \bmod 11$$

$$= 5$$

and

$$y_3 = 8(2 - 5) - 7 \bmod 11$$

$$= 2,$$

so $2\alpha = (5, 2)$.

The next multiple would be $3\alpha = 2\alpha + \alpha = (5, 2) + (2, 7)$. Again, we begin by computing λ, which in this situation is done as follows:

$$\lambda = (7 - 2)(2 - 5)^{-1} \bmod 11$$

$$= 5 \times 8^{-1} \bmod 11$$

$$= 5 \times 7 \bmod 11$$

$$= 2.$$

Then we have

$$x_3 = 2^2 - 5 - 2 \bmod 11$$
$$= 8$$

and

$$y_3 = 2(5 - 8) - 2 \bmod 11$$
$$= 3,$$

so $3\alpha = (8, 3)$.

Continuing in this fashion, the remaining multiples can be computed to be the following:

α	$=$	$(2,7)$	2α	$=$	$(5,2)$	3α	$=$	$(8,3)$
4α	$=$	$(10,2)$	5α	$=$	$(3,6)$	6α	$=$	$(7,9)$
7α	$=$	$(7,2)$	8α	$=$	$(3,5)$	9α	$=$	$(10,9)$
10α	$=$	$(8,8)$	11α	$=$	$(5,9)$	12α	$=$	$(2,4)$

Hence $\alpha = (2, 7)$ is indeed a primitive element. $\square$

An elliptic curve E defined over $\mathbb{Z}_p$ (p prime, $p > 3$) will have roughly p points on it. More precisely, a well-known theorem due to Hasse asserts that the number of points on E, which we denote by $\#E$, satisfies the following inequality

$$p + 1 - 2\sqrt{p} \leq \#E \leq p + 1 + 2\sqrt{p}.$$

Computing the exact value of $\#E$ is more difficult, but there is an efficient algorithm to do this, due to Schoof. (By "efficient" we mean that it has a running time that is polynomial in $\log p$. Schoof's algorithm has a running time of $O((\log p)^8)$ bit operations and is practical for primes p having several hundred digits.)

Now, given that we can compute $\#E$, we further want to find a cyclic subgroup of E in which the discrete log problem is intractible. So we would like to know something about the structure of the group E. The following theorem gives a considerable amount of information on the group structure of E.

THEOREM 5.1
Let E be an elliptic curve defined over $\mathbb{Z}_p$, where p is prime, $p > 3$. Then there exist integers n_1 and n_2 such that E is isomorphic to $\mathbb{Z}_{n_1} \times \mathbb{Z}_{n_2}$. Further, $n_2 \mid n_1$ and $n_2 \mid (p - 1)$.

Hence, if the integers n_1 and n_2 can be computed, then we know that E has a cyclic subgroup isomorphic to $\mathbb{Z}_{n_1}$ that can potentially be used as a setting for an **ElGamal Cryptosystem**.

Note that if $n_2 = 1$, then E is a cyclic group. Also, if $\#E$ is a prime, or the product of distinct primes, then E must be a cyclic group.

The Shanks and Pohlig-Hellman algorithms apply to the elliptic curve logarithm problem, but there is no known adaptation of the index calculus method to elliptic curves. However, there is a method of exploiting an explicit isomorphism between elliptic curves and finite fields that leads to efficient algorithms for certain classes of elliptic curves. This technique, due to Menezes, Okamoto and Vanstone, can be applied to some particular examples within a special class of elliptic curves called supersingular curves that were suggested for use in cryptosystems. If the supersingular curves are avoided, however, then it appears that an elliptic curve having a cyclic subgroup of size about 2^{160} will provide a secure setting for a cryptosystem, provided that the order of the subgroup is divisible by at least one large prime factor (again, to guard against a Pohlig-Hellman attack).

Let's now look at an example of **ElGamal** encryption using the elliptic curve of Example 5.7.

Example 5.8
Suppose that $\alpha = (2, 7)$ and Bob's secret "exponent" is $a = 7$, so

$$\beta = 7\alpha = (7, 2).$$

Thus the encryption operaton is

$$e_K(x, k) = (k(2, 7), x + k(7, 2)),$$

where $x \in E$ and $0 \le k \le 12$, and the decryption operation is

$$d_K(y_1, y_2) = y_2 - 7y_1.$$

Suppose that Alice wishes to encrypt the message $x = (10, 9)$ (which is a point on E). If she chooses the random value $k = 3$, then she will compute

$$y_1 = 3(2, 7)$$
$$= (8, 3)$$

and

$$y_2 = (10, 9) + 3(7, 2)$$
$$= (10, 9) + (3, 5)$$
$$= (10, 2).$$

Hence, $y = ((8, 3), (10, 2))$. Now, if Bob receives the ciphertext y, he decrypts it as follows:

$$x = (10, 2) - 7(8, 3)$$

$$= (10, 2) - (3, 5)$$
$$= (10, 2) + (3, 6)$$
$$= (10, 9).$$

Hence, the decryption yields the correct plaintext.   □

There are some practical difficulties in implementing an **ElGamal Cryptosystem** on an elliptic curve. This system, when implemented in $\mathbb{Z}_p$ (or in $GF(p^n)$ with $n > 1$) has a message expansion factor of two. An elliptic curve implementation has a message expansion factor of (about) four. This happens since there are approximately p plaintexts, but each ciphertext consists of four field elements. A more serious problem is that the plaintext space consists of the points on the curve E, and there is no convenient method known of deterministically generating points on E.

A more efficient variation has been found by Menezes and Vanstone. In this variation, the elliptic curve is used for "masking," and plaintexts and ciphertexts are allowed to be arbitrary ordered pairs of (nonzero) field elements (i.e., they are not required to be points on E). This yields a message expansion factor of two, the same as in the original **ElGamal Cryptosystem**. The **Menezes-Vanstone Cryptosystem** is presented in Figure 5.10.

If we return to the curve $y^2 = x^3 + x + 6$ over $\mathbb{Z}_{11}$, we see that the **Menezes-Vanstone Cryptosystem** allows $10 \times 10 = 100$ plaintexts, as compared to 13 in the original system. We illustrate encryption and decryption in this system using this same curve.

Example 5.9
As in the previous example, suppose that $\alpha = (2, 7)$ and Bob's secret "exponent" is $a = 7$, so
$$\beta = 7\alpha = (7, 2).$$
Suppose Alice wants to encrypt the plaintext
$$x = (x_1, x_2) = (9, 1)$$
(note that x is not a point on E), and she chooses the random value $k = 6$. First, she computes
$$y_0 = k\alpha = 6(2, 7) = (7, 9)$$
and
$$k\beta = 6(7, 2) = (8, 3),$$
so $c_1 = 8$ and $c_2 = 3$.
    Next, she calculates
$$y_1 = c_1 x_1 \bmod p = 8 \times 9 \bmod 11 = 6$$

FIGURE 5.10
Menezes-Vanstone Elliptic Curve Cryptosystem

Let E be an elliptic curve defined over $\mathbb{Z}_p$ ($p > 3$ prime) such that E contains a cyclic subgroup H in which the discrete log problem is intractible.
Let $\mathcal{P} = \mathbb{Z}_p{}^* \times \mathbb{Z}_p{}^*$, $\mathcal{C} = E \times \mathbb{Z}_p{}^* \times \mathbb{Z}_p{}^*$, and define

$$\mathcal{K} = \{(E, \alpha, a, \beta) : \beta = a\alpha\},$$

where $\alpha \in E$. The values α and β are public, and a is secret.

For $K = (E, \alpha, a, \beta)$, for a (secret) random number $k \in \mathbb{Z}_{|H|}$, and for $x = (x_1, x_2) \in \mathbb{Z}_p{}^* \times \mathbb{Z}_p{}^*$, define

$$e_K(x, k) = (y_0, y_1, y_2),$$

where

$$y_0 = k\alpha,$$
$$(c_1, c_2) = k\beta,$$
$$y_1 = c_1 x_1 \bmod p, \qquad \text{and}$$
$$y_2 = c_2 x_2 \bmod p.$$

For a ciphertext $y = (y_0, y_1, y_2)$, define

$$d_K(y) = (y_1 c_1{}^{-1} \bmod p, y_2 c_2{}^{-1} \bmod p),$$

where
$$a y_0 = (c_1, c_2).$$

and
$$y_2 = c_2 x_2 \bmod p = 3 \times 1 \bmod 11 = 3.$$

The ciphertext she sends to Bob is

$$y = (y_0, y_1, y_2) = ((7, 9), 6, 3).$$

When Bob receives the ciphertext y, he first computes

$$(c_1, c_2) = a y_0 = 7(7, 9) = (8, 3),$$

FIGURE 5.11
Subset sum problem

Problem Instance $I = (s_1, \ldots s_n, T)$, where $s_1, \ldots s_n$ and T are positive integers. The s_i's are called *sizes* and T is called the *target sum*.

Question   Is there a 0-1 vector $\mathbf{x} = (x_1, \ldots, x_n)$ such that

$$\sum_{i=1}^{n} x_i s_i = T?$$

---

and then

$$x = (y_1 c_1{}^{-1} \bmod p, y_2 c_2{}^{-1} \bmod p)$$
$$= (6 \times 8^{-1} \bmod 11, 3 \times 3^{-1} \bmod 11)$$
$$= (6 \times 7 \bmod 11, 3 \times 4 \bmod 11)$$
$$= (9, 1).$$

Hence, the decryption yields the correct plaintext.   □

5.3   The Merkle-Hellman Knapsack System

The well-known **Merkle-Hellman Knapsack Cryptosystem** was first described by Merkle and Hellman in 1978. Although this cryptosystem, and several variants of it, were broken in the early 1980's, it is still worth studying for its conceptual elegance and for the underlying design technique.

The term "knapsack" is actually a misnomer[2]; the system is based on the **Subset Sum** problem which is presented in Figure 5.11.

The **Subset Sum** problem, as phrased in Figure 5.11, is a *decision problem* (i.e., we are required only to answer "yes" or "no"). If we rephrase the problem slightly, so that in any instance where the answer is "yes" we are required to find the desired vector $\mathbf{x}$ (which may not be unique), then we have a *search problem*.

---

[2] The **Knapsack** problem, as it is usually defined, is a problem involving selecting objects with given weights and profits in such a way that a specified capacity is not exceeded and a specified target profit is attained.

FIGURE 5.12
Algorithm for solving a superincreasing instance of the subset sum problem

1.	**for** $i = n$ **downto** 1 **do**
2.	**if** $T \geq s_i$ **then**
3.	$T = T - s_i$
4.	$x_i = 1$
5.	**else**
6.	$x_i = 0$
7.	**if** $T = 0$ **then**
8.	$X = (x_1, \ldots, x_n)$ is the solution
9.	**else**
10.	there is no solution.

The **Subset Sum** (decision) problem is one of the so-called NP-complete problems. Among other things, this means that there is no known polynomial-time algorithm that solves it. This is also the case for the **Subset Sum** search problem. But even if a problem has no polynomial-time algorithm to solve it in general, this does not rule out the possibility that certain special cases can be solved in polynomial time. This is indeed the situation with the **Subset Sum** problem.

We define a list of sizes, $(s_1, \ldots, s_n)$ to be *superincreasing* if

$$s_j > \sum_{i=1}^{j-1} s_i$$

for $2 \leq j \leq n$. If the list of sizes is superincreasing, then the search version of the **Subset Sum** problem can be solved very easily in time $O(n)$, and a solution $\mathbf{x}$ (if it exists) must be unique. The algorithm to do this is presented in Figure 5.12.

Suppose $\mathbf{s} = (s_1, \ldots, s_n)$ is superincreasing, and consider the function

$$e_{\mathbf{s}} : \{0,1\}^n \rightarrow \left\{ 0, \ldots, \sum_{i=1}^{n} s_i \right\}$$

defined by the rule

$$e_{\mathbf{s}}(x_1, \ldots, x_n) = \sum_{i=1}^{n} x_i s_i.$$

Is $e_{\mathbf{s}}$ a possible candidate for an encryption rule? Since $\mathbf{s}$ is superincreasing, $e_{\mathbf{s}}$ is an injection, and the algorithm presented in Figure 5.12 would be the corresponding decryption algorithm. However, such a system would be completely insecure since anyone (including Oscar) can decrypt a message that is encrypted in this way.

The strategy therefore is to transform the list of sizes in such a way that it is no longer superincreasing. Bob will be able to apply an inverse transformation to restore the superincreasing list of sizes. On the other hand Oscar, who does not know the transformation that was applied, is faced with what looks like a general, apparently difficult, instance of the subset sum problem when he tries to decrypt a ciphertext.

One suitable type of transformation is a *modular transformation*. That is, a prime modulus p is chosen such that

$$p > \sum_{i=1}^{n} s_i,$$

as well as a multiplier a, where $1 \leq a \leq p - 1$. Then we define

$$t_i = a s_i \bmod p,$$

$1 \leq i \leq n$. The list of sizes $\mathbf{t} = (t_1, \ldots, t_n)$ will be the public key used for encryption. The values a, p used to define the modular transformation are secret. The complete description of the **Merkle-Hellman Knapsack Cryptosystem** is given in Figure 5.13.

The following small example illustrates the encryption and decryption operations in the **Merkle-Hellman Cryptosystem**.

Example 5.10
Suppose

$$\mathbf{s} = (2, 5, 9, 21, 45, 103, 215, 450, 946)$$

is the secret superincreasing list of sizes. Suppose $p = 2003$ and $a = 1289$. Then the public list of sizes is

$$\mathbf{t} = (575, 436, 1586, 1030, 1921, 569, 721, 1183, 1570).$$

Now, if Alice wants to encrypt the plaintext $x = (1, 0, 1, 1, 0, 0, 1, 1, 1)$, she computes

$$y = 575 + 1586 + 1030 + 721 + 1183 + 1570 = 6665.$$

When Bob receives the ciphertext y, he first computes

$$z = a^{-1} y \bmod p$$
$$= 317 \times 6665 \bmod 2003$$
$$= 1643.$$

Then Bob solves the instance $I = (\mathbf{s}, z)$ of the **Subset Sum** problem using the algorithm presented in Figure 5.12. The plaintext $(1, 0, 1, 1, 0, 0, 1, 1, 1)$ is obtained. ▯

FIGURE 5.13
Merkle-Hellman Knapsack Cryptosystem

Let $\mathbf{s} = (s_1, \ldots, s_n)$ be a superincreasing list of integers, let $p > \sum_{i=1}^{n} s_i$ be prime, and let $1 \le a \le p - 1$. For $1 \le i \le n$, define

$$t_i = as_i \bmod p,$$

and denote $\mathbf{t} = (t_1, \ldots, t_n)$. Let $\mathcal{P} = \{0,1\}^n$, $\mathcal{C} = \{0, \ldots, n(p-1)\}$, and let

$$\mathcal{K} = \{(\mathbf{s}, p, a, \mathbf{t})\},$$

where $\mathbf{s}$, p, a, and $\mathbf{t}$ are constructed as described above. $\mathbf{t}$ is public, and p, a and $\mathbf{s}$ are secret.

For $K = (\mathbf{s}, p, a, \mathbf{t})$, define

$$e_K(x_1, \ldots, x_n) = \sum_{i=1}^{n} x_i t_i.$$

For $0 \le y \le n(p-1)$, define $z = a^{-1} y \bmod p$ and solve the subset problem $(s_1, \ldots, s_n, z)$, obtaining $d_K(y) = (x_1, \ldots, x_n)$.

By the early 1980's, the **Merkle-Hellman Knapsack Cryptosystem** had been broken by Shamir. Shamir was able to use an integer programming algorithm of Lenstra to break the system. This allows Bob's trapdoor (or an equivalent trapdoor) to be discovered by Oscar, the cryptanalyst. Then Oscar can decrypt messages exactly as Bob does.

5.4 The McEliece System

The **McEliece Cryptosystem** uses the same design principle as the **Merkle-Hellman Cryptosystem**: decryption is an easy special case of an NP-complete problem, disguised so that it looks like a general instance of the problem. In this system, the NP-complete problem that is employed is decoding a general linear (binary) error-correcting code. However, for many special classes of codes, polynomial-time algorithms are known to exist. One such class of codes, the Goppa codes, are used as the basis of the **McEliece Cryptosystem**.

We begin with some essential definitions.

DEFINITION 5.4   *Let k, n be positive integers, $k \le n$. An $[n, k]$ code, **C**, is a k-dimensional subspace of $(\mathbb{Z}_2)^n$, the vector space of all binary n-tuples.*

*A generating matrix for an $[n, k]$ code, **C**, is a $k \times n$ binary matrix whose rows form a basis for **C**.*

Let $\mathbf{x}, \mathbf{y} \in (\mathbb{Z}_2)^n$, where $\mathbf{x} = (x_1, \ldots, x_n)$ and $\mathbf{y} = (y_1, \ldots, y_n)$. Define the Hamming distance

$$d(\mathbf{x}, \mathbf{y}) = |\{i : 1 \le i \le n, x_i \ne y_i\}|,$$

i.e., the number of coordinates in which $\mathbf{x}$ and $\mathbf{y}$ differ.

*Let **C** be an $[n, k]$ code. Define the distance of **C** to be the quantity*

$$d(\mathbf{C}) = \min\{d(\mathbf{x}, \mathbf{y}) : \mathbf{x}, \mathbf{y} \in \mathbf{C}, \mathbf{x} \ne \mathbf{y}\}.$$

An $[n, k]$ code with distance d is denoted as an $[n, k, d]$ code.

·The purpose of an error-correcting code is to correct random errors that occur in the transmission of (binary) data through a noisy channel. Briefly, this is done as follows. Let G be a generating matrix for an $[n, k, d]$ code. Suppose $\mathbf{x}$ is the binary k-tuple we wish to transmit. Then Alice *encodes* $\mathbf{x}$ as the n-tuple $\mathbf{y} = \mathbf{x}G$, and transmits $\mathbf{y}$ through the channel.

Now, suppose Bob receives the n-tuple $\mathbf{r}$, which may not be the same as $\mathbf{y}$. He will *decode* $\mathbf{r}$ using the strategy of *nearest neighbor decoding*. In nearest neighbor decoding, Bob finds the codeword $\mathbf{y}'$ that has minimum distance to $\mathbf{r}$. Then he decodes $\mathbf{r}$ to $\mathbf{y}'$, and, finally, determines the k-tuple $\mathbf{x}'$ such that $\mathbf{y}' = \mathbf{x}'G$. Bob is hoping that $\mathbf{y}' = \mathbf{y}$, so $\mathbf{x}' = \mathbf{x}$ (i.e., he is hoping that any transmission errors have been corrected).

It is fairly easy to show that if at most $(d - 1)/2$ errors occurred during transmission, then nearest neighbor decoding does in fact correct all the errors.

Let us think about how nearest neighbor decoding would be done in practice. $|\mathbf{C}| = 2^k$, so if Bob compares $\mathbf{r}$ to every codeword, he will have to examine 2^k vectors, which is an exponentially large number compared to k. In other words, this obvious algorithm is not a polynomial-time algorithm.

Another approach, ˙which forms the basis for many practical decoding algorithms, is based on the idea of a syndrome. A *parity-check matrix* for an $[n, k, d]$ code **C** having generating matrix G is an $(n - k) \times n$ 0 – 1 matrix, denoted by H, whose rows form a basis for the orthogonal complement of **C**, which is denoted by $\mathbf{C}^\perp$ and called the *dual code* to **C**. Stated another way, the rows of H are linearly independent vectors, and GH^T is a $k \times (n - k)$ matrix of zeroes.

Given a vector $\mathbf{r} \in (\mathbb{Z}_2)^n$, we define the *syndrome* of $\mathbf{r}$ to be Hr^T. A syndrome is a column vector with $n - k$ components.

The following basic results follow immediately from linear algebra.

THEOREM 5.2
Suppose $\mathbf{C}$ *is an* $[n, k]$ *code with generating matrix* G *and parity-check matrix* H. *Then* $\mathbf{x} \in (\mathbb{Z}_2)^n$ *is a codeword if and only if*

$$H\mathbf{x}^T = \begin{pmatrix} 0 \\ 0 \\ \vdots \\ 0 \end{pmatrix}.$$

Further, if $\mathbf{x} \in \mathbf{C}$, $\mathbf{e} \in (\mathbb{Z}_2)^n$, *and* $\mathbf{r} = \mathbf{x} + \mathbf{e}$, *then* $H\mathbf{r}^T = H\mathbf{e}^T$.

Think of $\mathbf{e}$ as being the vector of errors that occur during transmission of a codeword $\mathbf{x}$. Then $\mathbf{r}$ represents the vector that is received. The above theorem is saying that the syndrome depends only on the errors, and not on the particular codeword that was transmitted.

This suggests the following approach to decoding, known as *syndrome decoding*: First, compute $\mathbf{s} = H\mathbf{r}^T$. If $\mathbf{s}$ is a vector of zeroes, then decode $\mathbf{r}$ as $\mathbf{r}$. If not, then generate all possible error vectors or weight 1 in turn. For each such $\mathbf{e}$, compute $H\mathbf{e}^T$. If, for any of these vectors $\mathbf{e}$, it happens that $H\mathbf{e}^T = \mathbf{s}$, then decode $\mathbf{r}$ to $\mathbf{r} - \mathbf{e}$. Otherwise, continue on to generate all error vectors of weight $2, \ldots, \lfloor (d-1)/2 \rfloor$. If at any time $H\mathbf{e}^T = \mathbf{s}$, then we decode $\mathbf{r}$ to $\mathbf{r} - \mathbf{e}$ and quit. If this equation is never satisfied, then we conclude that more than $\lfloor (d-1)/2 \rfloor$ errors have occurred during transmission.

By this approach, we can decode a received vector in at most

$$1 + \binom{n}{1} + \cdots + \binom{n}{\lfloor (d-1)/2 \rfloor}$$

steps.

This method will work on any linear code. For certain specific types of codes, the decoding procedure can be speeded up. However, a decision version of nearest neighbor decoding is in fact an NP-complete problem. Thus no polynomial-time algorithm is known for the general problem of nearest neighbor decoding (when the number of errors is not bounded by $\lfloor (d-1)/2 \rfloor$).

As was the case with the subset sum problem, we can identify an "easy" special case, and then disguise it so that it looks like a "difficult" general case of the problem. It would take us too long to go into the theory here, so we will just summarize the results. The "easy" special case that was suggested by McEliece is to use a code from a class of codes known as the *Goppa codes*. These codes do in fact have efficient decoding algorithms. Also, they are easy to generate, and there are a large number of inequivalent Goppa codes with the same parameters.

The parameters of the Goppa codes have the form $n = 2^m$, $d = 2t + 1$ and $k = n - mt$. For a practical implementation of the public-key cryptosystem, McEliece suggested taking $m = 10$ and $t = 50$. This gives rise to a Goppa code

FIGURE 5.14
McEliece Cryptosystem

Let G be a generating matrix for an $[n, k, d]$ Goppa code **C**, where $n = 2^m$, $d = 2t + 1$ and $k = n - mt$. Let S be a $k \times k$ matrix that is invertible over $\mathbb{Z}_2$, let P be an $n \times n$ permutation matrix, and let $G' = SGP$. Let $\mathcal{P} = (\mathbb{Z}_2)^k$, $\mathcal{C} = (\mathbb{Z}_2)^n$, and let

$$\mathcal{K} = \{(G, S, P, G')\},$$

where G, S, P, and G' are constructed as described above. G' is public, and G, S, and P are secret.

For $K = (G, S, P, G')$, define

$$e_K(\mathbf{x}, \mathbf{e}) = \mathbf{x}G' + \mathbf{e},$$

where $\mathbf{e} \in (\mathbb{Z}_2)^n$ is a random vector of weight t.

Bob decrypts a ciphertext $\mathbf{y} \in (\mathbb{Z}_2)^n$ by means of the following operations:

1. Compute $\mathbf{y}_1 = \mathbf{y}P^{-1}$.
2. Decode $\mathbf{y}_1$, obtaining $\mathbf{y}_1 = \mathbf{x}_1 + \mathbf{e}_1$, where $\mathbf{x}_1 \in$ **C**.
3. Compute $\mathbf{x}_0 \in (\mathbb{Z}_2)^k$ such that $\mathbf{x}_0 G = \mathbf{x}_1$.
4. Compute $\mathbf{x} = \mathbf{x}_0 S^{-1}$.

---

that is a $[1024, 524, 101]$ code. Each plaintext is a binary 524-tuple, and each ciphertext is a binary 1024-tuple. The public key is a 524×1024 binary matrix.

A description of the **McEliece Cryptosystem** is given in Figure 5.14.

We present a ridiculously small example to illustrate the encoding and decoding procedures.

Example 5.11
The matrix

$$G = \begin{pmatrix} 1 & 0 & 0 & 0 & 1 & 1 & 0 \\ 0 & 1 & 0 & 0 & 1 & 0 & 1 \\ 0 & 0 & 1 & 0 & 0 & 1 & 1 \\ 0 & 0 & 0 & 1 & 1 & 1 & 1 \end{pmatrix}$$

is a generating matrix for a $[7, 4, 3]$ code, known as a *Hamming code*. Suppose

Bob chooses the matrices

$$S = \begin{pmatrix} 1 & 1 & 0 & 1 \\ 1 & 0 & 0 & 1 \\ 0 & 1 & 1 & 1 \\ 1 & 1 & 0 & 0 \end{pmatrix}$$

and

$$P = \begin{pmatrix} 0 & 1 & 0 & 0 & 0 & 0 & 0 \\ 0 & 0 & 0 & 1 & 0 & 0 & 0 \\ 0 & 0 & 0 & 0 & 0 & 0 & 1 \\ 1 & 0 & 0 & 0 & 0 & 0 & 0 \\ 0 & 0 & 1 & 0 & 0 & 0 & 0 \\ 0 & 0 & 0 & 0 & 0 & 1 & 0 \\ 0 & 0 & 0 & 0 & 1 & 0 & 0 \end{pmatrix}.$$

Then, the public generating matrix is

$$G' = \begin{pmatrix} 1 & 1 & 1 & 1 & 0 & 0 & 0 \\ 1 & 1 & 0 & 0 & 1 & 0 & 0 \\ 1 & 0 & 0 & 1 & 1 & 0 & 1 \\ 0 & 1 & 0 & 1 & 1 & 1 & 0 \end{pmatrix}.$$

Now, suppose Alice encrypts the plaintext $\mathbf{x} = (1, 1, 0, 1)$ using as the random error vector of weight 1 the vector $\mathbf{e} = (0, 0, 0, 0, 1, 0, 0)$. The ciphertext is computed to be

$$\mathbf{y} = \mathbf{x}G' + \mathbf{e}$$

$$= (1, 1, 0, 1) \begin{pmatrix} 1 & 1 & 1 & 1 & 0 & 0 & 0 \\ 1 & 1 & 0 & 0 & 1 & 0 & 0 \\ 1 & 0 & 0 & 1 & 1 & 0 & 1 \\ 0 & 1 & 0 & 1 & 1 & 1 & 0 \end{pmatrix} + (0, 0, 0, 0, 1, 0, 0)$$

$$= (0, 1, 1, 0, 0, 1, 0) + (0, 0, 0, 0, 1, 0, 0)$$

$$= (0, 1, 1, 0, 1, 1, 0).$$

When Bob receives the ciphertext $\mathbf{y}$, he first computes

$$\mathbf{y}_1 = \mathbf{y}P^{-1}$$

$$= (0, 1, 1, 0, 1, 1, 0) \begin{pmatrix} 0 & 0 & 0 & 1 & 0 & 0 & 0 \\ 1 & 0 & 0 & 0 & 0 & 0 & 0 \\ 0 & 0 & 0 & 0 & 1 & 0 & 0 \\ 0 & 1 & 0 & 0 & 0 & 0 & 0 \\ 0 & 0 & 0 & 0 & 0 & 0 & 1 \\ 0 & 0 & 0 & 0 & 0 & 1 & 0 \\ 0 & 0 & 1 & 0 & 0 & 0 & 0 \end{pmatrix}$$

$$= (1, 0, 0, 0, 1, 1, 1).$$

Next, he decrypts $\mathbf{y}_1$ to get $\mathbf{x}_1 = (1, 0, 0, 0, 1, 1, 0)$ (note that $\mathbf{e}_1 \neq \mathbf{e}$ due to the multiplication by P^{-1}).

Next, Bob forms $\mathbf{x}_0 = (1, 0, 0, 0)$ (the first four components of $\mathbf{x}_1$).

Finally, Bob calculates

$$
\begin{aligned}
\mathbf{x} &= S^{-1}\mathbf{x}_0 \\
&= \begin{pmatrix} 1\ 1\ 0\ 1 \\ 1\ 1\ 0\ 0 \\ 0\ 1\ 1\ 1 \\ 1\ 0\ 0\ 1 \end{pmatrix} (1, 0, 0, 0) \\
&= (1, 1, 0, 1).
\end{aligned}
$$

This is indeed the plaintext that Alice encrypted. $\square$

5.5   Notes and References

The **ElGamal Cryptosystem** was presented in [EL85]. The Pohlig-Hellman algorithm was published in [PH78], and the material concerning individual bits of the **Discrete Logarithm** problem is based on Peralta [PE86]. For further information on the **Discrete Logarithm** problem, we recommend the articles by LaMacchia and Odlyzko [LO91] and McCurley [MC90].

The main reference book for finite fields is Lidl and Niederreiter [LN83]. McEliece [MC87] is a good textbook on the subject, and a research monograph on applications of finite fields was published by Menezes *et al.* [MBGMVY93]. A recent article on the **Discrete Logarithm** problem in $GF(2^n)$ is Gordon and McCurley [GM93].

The idea of using elliptic curves for public-key cryptosystems is due to Koblitz [KO87] and Miller [MI86]. Menezes [ME93] is a monograph on elliptic curve cryptosystems. See also Menezes and Vanstone [MV93] and Koblitz [KO94]. For an elementary treatment of elliptic curves, see Silverman and Tate [ST92]. The Menezes-Okamoto-Vanstone reduction of discrete logarithms from elliptic curves to finite fields is given in [MOV94] (see also [ME93]).

The **Merkle-Hellman Cryptosystem** was presented in [MH78]. This system was broken by Shamir [SH84], and the "iterated" version of the system was broken by Brickell [BR85]. A different knapsack-type system, due to Chor and Rivest [CR88], has not been broken. For more information, see the survey article by Brickell and Odlyzko [BO92].

The most important reference book for coding theory is MacWilliams and Sloane [MS77]. There are many good textbooks on coding theory, e.g., Hoffman *et al.* [HLLPRW91] and Vanstone and van Oorschot [VV89]. The **McEliece**

Cryptosystem was first described in [MC78]. A recent article discussing the security of this cryptosystem is by Chabaud [CH95].

Exercises

5.1 Implement Shanks' algorithm for finding discrete logarithms in $\mathbb{Z}_p$, where p is prime and α is a primitive element. Use your program to find $\log_{106} 12375$ in $\mathbb{Z}_{24691}$ and $\log_6 248388$ in $\mathbb{Z}_{458009}$.

5.2 Implement the Pohlig-Hellman algorithm for finding discrete logarithms in $\mathbb{Z}_p$, where p is prime and α is a primitive element. Use your program to find $\log_5 8563$ in $\mathbb{Z}_{28703}$ and $\log_{10} 12611$ in $\mathbb{Z}_{31153}$.

5.3 Find $\log_5 896$ in $\mathbb{Z}_{1103}$ using the algorithm presented in Figure 5.6, given that $L_2(\beta) = 1$ for $\beta = 25$, 219 and 841, and $L_2(\beta) = 0$ for $\beta = 163$, 532, 625 and 656.

5.4 Decrypt the **ElGamal** ciphertext presented in Table 5.3. The parameters of the system are $p = 31847$, $\alpha = 5$, $a = 7899$ and $\beta = 18074$. Each element of $\mathbb{Z}_n$ represents three alphabetic characters as in Exercise 4.6.

The plaintext was taken from "The English Patient," by Michael Ondaatje, Alfred A. Knopf, Inc., New York, 1992.

5.5 Determine which of the following polynomials are irreducible over $\mathbb{Z}_2[x]$: $x^5 + x^4 + 1$, $x^5 + x^3 + 1$, $x^5 + x^4 + x^2 + 1$.

5.6 The field $GF(2^5)$ can be constructed as $\mathbb{Z}_2[x]/(x^5 + x^2 + 1)$. Perform the following computations in this field.

(a) Compute $(x^4 + x^2) \times (x^3 + x + 1)$.

(b) Using the extended Euclidean algorithm, compute $(x^3 + x^2)^{-1}$.

(c) Using the square-and-multiply algorithm, compute x^{25}.

5.7 We give an example of the **ElGamal Cryptosystem** implemented in $GF(3^3)$. The polynomial $x^3 + 2x^2 + 1$ is irreducible over $\mathbb{Z}_3[x]$ and hence $\mathbb{Z}_3[x]/(x^3 + 2x^2 + 1)$ is the field $GF(3^3)$. We can associate the 26 letters of the alphabet with the 26 nonzero field elements, and thus encrypt ordinary text in a convenient way. We will use a lexicographic ordering of the (nonzero) polynomials to set up the correspondence. This correspondence is as follows:

A	$\leftrightarrow$	1	B	$\leftrightarrow$	2	C	$\leftrightarrow$	x
D	$\leftrightarrow$	$x+1$	E	$\leftrightarrow$	$x+2$	F	$\leftrightarrow$	$2x$
G	$\leftrightarrow$	$2x+1$	H	$\leftrightarrow$	$2x+2$	I	$\leftrightarrow$	x^2
J	$\leftrightarrow$	x^2+1	K	$\leftrightarrow$	x^2+2	L	$\leftrightarrow$	x^2+x
M	$\leftrightarrow$	x^2+x+1	N	$\leftrightarrow$	x^2+x+2	O	$\leftrightarrow$	x^2+2x
P	$\leftrightarrow$	x^2+2x+1	Q	$\leftrightarrow$	x^2+2x+2	R	$\leftrightarrow$	$2x^2$
S	$\leftrightarrow$	$2x^2+1$	T	$\leftrightarrow$	$2x^2+2$	U	$\leftrightarrow$	$2x^2+x$
V	$\leftrightarrow$	$2x^2+x+1$	W	$\leftrightarrow$	$2x^2+x+2$	X	$\leftrightarrow$	$2x^2+2x$
Y	$\leftrightarrow$	$2x^x+2x+1$	Z	$\leftrightarrow$	$2x^x+2x+2$			

Suppose Bob uses $\alpha = x$ and $a = 11$ in an ElGamal system; then $\beta = x+2$. Show how Bob will decrypt the following string of ciphertext:

(K,H) (P,X) (N,K) (H,R) (T,F) (V,Y) (E,H) (F,A) (T,W) (J,D) (U,J)

TABLE 5.3
ElGamal Ciphertext

$(3781, 14409)$	$(31552, 3930)$	$(27214, 15442)$	$(5809, 30274)$
$(5400, 31486)$	$(19936, 721)$	$(27765, 29284)$	$(29820, 7710)$
$(31590, 26470)$	$(3781, 14409)$	$(15898, 30844)$	$(19048, 12914)$
$(16160, 3129)$	$(301, 17252)$	$(24689, 7776)$	$(28856, 15720)$
$(30555, 24611)$	$(20501, 2922)$	$(13659, 5015)$	$(5740, 31233)$
$(1616, 14170)$	$(4294, 2307)$	$(2320, 29174)$	$(3036, 20132)$
$(14130, 22010)$	$(25910, 19663)$	$(19557, 10145)$	$(18899, 27609)$
$(26004, 25056)$	$(5400, 31486)$	$(9526, 3019)$	$(12962, 15189)$
$(29538, 5408)$	$(3149, 7400)$	$(9396, 3058)$	$(27149, 20535)$
$(1777, 8737)$	$(26117, 14251)$	$(7129, 18195)$	$(25302, 10248)$
$(23258, 3468)$	$(26052, 20545)$	$(21958, 5713)$	$(346, 31194)$
$(8836, 25898)$	$(8794, 17358)$	$(1777, 8737)$	$(25038, 12483)$
$(10422, 5552)$	$(1777, 8737)$	$(3780, 16360)$	$(11685, 133)$
$(25115, 10840)$	$(14130, 22010)$	$(16081, 16414)$	$(28580, 20845)$
$(23418, 22058)$	$(24139, 9580)$	$(173, 17075)$	$(2016, 18131)$
$(19886, 22344)$	$(21600, 25505)$	$(27119, 19921)$	$(23312, 16906)$
$(21563, 7891)$	$(28250, 21321)$	$(28327, 19237)$	$(15313, 28649)$
$(24271, 8480)$	$(26592, 25457)$	$(9660, 7939)$	$(10267, 20623)$
$(30499, 14423)$	$(5839, 24179)$	$(12846, 6598)$	$(9284, 27858)$
$(24875, 17641)$	$(1777, 8737)$	$(18825, 19671)$	$(31306, 11929)$
$(3576, 4630)$	$(26664, 27572)$	$(27011, 29164)$	$(22763, 8992)$
$(3149, 7400)$	$(8951, 29435)$	$(2059, 3977)$	$(16258, 30341)$
$(21541, 19004)$	$(5865, 29526)$	$(10536, 6941)$	$(1777, 8737)$
$(17561, 11884)$	$(2209, 6107)$	$(10422, 5552)$	$(19371, 21005)$
$(26521, 5803)$	$(14884, 14280)$	$(4328, 8635)$	$(28250, 21321)$
$(28327, 19237)$	$(15313, 28649)$		

5.8 Let E be the elliptic curve $y^2 = x^3 + x + 28$ defined over $\mathbb{Z}_{71}$.

   (a) Determine the number of points on E.

   (b) Show that E is not a cyclic group.

   (c) What is the maximum order of an element in E? Find an element having this order.

5.9 Let E be the elliptic curve $y^2 = x^3 + x + 13$ defined over $\mathbb{Z}_{31}$. It can be shown that $\#E = 34$ and $(9, 10)$ is an element of order 34 in E. The **Menezes-Vanstone Cryptosystem** defined on E will have as its plaintext space $\mathbb{Z}_{31}^* \times \mathbb{Z}_{31}^*$. Suppose Bob's secret exponent is $a = 25$.

   (a) Compute $\beta = a\alpha$.

   (b) Decrypt the following string of ciphertext:

$$((4, 9), 28, 7), ((19, 28), 9, 13), ((5, 22), 20, 17), ((25, 16), 12, 27).$$

   (c) Assuming that each plaintext represents two alphabetic characters, convert the plaintext into an English word. (Here we will use the correspondence $A \leftrightarrow 1, \ldots, Z \leftrightarrow 26$, since 0 is not allowed in a (plaintext) ordered pair.)

5.10 Suppose the **Merkle-Hellman Cryptosystem** has as its public list of sizes the vector

$$\mathbf{t} = (1394, 1256, 1508, 1987, 439, 650, 724, 339, 2303, 810).$$

Suppose Oscar discovers that $p = 2503$.

(a) By trial and error, determine the value a such that the list $a^{-1}\mathbf{t} \bmod p$ is a permutation of a superincreasing list.

(b) Show how the ciphertext 5746 would be decrypted.

5.11 It can be shown that the matrix H shown below is a parity-check matrix for a $[15, 7, 5]$ code called a BCH code.

$$H = \begin{pmatrix} 1 & 0 & 0 & 0 & 1 & 0 & 0 & 1 & 1 & 0 & 1 & 0 & 1 & 1 & 1 \\ 0 & 1 & 0 & 0 & 1 & 1 & 0 & 1 & 0 & 1 & 1 & 1 & 1 & 0 & 0 \\ 0 & 0 & 1 & 0 & 0 & 1 & 1 & 0 & 1 & 0 & 1 & 1 & 1 & 1 & 0 \\ 0 & 0 & 0 & 1 & 0 & 0 & 1 & 1 & 0 & 1 & 0 & 1 & 1 & 1 & 1 \\ 1 & 0 & 0 & 0 & 1 & 1 & 0 & 0 & 0 & 1 & 1 & 0 & 0 & 0 & 1 \\ 0 & 0 & 0 & 1 & 1 & 0 & 0 & 0 & 1 & 1 & 0 & 0 & 0 & 1 & 1 \\ 0 & 0 & 1 & 0 & 1 & 0 & 0 & 1 & 0 & 1 & 0 & 0 & 1 & 0 & 1 \\ 0 & 1 & 1 & 1 & 1 & 0 & 1 & 1 & 1 & 1 & 0 & 1 & 1 & 1 & 1 \end{pmatrix}.$$

Decode, if possible, each of the following received vectors $\mathbf{r}$ using the syndrome decoding method.

(a) $\mathbf{r} = (1, 1, 0, 0, 0, 0, 0, 0, 0, 0, 0, 0, 0, 0, 0)$.

(b) $\mathbf{r} = (1, 1, 0, 1, 1, 1, 1, 0, 1, 0, 1, 1, 0, 0, 0)$.

(c) $\mathbf{r} = (1, 0, 1, 0, 1, 0, 0, 1, 0, 1, 1, 0, 0, 0, 0)$.

6
Signature Schemes

6.1 Introduction

In this chapter, we study *signature schemes*, which are also called digital signatures. A "conventional" handwritten signature attached to a document is used to specify the person responsible for it. A signature is used in everyday situations such as writing a letter, withdrawing money from a bank, signing a contract, etc.

A signature scheme is a method of signing a message stored in electronic form. As such, a signed message can be transmitted over a computer network. In this chapter, we will study several signature schemes, but first we discuss some fundamental differences between conventional and digital signatures.

First is the question of signing a document. With a conventional signature, a signature is physically part of the document being signed. However, a digital signature is not attached physically to the message that is signed, so the algorithm that is used must somehow "bind" the signature to the message.

Second is the question of verification. A conventional signature is verified by comparing it to other, authentic signatures. For example, when someone signs a credit card purchase, the salesperson is supposed to compare the signature on the sales slip to the signature on the back of the credit card in order to verify the signature. Of course, this is not a very secure method as it is relatively easy to forge someone else's signature. Digital signatures, on the other hand, can be verified using a publicly known verification algorithm. Thus, "anyone" can verify a digital signature. The use of a secure signature scheme will prevent the possibility of forgeries.

Another fundamental difference between conventional and digital signatures is that a "copy" of a signed digital message is identical to the original. On the other hand, a copy of a signed paper document can usually be distinguished from an original. This feature means that care must be taken to prevent a signed digital message from being reused. For example, if Bob signs a digital message authorizing Alice to withdraw $100 from his bank account (i.e., a check), he only wants Alice to be able to do so once. So the message itself should contain information, such as a date, that prevents it from being reused.

A signature scheme consists of two components: a *signing algorithm* and a *verification algorithm*. Bob can sign a message x using a (secret) signing algorithm sig. The resulting signature $sig(x)$ can subsequently be verified using a public verification algorithm ver. Given a pair (x, y), the verification algorithm returns an answer "true" or "false" depending on whether the signature is authentic.

Here is a formal defintion of a signature scheme.

DEFINITION 6.1 *A signature scheme is a five-tuple* $(\mathcal{P}, \mathcal{A}, \mathcal{K}, \mathcal{S}, \mathcal{V})$, *where the following conditions are satisfied:*

1. $\mathcal{P}$ *is a finite set of possible messages*

2. $\mathcal{A}$ *is a finite set of possible signatures*

3. $\mathcal{K}$, *the keyspace, is a finite set of possible keys*

4. *For each* $K \in \mathcal{K}$, *there is a signing algorithm* $sig_K \in \mathcal{S}$ *and a corresponding verification algorithm* $ver_K \in \mathcal{V}$. *Each* $sig_K : \mathcal{P} \to \mathcal{A}$ *and* $ver_K : \mathcal{P} \times \mathcal{A} \to \{\text{true}, \text{false}\}$ *are functions such that the following equation is satisfied for every message* $x \in \mathcal{P}$ *and for every signature* $y \in \mathcal{A}$:

$$ver(x, y) = \begin{cases} \text{true} & \text{if } y = sig(x) \\ \text{false} & \text{if } y \neq sig(x). \end{cases}$$

For every $K \in \mathcal{K}$, the functions sig_K and ver_K should be polynomial-time functions. ver_K will be a public function and sig_K will be secret. It should be computationally infeasible for Oscar to "forge" Bob's signature on a message x. That is, given x, only Bob should be able to compute the signature y such that $ver(x, y) = \text{true}$. A signature scheme cannot be unconditionally secure, since Oscar can test all possible signatures y for a message x using the public algorithm ver, until he finds the right signature. So, given sufficient time, Oscar can always forge Bob's signature. Thus, as was the case with public-key cryptosystems, our goal is to find signature schemes that are computationally secure.

As our first example of a signature scheme, we observe that the **RSA** public-key cryptosystem can be used to provide digital signatures. See Figure 6.1.

Thus, Bob signs a message x using the **RSA** decryption rule d_K. Bob is the only person that can create the signature since $d_K = sig_K$ is secret. The verification algorithm uses the **RSA** encryption rule e_K. Anyone can verify a signature since e_K is public.

Note that anyone can forge Bob's signature on a "random" message x by computing $x = e_K(y)$ for some y; then $y = sig_K(x)$. One way around this difficulty is to require that messages contain sufficient redundancy that a forged signature of this type does not correspond to a "meaningful" message x except with a very small probability. Alternatively, the use of hash functions in conjunction with signature schemes will eliminate this method of forging (cryptographic hash functions will be discussed in Chapter 7).

FIGURE 6.1
RSA Signature Scheme

Let $n = pq$, where p and q are primes. Let $\mathcal{P} = \mathcal{A} = \mathbb{Z}_n$, and define

$$K = \{(n, p, q, a, b) : n = pq, p, q \text{ prime}, ab \equiv 1 \pmod{\phi(n)}\}.$$

The values n and b are public, and the values p, q, a are secret.

For $K = (n, p, q, a, b)$, define

$$sig_K(x) = x^a \bmod n$$

and

$$ver_K(x, y) = \text{true} \Leftrightarrow x \equiv y^b \pmod{n}$$

$(x, y \in \mathbb{Z}_n)$.

Finally, let's look briefly at how we would combine signing and public-key encryption. Suppose Alice wishes to send a signed, encrypted message to Bob. Given a plaintext x, Alice would compute her signature $y = sig_{\text{Alice}}(x)$, and then encrypt both x and y using Bob's public encryption function e_{Bob}, obtaining $z = e_{\text{Bob}}(x, y)$. The ciphertext z would be transmitted to Bob. When Bob receives z, he first decrypts it with his decryption function d_{Bob} to get (x, y). Then he uses Alice's public verification function to check that $ver_{\text{Alice}}(x, y) = \text{true}$.

What if Alice first encrypted x, and then signed the result? Then she would compute

$$z = e_{\text{Bob}}(x) \text{ and } y = sig_{\text{Alice}}(z).$$

Alice would transmit the pair (z, y) to Bob. Bob would decrypt z, obtaining x, and then verify the signature y on x using ver_{Alice}. One potential problem with this approach is that if Oscar obtains a pair (z, y) of this type, he could replace Alice's signature y by his own signature

$$y' = sig_{\text{Oscar}}(z).$$

(Note that Oscar can sign the ciphertext $z = e_{\text{Bob}}(x)$ even though he doesn't know the plaintext x.) Then, if Oscar transmits (z, y') to Bob, Oscar's signature will be verified by Bob using ver_{Oscar}, and Bob may infer that the plaintext x originated with Oscar. Because of this potential difficulty, most people recommend signing before encrypting.

FIGURE 6.2
ElGamal Signature Scheme

Let p be a prime such that the discrete log problem in $\mathbb{Z}_p$ is intractable, and let $\alpha \in \mathbb{Z}_p^*$ be a primitive element. Let $\mathcal{P} = \mathbb{Z}_p^*$, $\mathcal{A} = \mathbb{Z}_p^* \times \mathbb{Z}_{p-1}$, and define

$$\mathcal{K} = \{(p, \alpha, a, \beta) : \beta \equiv \alpha^a \pmod{p}\}.$$

The values p, α and β are public, and a is secret.

For $K = (p, \alpha, a, \beta)$, and for a (secret) random number $k \in \mathbb{Z}_{p-1}^*$, define

$$sig_K(x, k) = (\gamma, \delta),$$

where

$$\gamma = \alpha^k \bmod p$$

and

$$\delta = (x - a\gamma)k^{-1} \bmod (p - 1).$$

For $x, \gamma \in \mathbb{Z}_p^*$ and $\delta \in \mathbb{Z}_{p-1}$, define

$$ver_K(x, \gamma, \delta) = \text{true} \Leftrightarrow \beta^\gamma \gamma^\delta \equiv \alpha^x \pmod{p}.$$

6.2 The ElGamal Signature Scheme

We now describe the **ElGamal Signature Scheme**, which was described in a 1985 paper. A modification of this scheme has been adopted as a digital signature standard by the National Institute of Standards and Technology (NIST). The **El-Gamal Scheme** is designed specifically for the purpose of signatures, as opposed to **RSA**, which can be used both as a public-key cryptosystem and a signature scheme.

The **ElGamal Signature Scheme** is non-deterministic, as was the **ElGamal Public-key Cryptosystem**. This means that there are many valid signatures for any given message. The verification algorithm must be able to accept any of the valid signatures as authentic. The description of the **ElGamal Signature Scheme** is given in Figure 6.2.

If the signature was constructed correctly, then the verification will succeed, since

$$\beta^\gamma \gamma^\delta \equiv \alpha^{a\gamma} \alpha^{k\delta} \pmod{p}$$

$$\equiv \alpha^x \pmod{p},$$

where we use the fact that

$$a\gamma + k\delta \equiv x \ (\mathrm{mod} \ p - 1).$$

Bob computes a signature using both the secret value a (which is part of the key) and the secret random number k (which is used to sign one message, x). The verification can be accomplished using only public information.

Let's do a small example to illustrate the arithmetic.

Example 6.1
Suppose we take $p = 467$, $\alpha = 2$, $a = 127$; then

$$\beta = \alpha^a \ \mathrm{mod} \ p$$
$$= 2^{127} \ \mathrm{mod} \ 467$$
$$= 132.$$

Suppose Bob wants to sign the message $x = 100$ and he chooses the random value $k = 213$ (note that $\gcd(213, 466) = 1$ and $213^{-1} \ \mathrm{mod} \ 466 = 431$). Then

$$\gamma = 2^{213} \ \mathrm{mod} \ 467 = 29$$

and

$$\delta = (100 - 127 \times 29)431 \ \mathrm{mod} \ 466 = 51.$$

Anyone can verify this signature by checking that

$$132^{29} 29^{51} \equiv 189 \ (\mathrm{mod} \ 467)$$

and

$$2^{100} \equiv 189 \ (\mathrm{mod} \ 467).$$

Hence, the signature is valid.   ▯

Let's look at the security of the **ElGamal Signature Scheme**. Suppose Oscar tries to forge a signature for a given message x, without knowing a. If Oscar chooses a value γ and then tries to find the corresponding δ, he must compute the discrete logarithm $\log_\gamma \alpha^x \beta^{-\gamma}$. On the other hand, if he first chooses δ and then tries to find γ, he is trying to "solve" the equation

$$\beta^\gamma \gamma^\delta \equiv \alpha^x \ (\mathrm{mod} \ p)$$

for the "unknown" γ. This is a problem for which no feasible solution is known; however, it does not seem to be related to any well-studied problem such as the **Discrete Logarithm** problem. There also remains the possibility that there might be some way to compute γ and δ simultaneously in such a way that (γ, δ) will be a signature. No one has discovered a way to do this, but conversely, no one has proved that it cannot be done.

If Oscar chooses γ and δ and then tries to solve for x, he is again faced with an instance of the **Discrete Logarithm** problem, namely the computation of $\log_\alpha \beta^\gamma \gamma^\delta$. Hence, Oscar cannot sign a "random" message using this approach. However, there is a method by which Oscar can sign a random message by choosing γ, δ and x simultaneously: Suppose i and j are integers, $0 \le i \le p-2$, $0 \le j \le p-2$, and $\gcd(j, p-1) = 1$. Then perform the following computations:

$$\gamma = \alpha^i \beta^j \bmod p$$
$$\delta = -\gamma j^{-1} \bmod (p-1)$$
$$x = -\gamma i j^{-1} \bmod (p-1),$$

where j^{-1} is computed modulo $(p-1)$ (this is where we require that j be relatively prime to $p-1$).

We claim that (γ, δ) is a valid signature for the message x. This is proved by checking the verification condition:

$$\beta^\gamma \gamma^\delta \equiv \beta^\gamma (\alpha^i \beta^j)^{-\gamma j^{-1}} \pmod{p}$$
$$\equiv \beta^\gamma \alpha^{-i\gamma j^{-1}} \beta^{-\gamma} \pmod{p}$$
$$\equiv \alpha^{-i\gamma j^{-1}} \pmod{p}$$
$$\equiv \alpha^x \pmod{p}.$$

We illustrate with an example.

Example 6.2
As in the previous example, suppose $p = 467$, $\alpha = 2$ and $\beta = 132$. Suppose Oscar chooses $i = 99$ and $j = 179$; then $j^{-1} \bmod (p-1) = 151$. He would compute the following:

$$
\begin{aligned}
\gamma &= 2^{99} 132^{179} \bmod 467 &&= 117 \\
\delta &= -117 \times 151 \bmod 466 &&= 41 \\
x &= 99 \times 41 \bmod 466 &&= 331.
\end{aligned}
$$

Then $(117, 41)$ is a valid signature for the message 331, as may be verified by checking that
$$132^{117} 117^{41} \equiv 303 \pmod{467}$$
and
$$2^{331} \equiv 303 \pmod{467}.$$

Hence, the signature is valid. $\square$

Here is a second type of forgery, in which Oscar begins with a message previously signed by Bob. Suppose (γ, δ) is a valid signature for a message x. Then it is possible for Oscar to sign various other messages. Suppose h, i and j are integers, $0 \leq h, i, j \leq p - 2$, and $\gcd(h\gamma - j\delta, p - 1) = 1$. Compute the following:

$$\lambda = \gamma^h \alpha^i \beta^j \bmod p$$

$$\mu = \delta\lambda(h\gamma - j\delta)^{-1} \bmod (p - 1)$$

$$x' = \lambda(hx + i\delta)(h\gamma - j\delta)^{-1} \bmod (p - 1),$$

where $(h\gamma - j\delta)^{-1}$ is computed modulo $(p - 1)$. Then, it is tedious but straightforward to check the verification condition:

$$\beta^\lambda \lambda^\mu \equiv \alpha^{x'} \pmod{p}.$$

Hence (λ, μ) is a valid signature for x'.

Both of these methods produce valid forged signatures, but they do not appear to enable an opponent to forge a signature on a message of his own choosing without first solving a discrete logarithm problem. Hence, they do not seem to represent a threat to the security of the **ElGamal Signature Scheme**.

Finally, we mention a couple of ways in which the **ElGamal Scheme** can be broken if it is used carelessly (these are further examples of protocol failures, some of which were discussed in the exercises of Chapter 4). First, the random value k used in computing a signature should not be revealed. For, if k is known, it is a simple matter to compute

$$a = (x - k\delta)\gamma^{-1} \bmod (p - 1).$$

Of course, once a is known, then the system is broken and Oscar can forge signatures at will.

Another misuse of the system is to use the same value k in signing two different messages. This also makes it easy for Oscar to compute a and hence break the system. This can be done as follows. Suppose (γ, δ_1) is a signature on x_1 and (γ, δ_2) is a signature on x_2. Then we have

$$\beta^\gamma \gamma^{\delta_1} \equiv \alpha^{x_1} \pmod{p}$$

and

$$\beta^\gamma \gamma^{\delta_2} \equiv \alpha^{x_2} \pmod{p}.$$

Thus

$$\alpha^{x_1 - x_2} \equiv \gamma^{\delta_1 - \delta_2} \pmod{p}.$$

Writing $\gamma = \alpha^k$, we obtain the following equation in the unknown k:

$$\alpha^{x_1 - x_2} \equiv \alpha^{k(\delta_1 - \delta_2)} \pmod{p},$$

which is equivalent to

$$x_1 - x_2 \equiv k(\delta_1 - \delta_2) \pmod{p - 1}.$$

Now let $d = \gcd(\delta_1 - \delta_2, p - 1)$. Since $d \mid (p - 1)$ and $d \mid (\delta_1 - \delta_2)$, it follows that $d \mid (x_1 - x_2)$. Define

$$x' = \frac{x_1 - x_2}{d}$$

$$\delta' = \frac{\delta_1 - \delta_2}{d}$$

$$p' = \frac{p - 1}{d}.$$

Then the congruence becomes:

$$x' \equiv k\delta' \pmod{p'}.$$

Since $\gcd(\delta', p') = 1$, we can compute

$$\epsilon = (\delta')^{-1} \bmod p'.$$

Then value of k is determined modulo p' to be

$$k = x'\epsilon \bmod p'.$$

This yields d candidate values for k:

$$k = x'\epsilon + ip' \bmod (p - 1)$$

for some i, $0 \le i \le d - 1$. Of these d candidate values, the (unique) correct one can be determined by testing the condition

$$\gamma \equiv \alpha^k \pmod{p}.$$

6.3 The Digital Signature Standard

The **Digital Signature Standard** (or **DSS**) is a modification of the **ElGamal Signature Scheme**. It was published in the Federal Register on May 19, 1994 and adopted as a standard on December 1, 1994 (however, it was first proposed in August, 1991). First, we want to motivate the changes that are made to **ElGamal**, and then we will describe how they are accomplished.

In many situations, a message might be encrypted and decrypted only once, so it suffices to use any cryptosystem which is known to be secure at the time the message is encrypted. On the other hand, a signed message could function as a legal document such as a contract or will, so it is very likely that it would be necessary to verify a signature many years after the message is signed. So it is important to take even more precautions regarding the security of a signature scheme as opposed to a cryptosystem. Since the **ElGamal Scheme** is no more secure than the **Discrete Logarithm** problem, this necessitates the use of a large modulus p. Certainly p should have at least 512 bits, and many people would argue that the length of p should be 1024 bits in order to provide security into the foreseeable future.

However, even a 512 bit modulus leads to a signature having 1024 bits. For potential applications, many of which involve the use of smart cards, a shorter signature is desirable. **DSS** modifies the **ElGamal Scheme** in an ingenious way so that a 160-bit message is signed using a 320-bit signature, but the computations are done using a 512-bit modulus p. The way that this done is to work in a subgroup of $\mathbb{Z}_p{}^*$ of size 2^{160}. The assumed security of the scheme is based on the belief that finding discrete logarithms in this specified subgroup of $\mathbb{Z}_p{}^*$ is secure.

The first change we make is to change the "$-$" to a "$+$" in the definition of δ, so

$$\delta = (x + a\gamma)k^{-1} \bmod (p - 1).$$

This changes the verification condition to the following:

$$\alpha^x \beta^\gamma \equiv \gamma^\delta \; (\bmod \; p). \tag{6.1}$$

If $\gcd(x + a\gamma, p - 1) = 1$, then $\delta^{-1} \bmod (p - 1)$ exists, and we can modify condition (6.1), producing the following:

$$\alpha^{x\delta^{-1}} \beta^{\gamma\delta^{-1}} \equiv \gamma \; (\bmod \; p). \tag{6.2}$$

Now here is the major innovation in the **DSS**. We suppose that q is a 160-bit prime such that $q \mid (p - 1)$, and α is a qth root of 1 modulo p. (It is easy to construct such an α: Let α_0 be a primitive element of $\mathbb{Z}_p$, and define $\alpha = \alpha_0{}^{(p-1)/q} \bmod p$.) Then β and γ will also be qth roots of 1. Hence, any exponents of α, β and γ can be reduced modulo q without affecting verification condition (6.2). The tricky point is that γ appears as an exponent on the left side of (6.2), and again — but not as an exponent — on the right side of (6.2). So if γ is reduced modulo q, then we must also reduce the entire left side of (6.2) modulo q in order to perform the verification. Observe that (6.1) will not work if the extra reductions modulo q are done. The complete description of the **DSS** is given in Figure 6.3.

Notice that is necessary that $\delta \not\equiv 0 \; (\bmod \; q)$ since the value $\delta^{-1} \bmod q$ is needed to verify the signature (this is analogous to the requirement that $\gcd(\delta, p - 1) = 1$ when we modified (6.1) to obtain (6.2)). If Bob computes a value $\delta \equiv 0$ $(\bmod \; q)$ in the signing algorithm, he should reject it and construct a new signature with a new random k. We should point out that this is not likely to cause a problem in practice: the probability that $\delta \equiv 0 \; (\bmod \; q)$ is likely to be on the order of 2^{-160}, so for all intents and purposes it will almost never happen.

FIGURE 6.3
Digital Signature Standard

Let p be a 512-bit prime such that the discrete log problem in $\mathbb{Z}_p$ is intractible, and let q be a 160-bit prime that divides $p - 1$. Let $\alpha \in \mathbb{Z}_p^*$ be a qth root of 1 modulo p. Let $\mathcal{P} = \mathbb{Z}_q^*$, $\mathcal{A} = \mathbb{Z}_q \times \mathbb{Z}_q$, and define

$$\mathcal{K} = \{(p, q, \alpha, a, \beta) : \beta \equiv \alpha^a \pmod{p}\}.$$

The values p, q, α and β are public, and a is secret.

For $K = (p, q, \alpha, a, \beta)$, and for a (secret) random number k, $1 \le k \le q - 1$, define

$$sig_K(x, k) = (\gamma, \delta),$$

where

$$\gamma = (\alpha^k \bmod p) \bmod q$$

and

$$\delta = (x + a\gamma)k^{-1} \bmod q.$$

For $x \in \mathbb{Z}_q^*$ and $\gamma, \delta \in \mathbb{Z}_q$, verification is done by performing the following computations:

$$e_1 = x\delta^{-1} \bmod q$$
$$e_2 = \gamma\delta^{-1} \bmod q$$
$$ver_K(x, \gamma, \delta) = \text{true} \Leftrightarrow (\alpha^{e_1}\beta^{e_2} \bmod p) \bmod q = \gamma.$$

Here is a small example to illustrate.

Example 6.3
Suppose we take $q = 101$ and $p = 78q + 1 = 7879$. 3 is a primitive element in $\mathbb{Z}_{7879}$, so we can take

$$\alpha = 3^{78} \bmod 7879 = 170.$$

Suppose $a = 75$; then

$$\beta = \alpha^a \bmod 7879 = 4567.$$

Now, suppose Bob wants to sign the message $x = 22$ and he chooses the random value $k = 50$, so

$$k^{-1} \bmod 101 = 99.$$

Then

$$\gamma = (170^{50} \bmod 7879) \bmod 101$$
$$= 2518 \bmod 101$$
$$= 94$$

and

$$\delta = (22 + 75 \times 94)99 \bmod 101$$
$$= 97.$$

The signature $(94, 97)$ on the message 22 is verified by the following computations:

$$\delta^{-1} = 97^{-1} \bmod 101 = 25$$
$$e_1 = 22 \times 25 \bmod 101 = 45$$
$$e_2 = 94 \times 25 \bmod 101 = 27$$
$$(170^{45} 4567^{27} \bmod 7879) \bmod 101 = 2518 \bmod 101 = 94.$$

Hence, the signature is valid.   ▯

When the **DSS** was proposed in 1991, there were several criticisms put forward. One complaint was that the selection process by NIST was not public. The standard was developed by the National Security Agency (NSA) without the input of U. S. industry. Regardless of the merits of the resulting scheme, many people resented the "closed-door" approach.

Of the technical criticisms put forward, the most serious was that the size of the modulus p was fixed at 512 bits. Many people would prefer that the modulus size not be fixed, so that larger modulus sizes could be used if desired. In reponse to these comments, NIST altered the description of the standard so that a variety of modulus sizes are allowed, namely, any modulus size divisible by 64, in the range from 512 to 1024 bits.

Another complaint about the **DSS** was that signatures can be generated considerably faster than they can be verified. In contrast, if **RSA** is used as a signature scheme and the public verification exponent is very small (say 3, for example), then verification can be performed much more quickly than signing. This leads to a couple of considerations concerning the potential applications of the signature scheme:

1. A message will only be signed once. On the other hand, it might be necessary to verify the signature many times over a period of years. This suggests that a faster verification algorithm would be desirable.

FIGURE 6.4
Lamport Signature Scheme

Let k be a positive integer and let $\mathcal{P} = \{0,1\}^k$. Suppose $f : Y \to Z$ is a one-way function, and let $\mathcal{A} = Y^k$. Let $y_{i,j} \in Y$ be chosen at random, $1 \leq i \leq k, j = 0, 1$, and let $z_{i,j} = f(y_{i,j}), 1 \leq i \leq k, j = 0, 1$. The key K consists of the $2k$ y's and the $2k$ z's. The y's are secret while the z's are public.

For $K = (y_{i,j}, z_{i,j} : 1 \leq i \leq k, j = 0, 1)$, define

$$sig_K(x_1, \ldots, x_k) = (y_{1,x_1}, \ldots, y_{k,x_k})$$

and

$$ver_K(x_1, \ldots, x_k, a_1, \ldots a_k) = \text{true} \Leftrightarrow f(a_i) = z_{i,x_i}, 1 \leq i \leq k.$$

---

2. What types of computers are likely to be doing the signing and verifying? Many potential applications involve smart cards, with limited processing power, communicating with a more powerful computer. So one might try to design a scheme so that fewer computations are likely to be done by a card. But one can imagine situations where a smart card would generate a signature, and other situations where a smart card would verify a signature, so it is difficult to give a definitive answer here.

The response of NIST to the question of signature generation/verification times is that it does not really matter which is faster, provided that both can be done sufficiently quickly.

6.4 One-time Signatures

In this section, we describe a conceptually simple way to construct a one-time signature scheme from any one-way function. The term "one-time" means that only one message can be signed. (The signature can be verified an arbitrary number of times, of course.) The description of the scheme, known as the **Lamport Signature Scheme**, is given in Figure 6.4.

Informally, this is how the system works. A message to be signed is a binary k-tuple. Each bit is signed individually: the value $z_{i,j}$ corresponds to the ith bit of the message having the value j $(j = 0, 1)$. Each $z_{i,j}$ is the image of $y_{i,j}$ under the one-way function f. The ith bit of the message is signed using the preimage

$y_{i,j}$ of the $z_{i,j}$ corresponding to the ith bit of the message. The verification consists simply of checking that each element in the signature is the preimage of the appropriate public key element.

We illustrate the scheme by considering one possible implementation using the exponentiation function $f(x) = \alpha^x \bmod p$, where α is a primitive element modulo p.

Example 6.4
7879 is prime and 3 is a primitive element in $\mathbb{Z}_{7879}$. Define

$$f(x) = 3^x \bmod 7879.$$

Suppose Bob wishes to sign a message of three bits, and he chooses the six (secret) random numbers

$$y_{1,0} = 5831$$
$$y_{1,1} = 735$$
$$y_{2,0} = 803$$
$$y_{2,1} = 2467$$
$$y_{3,0} = 4285$$
$$y_{3,1} = 6449.$$

Then he computes the images of the y's under the function f:

$$z_{1,0} = 2009$$
$$z_{1,1} = 3810$$
$$z_{2,0} = 4672$$
$$z_{2,1} = 4721$$
$$z_{3,0} = 268$$
$$z_{3,1} = 5731.$$

These z's are published. Now, suppose Bob wants to sign the message

$$x = (1, 1, 0).$$

The signature for x is

$$(y_{1,1}, y_{2,1}, y_{3,0}) = (735, 2467, 4285).$$

To verify this signature, it suffices to compute the following:

$$3^{735} \bmod 7879 = 3810$$
$$3^{2467} \bmod 7879 = 4721$$
$$3^{4285} \bmod 7879 = 268.$$

Hence, the signature is valid. $\square$

Oscar cannot forge a signature because he is unable to invert the one-way function f to obtain the secret y's. However, the signature scheme can be used to sign only one message. For, given signatures for two different messages, it is (usually) an easy matter for Oscar to construct signatures for further messages (different from the first two).

For example, suppose the messages $(0, 1, 1)$ and $(1, 0, 1)$ are both signed using the same scheme. The message $(0, 1, 1)$ would have as its signature the triple $(y_{1,0}, y_{2,1}, y_{3,1})$, and the message $(1, 0, 1)$ would be signed with $(y_{1,1}, y_{2,0}, y_{3,1})$. Given these two signatures, Oscar can manufacture signatures for the messages $(1, 1, 1)$ (namely, $(y_{1,1}, y_{2,1}, y_{3,1})$) and $(0, 0, 1)$ (namely, $(y_{1,0}, y_{2,0}, y_{3,1})$).

Even though this scheme is quite elegant, it is not of great practical use due to the size of the signatures it produces. For example, if we use the modular exponentiation function, as in the example above, then a secure implementation would require that p be at least 512 bits in length. This means that each bit of the message is signed using 512 bits. Consequently, the signature is 512 times as long as the message!

We now look at a modification due to Bos and Chaum that allows the signatures to be made somewhat shorter, with no loss of security. In the **Lamport Scheme**, the reason that Oscar cannot forge a signature on a (second) message, given a signature on one message, is that the y's corresponding to one message are never a subset of the y's corresponding to another (distinct) message.

Suppose we have a set $\mathcal{B}$ of subsets of a set B such that $B_1 \subseteq B_2$ only if $B_1 = B_2$, for all $B_1, B_2 \in \mathcal{B}$. Then $\mathcal{B}$ is said to satisfy the *Sperner property*. Given a set B of even cardinality $2n$, it is known that the maximum size of a set $\mathcal{B}$ of subsets of B having the Sperner property is $\binom{2n}{n}$. This can easily be obtained by taking all the n-subsets of B: clearly no n-subset is contained in another n-subset.

Now suppose we want to sign a k-bit message, as before, and we choose n large enough so that

$$2^k \le \binom{2n}{n}.$$

Let $|B| = 2n$ and let $\mathcal{B}$ denote the set of n-subsets of B. Let $\phi : \{0, 1\}^k \to \mathcal{B}$ be a publicly known injection. Then we can associate each possible message with an n-subset in $\mathcal{B}$. We will have $2n$ y's and $2n$ z's, and each message will be signed with n y's. The complete description of the **Bos-Chaum Scheme** is given in Figure 6.5.

The advantage of the **Bos-Chaum Scheme** is that signatures are shorter than with the **Lamport Scheme**. For example, suppose we wish to sign a message of six bits (i.e., $k = 6$). Since $2^6 = 64$ and $\binom{8}{4} = 70$, we can take $n = 4$. This allows a six-bit message to be signed with four y's, as opposed to six with

FIGURE 6.5
Bos-Chaum Signature Scheme

Let k be a positive integer and let $\mathcal{P} = \{0,1\}^k$. Let n be an integer such that $2^k \leq \binom{2n}{n}$, let B be a set of cardinality $2n$, and let

$$\phi : \{0,1\}^k \to \mathcal{B}$$

be an injection, where $\mathcal{B}$ is the set of all n-subsets of B. Suppose $f : Y \to Z$ is a one-way function, and let $\mathcal{A} = Y^n$. Let $y_i \in Y$ be chosen at random, $1 \leq i \leq 2n$, and let $z_i = f(y_i)$, $1 \leq i \leq 2n$. The key K consists of the $2n$ y's and the $2n$ z's. The y's are secret while the z's are public.

For $K = (y_i, z_i, 1 \leq i \leq 2n)$, define

$$sig_K(x_1, \ldots, x_k) = \{y_j : j \in \phi(x_1, \ldots, x_k)\}$$

and

$$ver_K(x_1, \ldots, x_k, a_1, \ldots a_n) = \text{true}$$
$$\Leftrightarrow \{f(a_i) : 1 \leq i \leq n\} = \{z_j : j \in \phi(x_1, \ldots, x_k)\}.$$

Lamport. As well, the key is shorter, consisting of eight z's as opposed to twelve with **Lamport**.

The **Bos-Chaum Scheme** requires an injective function ϕ that associates an n-subset of a $2n$-set with each possible binary k-tuple $x = (x_1, \ldots, x_k)$. We present one simple algorithm to do this in Figure 6.6. Applying this algorithm with $x = (0, 1, 0, 0, 1, 1)$, for example, yields

$$\phi(x) = \{2, 4, 6, 8\}.$$

In general, how big is n in the **Bos-Chaum Scheme** as compared to k? We need to satisfy the inequality $2^k \leq \binom{2n}{n}$. If we estimate the binomial coefficient

$$\binom{2n}{n} = \frac{(2n)!}{(n!)^2}$$

using Stirling's formula, we obtain the quantity $2^{2n}/\sqrt{\pi n}$. After some simplification, the inequality becomes

$$k \leq 2n - \frac{\log_2(n\pi)}{2}.$$

FIGURE 6.6
Computation of ϕ in the Bos-Chaum Scheme

1.	$x = \sum_{i=1}^{k} x_i 2^{i-1}$
2.	$\phi(x) = \emptyset$
3.	$t = 2n$
4.	$e = n$
5.	**while** $t > 0$ **do**
6.	$\quad t = t - 1$
7.	$\quad$ **if** $x > \binom{t}{e}$ **then**
8.	$\quad\quad x = x - \binom{t}{e}$
9.	$\quad\quad e = e - 1$
10.	$\quad\quad \phi(x) = \phi(x) \cup \{t + 1\}.$

Asymptotically, n is about $k/2$, so we obtain an almost 50% reduction in signature size by using the **Bos-Chaum Scheme**.

6.5 Undeniable Signatures

Undeniable signatures were introduced by Chaum and van Antwerpen in 1989. They have several novel features. Primary among these is that a signature cannot be verified without the cooperation of the signer, Bob. This protects Bob against the possibility that documents signed by him are duplicated and distributed electronically without his approval. The verification will be accomplished by means of a *challenge-and-response protocol*.

But if Bob's cooperation is required to verify a signature, what is to prevent Bob from disavowing a signature he made at an earlier time? Bob might claim that a valid signature is a forgery, and either refuse to verify it, or carry out the protocol in such a way that the signature will not be verified. To prevent this from happening, an undeniable signature scheme incorporates a *disavowal protocol* by which Bob can prove that a signature is a forgery. Thus, Bob will be able to prove in court that a given forged signature is in fact a forgery. (If he refuses to take part in the disavowal protocol, this would be regarded as evidence that the signature is, in fact, genuine.)

Thus, an undeniable signature scheme consists of three components: a signing

FIGURE 6.7
Chaum-van Antwerpen Undeniable Signature Scheme

Let $p = 2q + 1$ be a prime such that q is prime and the discrete log problem in $\mathbb{Z}_p$ is intractible. Let $\alpha \in \mathbb{Z}_p{}^*$ be an element of order q. Let $1 \le a \le q-1$ and define $\beta = \alpha^a \bmod p$. Let G denote the multiplicative subgroup of $\mathbb{Z}_p{}^*$ of order q (G consists of the quadratic residues modulo p). Let $\mathcal{P} = \mathcal{A} = G$, and define

$$\mathcal{K} = \{(p, \alpha, a, \beta) : \beta \equiv \alpha^a \ (\bmod\ p)\}.$$

The values p, α and β are public, and a is secret.

For $K = (p, \alpha, a, \beta)$ and $x \in G$, define

$$y = sig_K(x) = x^a \bmod p.$$

For $x, y \in G$, verification is done by executing the following protocol:

1. Alice chooses e_1, e_2 at random, $e_1, e_2 \in \mathbb{Z}_q{}^*$.
2. Alice computes $c = y^{e_1} \beta^{e_2} \bmod p$ and sends it to Bob.
3. Bob computes $d = c^{a^{-1} \bmod q} \bmod p$ and sends it to Alice.
4. Alice accepts y as a valid signature if and only if

$$d \equiv x^{e_1} \alpha^{e_2} \ (\bmod\ p).$$

---

algorithm, a verification protocol, and a disavowal protocol. First, we present the signing algorithm and verification protocol of the **Chaum-van Antwerpen Undeniable Signature Scheme** in Figure 6.7.

We should explain the roles of p and q in this scheme. The scheme lives in $\mathbb{Z}_p$; however, we need to be able to do computations in a multiplicative subgroup G of $\mathbb{Z}_p{}^*$ of prime order. In particular, we need to be able to compute inverses modulo $|G|$, which is why $|G|$ should be prime. It is convenient to take $p = 2q + 1$ where q is prime. In this way, the subgroup G is as large as possible, which is desirable since messages and signatures are both elements of G.

We first prove that Alice will accept a valid signature. In the following computations, all exponents are to be reduced modulo q. First, observe that

$$d \equiv c^{a^{-1}} \ (\bmod\ p)$$
$$\equiv y^{e_1 a^{-1}} \beta^{e_2 a^{-1}} \ (\bmod\ p).$$

Since
$$\beta \equiv \alpha^a \pmod{p},$$

we have that
$$\beta^{a^{-1}} \equiv \alpha \pmod{p}.$$

Similarly,
$$y = x^a \pmod{p}$$

implies that
$$y^{a^{-1}} \equiv x \pmod{p}.$$

Hence,
$$d \equiv x^{e_1} \alpha^{e_2} \pmod{p},$$

as desired.

Here is a small example.

Example 6.5

Suppose we take $p = 467$. Since 2 is a primitive element, $2^2 = 4$ is a generator of G, the quadratic residues modulo 467. So we can take $\alpha = 4$. Suppose $a = 101$; then
$$\beta = \alpha^a \bmod 467 = 449.$$

Bob will sign the message $x = 119$ with the signature
$$y = 119^{101} \bmod 467 = 129.$$

Now, suppose Alice wants to verify the signature y. Suppose she chooses the random values $e_1 = 38$, $e_2 = 397$. She will compute $c = 13$, whereupon Bob will respond with $d = 9$. Alice checks the response by verifying that
$$119^{38} 4^{397} \equiv 9 \pmod{467}.$$

Hence, Alice accepts the signature as valid. $\quad\Box$

We next prove that Bob cannot fool Alice into accepting a fradulent signature as valid, except with a very small probability. This result does not depend on any computational assumptions, i.e., the security is unconditional.

THEOREM 6.1
If $y \not\equiv x^a \pmod{p}$, then Alice will accept y as a valid signature for x with probability $1/q$.

PROOF First, we observe that each possible challenge c corresponds to exactly q ordered pairs (e_1, e_2) (this is because y and β are both elements of the multiplicative group G of prime order q). Now, when Bob receives the challenge c, he has no way of knowing which of the q possible ordered pairs (e_1, e_2) Alice used to construct c. We claim that, if $y \not\equiv x^a \pmod{p}$, then any possible response $d \in G$ that Bob might make is consistent with exactly one of the q possible ordered pairs (e_1, e_2).

Since α generates G, we can write any element of G as a power of α, where the exponent is defined uniquely modulo q. So write $c = \alpha^i$, $d = \alpha^j$, $x = \alpha^k$, and $y = \alpha^\ell$, where $i, j, k, \ell \in \mathbb{Z}_q$ and all arithmetic is modulo p. Consider the following two congruences:

$$c \equiv y^{e_1} \beta^{e_2} \pmod{p}$$
$$d \equiv x^{e_1} \alpha^{e_2} \pmod{p}.$$

This system is equivalent to the following system:

$$i \equiv \ell e_1 + a e_2 \pmod{q}$$
$$j \equiv k e_1 + e_2 \pmod{q}.$$

Now, we are assuming that

$$y \not\equiv x^a \pmod{p},$$

so it follows that

$$\ell \not\equiv a k \pmod{q}.$$

Hence, the coefficient matrix of this system of congruences modulo q has non-zero determinant, and thus there is a unique solution to the system. That is, every $d \in G$ is the correct response for exactly one of the q possible ordered pairs (e_1, e_2). Consequently, the probability that Bob gives Alice a response d that will be verified is exactly $1/q$, and the theorem is proved. ∎

We now turn to the disavowal protocol. This protocol consists of two runs of the verification protocol and is presented in Figure 6.8.

Steps 1–4 and steps 5–8 comprise two unsuccessful runs of the verification protocol. Step 9 is a "consistency check" that enables Alice to determine if Bob is forming his responses in the manner specified by the protocol.

The following example illustrates the disavowal protocol.

Example 6.6
As before, suppose $p = 467$, $\alpha = 4$, $a = 101$ and $\beta = 449$. Suppose the message $x = 286$ is signed with the (bogus) signature $y = 83$, and Bob wants to convince Alice that the signature is invalid.

FIGURE 6.8
Disavowal protocol

---

1. Alice chooses e_1, e_2 at random, $e_1, e_2 \in \mathbb{Z}_q{}^*$
2. Alice computes $c = y^{e_1} \beta^{e_2} \bmod p$ and sends it to Bob
3. Bob computes $d = c^{a^{-1} \bmod q} \bmod p$ and sends it to Alice
4. Alice verifies that $d \not\equiv x^{e_1} \alpha^{e_2} \pmod{p}$
5. Alice chooses f_1, f_2 at random, $f_1, f_2 \in \mathbb{Z}_q{}^*$
6. Alice computes $C = y^{f_1} \beta^{f_2} \bmod p$ and sends it to Bob
7. Bob computes $D = C^{a^{-1} \bmod q} \bmod p$ and sends it to Alice
8. Alice verifies that $D \not\equiv x^{f_1} \alpha^{f_2} \pmod{p}$
9. Alice concludes that y is a forgery if and only if

$$(d\alpha^{-e_2})^{f_1} \equiv (D\alpha^{-f_2})^{e_1} \pmod{p}.$$

Suppose Alice begins by choosing the random values $e_1 = 45, e_2 = 237$. Alice computes $c = 305$ and Bob responds with $d = 109$. Then Alice computes

$$286^{45} 4^{237} \bmod 467 = 149.$$

Since $149 \neq 109$, Alice proceeds to step 5 of the protocol.

Now suppose Alice chooses the random values $f_1 = 125$, $f_2 = 9$. Alice computes $C = 270$ and Bob responds with $D = 68$. Alice computes

$$286^{125} 4^9 \bmod 467 = 25.$$

Since $25 \neq 68$, Alice proceeds to step 9 of the protocol and performs the consistency check. This check succeeds, since

$$(109 \times 4^{-237})^{125} \equiv 188 \pmod{467}$$

and

$$(68 \times 4^{-9})^{45} \equiv 188 \pmod{467}.$$

Hence, Alice is convinced that the signature is invalid. ◻

We have to prove two things at this point:

1. Bob can convince Alice that an invalid signature is a forgery.

2. Bob cannot make Alice believe that a valid signature is a forgery except with a very small probability.

THEOREM 6.2
If $y \not\equiv x^a \pmod p$, and Alice and Bob follow the disavowal protocol, then

$$(d\alpha^{-e_2})^{f_1} \equiv (D\alpha^{-f_2})^{e_1} \pmod p.$$

PROOF   Using the facts that

$$d \equiv c^{a^{-1}} \pmod p,$$

$$c \equiv y^{e_1} \beta^{e_2} \pmod p$$

and

$$\beta \equiv \alpha^a \pmod p,$$

we have that

$$(d\alpha^{-e_2})^{f_1} \equiv \left((y^{e_1} \beta^{e_2})^{a^{-1}} \alpha^{-e_2} \right)^{f_1} \pmod p$$

$$\equiv y^{e_1 a^{-1} f_1} \beta^{e_2 a^{-1} f_1} \alpha^{-e_2 f_1} \pmod p$$

$$\equiv y^{e_1 a^{-1} f_1} \alpha^{e_2 f_1} \alpha^{-e_2 f_1} \pmod p$$

$$\equiv y^{e_1 a^{-1} f_1} \pmod p.$$

A similar computation, using the facts that $D \equiv C^{a^{-1}} \pmod p$, $C \equiv y^{f_1} \beta^{f_2} \pmod p$ and $\beta \equiv \alpha^a \pmod p$, establishes that

$$(D\alpha^{-f_2})^{e_1} \equiv y^{e_1 a^{-1} f_1} \pmod p,$$

so the consistency check in step 9 succeeds.   ∎

Now we look at the possibility that Bob might attempt to disavow a valid signature. In this situation, we do not assume that Bob follows the protocol. That is, Bob might not construct d and D as specified by the protocol. Hence, in the following theorem, we assume only that Bob is able to produce values d and D which satisfy the conditions in steps 4, 8, and 9 of the protocol presented in Figure 6.8.

THEOREM 6.3
Suppose $y \equiv x^a \pmod p$ and Alice follows the disavowal protocol. If

$$d \not\equiv x^{e_1} \alpha^{e_2} \pmod p$$

and

$$D \not\equiv x^{f_1} \alpha^{f_2} \pmod p,$$

then the probability that

$$(d\alpha^{-e_2})^{f_1} \not\equiv (D\alpha^{-f_2})^{e_1} \pmod{p}$$

is $1 - 1/q$.

PROOF Suppose that the following congruences are satisfied:

$$y \equiv x^a \pmod{p}$$
$$d \not\equiv x^{e_1}\alpha^{e_2} \pmod{p}$$
$$D \not\equiv x^{f_1}\alpha^{f_2} \pmod{p}$$
$$(d\alpha^{-e_2})^{f_1} \equiv (D\alpha^{-f_2})^{e_1} \pmod{p}.$$

We will derive a contradiction.

The consistency check (step 9) can be rewritten in the following form:

$$D \equiv d_0{}^{f_1}\alpha^{f_2} \pmod{p},$$

where

$$d_0 = d^{1/e_1}\alpha^{-e_2/e_1} \bmod p$$

is a value that depends only on steps 1–4 of the protocol.

Applying Theorem 6.1, we conclude that y is a valid signature for d_0 with probability $1 - 1/q$. But we are assuming that y is a valid signature for x. That is, with high probability we have

$$x^a \equiv d_0{}^a \pmod{p},$$

which implies that $x = d_0$.

However, the fact that

$$d \not\equiv x^{e_1}\alpha^{e_2} \pmod{p}$$

means that

$$x \not\equiv d^{1/e_1}\alpha^{-e_2/e_1} \pmod{p}.$$

Since

$$d_0 \equiv d^{1/e_1}\alpha^{-e_2/e_1} \pmod{p},$$

we conclude that $x \neq d_0$ and we have a contradiction.

Hence, Bob can fool Alice in this way with probability $1/q$. ∎

6.6   Fail-stop Signatures

A fail-stop signature scheme provides enhanced security against the possibility that a very powerful adversary might be able to forge a signature. In the event that Oscar is able to forge Bob's signature on a message, Bob will (with high probability) subsequently be able to prove that Oscar's signature is a forgery.

In this section, we describe a fail-stop signature scheme constructed by van Heyst and Pedersen in 1992. This is a one-time scheme (only one message can be signed with a given key). The system consists of signing and verification algorithms, as well as a "proof of forgery" algorithm. The description of the signing and verification algorithms of the **van Heyst and Pedersen Fail-stop Signature Scheme** is presented in Figure 6.9.

It is straightforward to see that a signature produced by Bob will satisfy the verification condition, so let's turn to the security aspects of this scheme and how the fail-stop property works. First we establish some important facts relating to the keys of the scheme. We begin with a definition. Two keys $(\gamma_1, \gamma_2, a_1, a_2, b_1, b_2)$ and $(\gamma_1', \gamma_2', a_1', a_2', b_1', b_2')$ are said to be *equivalent* if $\gamma_1 = \gamma_1'$ and $\gamma_2 = \gamma_2'$. It is easy to see that there are exactly q^2 keys in any equivalence class.

We establish several lemmas.

LEMMA 6.4
Suppose K and K' are equivalent keys and suppose that $ver_K(x, y) =$ true. Then $ver_{K'}(x, y) =$ true.

PROOF   Suppose $K = (\gamma_1, \gamma_2, a_1, a_2, b_1, b_2)$ and $K' = (\gamma_1, \gamma_2, a_1', a_2', b_1', b_2')$, where

$$\gamma_1 = \alpha^{a_1} \beta^{a_2} \bmod p = \alpha^{a_1'} \beta^{a_2'} \bmod p$$

and

$$\gamma_2 = \alpha^{b_1} \beta^{b_2} \bmod p = \alpha^{b_1'} \beta^{b_2'} \bmod p.$$

Suppose x is signed using K, producing the signature $y = (y_1, y_2)$, where

$$y_1 = a_1 + xb_1 \bmod q,$$

$$y_2 = a_2 + xb_2 \bmod q.$$

Now suppose that we verify y using K':

$$\alpha^{y_1} \beta^{y_2} \equiv \alpha^{a_1' + xb_1'} \beta^{a_2' + xb_2'} \pmod{p}$$

$$\equiv \alpha^{a_1'} \beta^{a_2'} (\alpha^{b_1'} \beta^{b_2'})^x \pmod{p}$$

$$\equiv \gamma_1 \gamma_2{}^x \pmod{p}.$$

Thus, y will also be verified using K'.   ∎

FIGURE 6.9
van Heyst and Pedersen Fail-stop Signature Scheme

Let $p = 2q + 1$ be a prime such that q is prime and the discrete log problem in $\mathbb{Z}_p$ is intractible, Let $\alpha \in \mathbb{Z}_p^*$ be an element of order q. Let $1 \leq a_0 \leq q-1$ and define $\beta = \alpha^{a_0} \bmod p$. The values p, q, α, β, and a_0 are chosen by a central (trusted) authority. p, q, α, and β are public and will be regarded as fixed. The value of a_0 is kept secret from everyone (even Bob).

Let $\mathcal{P} = \mathbb{Z}_q$ and $\mathcal{A} = \mathbb{Z}_q \times \mathbb{Z}_q$. A key has the form

$$K = (\gamma_1, \gamma_2, a_1, a_2, b_1, b_2),$$

where $a_1, a_2, b_1, b_2 \in \mathbb{Z}_q$,

$$\gamma_1 = \alpha^{a_1} \beta^{a_2} \bmod p,$$

and

$$\gamma_2 = \alpha^{b_1} \beta^{b_2} \bmod p.$$

For $K = (\gamma_1, \gamma_2, a_1, a_2, b_1, b_2)$ and $x \in \mathbb{Z}_q$, define

$$sig_K(x) = (y_1, y_2),$$

where

$$y_1 = a_1 + xb_1 \bmod q$$

and

$$y_2 = a_2 + xb_2 \bmod q.$$

For $y = (y_1, y_2) \in \mathbb{Z}_q \times \mathbb{Z}_q$, we have

$$ver_K(x, y) = \text{true} \Leftrightarrow \gamma_1 \gamma_2^x \equiv \alpha^{y_1} \beta^{y_2} \pmod{p}.$$

LEMMA 6.5

Suppose K is a key and $y = sig_K(x)$. Then there are exactly q keys K' equivalent to K such that $y = sig_{K'}(x)$.

PROOF　Suppose γ_1 and γ_2 are the public components of K. We want to determine the number of 4-tuples (a_1, a_2, b_1, b_2) such that the following congruences are satisfied:

$$\gamma_1 \equiv \alpha^{a_1} \beta^{a_2} \pmod{p}$$

$$\gamma_2 \equiv \alpha^{b_1} \beta^{b_2} \pmod{p}$$

$$y_1 \equiv a_1 + x b_1 \pmod{q}$$

$$y_2 \equiv a_2 + x b_2 \pmod{q}.$$

Since α generates G, there exist unique exponents $c_1, c_2, a_0 \in \mathbb{Z}_q$ such that

$$\gamma_1 \equiv \alpha^{c_1} \pmod{p},$$

$$\gamma_2 \equiv \alpha^{c_2} \pmod{p}$$

and

$$\beta \equiv \alpha^{a_0} \pmod{p}.$$

Hence, it is necessary and sufficient that the following system of congruences be satisfied:

$$c_1 \equiv a_1 + a_0 a_2 \pmod{q}$$

$$c_2 \equiv b_1 + a_0 b_2 \pmod{q}$$

$$y_1 \equiv a_1 + x b_1 \pmod{q}$$

$$y_2 \equiv a_2 + x b_2 \pmod{q}.$$

This system can, in turn, be written as a matrix equation in $\mathbb{Z}_q$, as follows:

$$\begin{pmatrix} 1 & a_0 & 0 & 0 \\ 0 & 0 & 1 & a_0 \\ 1 & 0 & x & 0 \\ 0 & 1 & 0 & x \end{pmatrix} \begin{pmatrix} a_1 \\ a_2 \\ b_1 \\ b_2 \end{pmatrix} = \begin{pmatrix} c_1 \\ c_2 \\ y_1 \\ y_2 \end{pmatrix}.$$

Now, the coefficient matrix of this system can be seen to have rank [1] three: Clearly, the rank is at least three since rows 1, 2 and 4 are linearly independent over $\mathbb{Z}_q$. And the rank is at most three since

$$r_1 + x r_2 - r_3 - a_0 r_4 = (0, 0, 0, 0),$$

---

[1] the *rank* of a matrix is the maximum number of linearly independent rows it contains

where r_i denotes the ith row of the matrix.

Now, this system of equations has at least one solution, obtained by using the key K. Since the rank of the coefficient matrix is three, it follows that the dimension of the solution space is $4 - 3 = 1$, and there are exactly q solutions. The result follows. ∎

By similar reasoning, the following result can be proved. We omit the proof.

LEMMA 6.6
Suppose K is a key, $y = sig_K(x)$, and $ver_K(x', y') = $ true, where $x' \neq x$. Then there is at most one key K' equivalent to K such that $y = sig_{K'}(x)$ and $y' = sig_{K'}(x')$.

Let's interpret what the preceding two lemmas say about the security of the scheme. Given that y is a valid signature for message x, there are q possible keys that would have signed x with y. But for any message $x' \neq x$, these q keys will produce q different signatures on x'. Thus, the following theorem results.

THEOREM 6.7
Given that $sig_K(x) = y$ and $x' \neq x$, Oscar can compute $sig_K(x')$ with probablity $1/q$.

Note that this theorem does not depend on the computational power of Oscar: the stated level of security is obtained because Oscar cannot tell which of q possible keys is being used by Bob. So the security is unconditional.

We now go on to look at the fail-stop concept. What we have said so far is that, given a signature y on message x, Oscar cannot compute Bob's signature y' on a different message x'. It is still conceivable that Oscar can compute a forged signature $y'' \neq sig_K(x')$ which will still be verified. However, if Bob is given a valid forged signature, then with probability $1 - 1/q$ he can produce a "proof of forgery." The proof of forgery is the value $a_0 = \log_\alpha \beta$, which is known only to the central authority.

So we assume that Bob possesses a pair (x', y'') such that $ver_K(x', y'') = $ true and $y'' \neq sig_K(x')$. That is,

$$\gamma_1 \gamma_2^{x'} \equiv \alpha^{y_1''} \beta^{y_2''} \pmod{p},$$

where $y'' = (y_1'', y_2'')$. Now, Bob can compute his own signature on x', namely $y' = (y_1', y_2')$, and it will be the case that

$$\gamma_1 \gamma_2^{x'} \equiv \alpha^{y_1'} \beta^{y_2'} \pmod{p}.$$

Hence,

$$\alpha^{y_1''} \beta^{y_2''} \equiv \alpha^{y_1'} \beta^{y_2'} \pmod{p}.$$

Writing $\beta = \alpha^{a_0} \bmod p$, we have that

$$\alpha^{y_1'' + a_0 y_2''} \equiv \alpha^{y_1' + a_0 y_2'} \pmod{p},$$

or

$$y_1'' + a_0 y_2'' \equiv y_1' + a_0 y_2' \pmod{q}.$$

This simplifies to give

$$y_1'' - y_1' \equiv a_0 (y_2' - y_2'') \pmod{q}.$$

Now, $y_2' \not\equiv y_2'' \pmod{q}$ since y' is a forgery. Hence, $(y_2' - y_2'')^{-1} \bmod q$ exists, and

$$a_0 = \log_\alpha \beta = (y_1'' - y_1')(y_2' - y_2'')^{-1} \bmod q.$$

Of course, by accepting such a proof of forgery, we assume that Bob cannot compute the discrete logarithm $\log_\alpha \beta$ by himself. This is a computational assumption.

Finally, we remark that the scheme is a one-time scheme since Bob's key K can easily be computed if two messages are signed using K.

We close with an example illustrating how Bob can produce a proof of forgery.

Example 6.7
Suppose $p = 3467 = 2 \times 1733 + 1$. The element $\alpha = 4$ has order 1733 in $\mathbb{Z}_{3467}{}^*$. Suppose that $a_0 = 1567$, so

$$\beta = 4^{1567} \bmod 3467 = 514.$$

(Recall that Bob knows the values of α and β, but not a_0.) Suppose Bob forms his key using $a_1 = 888$, $a_2 = 1024$, $b_1 = 786$ and $b_2 = 999$, so

$$\gamma_1 = 4^{888} 514^{1024} \bmod 3467 = 3405$$

and

$$\gamma_2 = 4^{786} 514^{999} \bmod 3467 = 2281.$$

Now, suppose Bob is presented with the forged signature $(822, 55)$ on the message 3383. This is a valid signature since the verification condition is satisfied:

$$3405 \times 2281^{3383} \equiv 2282 \pmod{3467}$$

and

$$4^{822} 514^{55} \equiv 2282 \pmod{3467}.$$

On the other hand, this is not the signature Bob would have constructed. Bob can compute his own signature to be

$$(888 + 3383 \times 786 \bmod 1733, 1024 + 3383 \times 999 \bmod 1733) = (1504, 1291).$$

Then, he proceeds to calculate the secret discrete log

$$a_0 = (822 - 1504)(1291 - 55)^{-1} \bmod 1733 = 1567.$$

This is the proof of forgery. ▯

6.7 Notes and References

For a nice survey of signature schemes, we recommend Mitchell, Piper, and Wild [MPW92]. This paper also contains the two methods of forging **ElGamal** signatures that we presented in Section 6.2.

The **ElGamal Signature Scheme** was presented in ElGamal [EL85]. The **Digital Signature Standard** was first published by NIST in August 1991, and it was adopted as a standard in December 1994 [NBS94]. There is a lengthy discussion of **DSS** and the controversy surrounding it in the July 1992 issue of the *Communications of the ACM*. For a response by NIST to some of the questions raised, see [SB93].

The **Lamport Scheme** is described in the 1976 paper by Diffie and Hellman [DH76]; the modification by Bos and Chaum is in [BC93]. The undeniable signature scheme presented in Section 6.5 is due to Chaum and van Antwerpen [CVA90]. The fail-stop signature scheme from Section 6.6 is due to van Heyst and Pedersen [VHP93].

Some examples of well-known "broken" signature schemes include the **Ong-Schnorr-Shamir Scheme** [OSS85] (broken by Estes *et al.* [EAKMM86]); and the **Birational Permutation Scheme** of Shamir [SH94] (broken by Coppersmith, Stern, and Vaudenay [CSV94]). Finally, **ESIGN** is a signature scheme due to Fujioka, Okamoto, and Miyaguchi [FOM91]. Some versions of the scheme were broken, but the variation in [FOM91] has not been broken.

Exercises

6.1 Suppose Bob is using the **ElGamal Signature Scheme**, and he signs two messages x_1 and x_2 with signatures (γ, δ_1) and (γ, δ_2), respectively. (The same value for γ occurs in both signatures.) Suppose also that $\gcd(\delta_1 - \delta_2, p - 1) = 1$.

   (a) Describe how k can be computed efficiently given this information.

   (b) Describe how the signature scheme can then be broken.

   (c) Suppose $p = 31847$, $\alpha = 5$ and $\beta = 25703$. Perform the computation of k and a, given the signature $(23972, 31396)$ for the message $x = 8990$ and the signature $(23972, 20481)$ for the message $x = 31415$.

6.2 Suppose I implement the **ElGamal Signature Scheme** with $p = 31847$, $\alpha = 5$ and $\beta = 26379$. Write a computer program which does the following.

    (a) Verify the signature $(20679, 11082)$ on the message $x = 20543$.

    (b) Determine my secret exponent, a, using the Shanks time-memory tradeoff. Then determine the random value k used in signing the message x.

6.3 Suppose Bob is using the **ElGamal Signature Scheme** as implemented in Example 6.1: $p = 467$, $\alpha = 2$ and $\beta = 132$. Suppose Bob has signed the message $x = 100$ with the signature $(29, 51)$. Compute the forged signature that Oscar can then form by using $h = 102$, $i = 45$ and $j = 293$. Check that the resulting signature satisfies the verification condition.

6.4 Prove that the second method of forgery on the **ElGamal Signature Scheme**, described in Section 6.2, also yields a signature that satisfies the verification condition.

6.5 Here is a variation of the **ElGamal Signature Scheme**. The key is constructed in a similar manner as before: Bob chooses $\alpha \in \mathbb{Z}_p{}^*$ to be a primitive element, a is a secret exponent $(0 \leq a \leq p - 2)$ such that $\gcd(a, p - 1) = 1$, and $\beta = \alpha^a \bmod p$. The key $K = (\alpha, a, \beta)$, where α and β are public and a is secret. Let $x \in \mathbb{Z}_p$ be a message to be signed. Bob computes the signature $sig(x) = (\gamma, \delta)$, where

$$\gamma = \alpha^k \bmod p$$

and

$$\delta = (x - k\gamma)a^{-1} \bmod (p - 1).$$

The only difference from the original **ElGamal Scheme** is in the computation of δ. Answer the following questions concerning this modified scheme.

    (a) Describe how a signature (γ, δ) on a message x would be verified using Bob's public key.

    (b) Describe a computational advantage of the modified scheme over the original scheme.

    (c) Briefly compare the security of the original and modified scheme.

6.6 Suppose Bob uses the **DSS** with $q = 101$, $p = 7879$, $\alpha = 170$, $a = 75$ and $\beta = 4567$, as in Example 6.3. Determine Bob's signature on the message $x = 52$ using the random value $k = 49$, and show how the resulting signature is verified.

6.7 In the **Lamport Scheme**, suppose that two k−tuples, x and x', are signed by Bob. Let $\ell = d(x, x')$ denote the number of coordinates in which x and x' differ. Show that Oscar can now sign $2^\ell - 2$ new messages.

6.8 In the **Bos-Chaum Scheme** with $k = 6$ and $n = 4$, suppose that the messages $x = (0, 1, 0, 0, 1, 1)$ and $x' = (1, 1, 0, 1, 1, 1)$ are signed. Determine the new messages that be signed by Oscar, knowing the signatures on x and x'.

6.9 In the **Bos-Chaum Scheme**, suppose that two k−tuples x and x' are signed by Bob. Let $\ell = |\phi(x) \cup \phi(x')|$. Show that Oscar can now sign $\binom{\ell}{n} - 2$ new messages.

6.10 Suppose Bob is using the **Chaum-van Antwerpen Undeniable Signature Scheme** as in Example 6.5. That is, $p = 467$, $\alpha = 4$, $a = 101$ and $\beta = 449$. Suppose Bob is presented with a signature $y = 25$ on the message $x = 157$ and he wishes to prove it is a forgery. Suppose Alice's random numbers are $e_1 = 46$, $e_2 = 123$, $f_1 = 198$ and $f_2 = 11$ in the disavowal protocol. Compute Alice's challenges, c and d, and Bob's responses, C and D, and show that Alice's consistency check will succeed.

6.11 Prove that each equivalence class of keys in the **Pedersen-van Heyst Fail-stop Signature Scheme** contains q^2 keys.

6.12 Suppose Bob is using the **Pedersen-van Heyst Fail-stop Signature Scheme**, where $p = 3467$, $\alpha = 4$, $a_0 = 1567$ and $\beta = 514$ (of course, the value of a_0 is not known to Bob).

(a) Using the fact that $a_0 = 1567$, determine all possible keys

$$K = (\gamma_1, \gamma_2, a_1, a_2, b_1, b_2)$$

such that $sig_K(42) = (1118, 1449)$.

(b) Suppose that $sig_K(42) = (1118, 1449)$ and $sig_K(969) = (899, 471)$. Without using the fact that $a_0 = 1567$, determine the value of K (this shows that the scheme is a one-time scheme).

6.13 Suppose Bob is using the **Pedersen-van Heyst Fail-stop Signature Scheme** with $p = 5087$, $\alpha = 25$ and $\beta = 1866$. Suppose the key is

$$K = (5065, 5076, 144, 874, 1873, 2345).$$

Now, suppose Bob finds the signature $(2219, 458)$ has been forged on the message 4785.

(a) Prove that this forgery satisfies the verification condition, so it is a valid signature.

(b) Show how Bob will compute the proof of forgery, a_0, given this forged signature.

7

Hash Functions

7.1 Signatures and Hash Functions

The reader might have noticed that the signature schemes described in Chapter 6 allow only "small" messages to be signed. For example, when using the **DSS**, a 160-bit message is signed with a 320-bit signature. In general, we will want to sign much longer messages. A legal document, for example, might be many megabytes in size.

A naive attempt to solve this problem would be to break a long message into 160-bit chunks, and then to sign each chunk independently. This is analogous to encrypting a long string of plaintext by encrypting each plaintext character independently using the same key (e.g., ECB mode in the **DES**).

But there are several problems with this approach in creating digital signatures. First of all, for a long message, we will end up with an enormous signature (twice as long as the original message in the case of the **DSS**). Another disadvantage is that most "secure" signature schemes are slow since they typically use complicated arithmetic operations such as modular exponentiation. But an even more serious problem with this approach is that the various chunks of a signed message could be rearranged, or some of them removed, and the resulting message would still be verified. We need to protect the integrity of the entire message, and this cannot be accomplished by independently signing little pieces of it.

The solution to all of these problems is to use a very fast public *cryptographic hash function*, which will take a message of arbitrary length and produce a *message digest* of a specified size (160 bits if the **DSS** is to be used). The message digest will then be signed. For the **DSS**, the use of a hash function h is depicted diagramatically in Figure 7.1

When Bob wants to sign a message x, he first constructs the message digest $z = h(x)$, and then computes the signature $y = sig_K(z)$. He transmits the ordered pair (x, y) over the channel. Now the verification can be performed (by anyone) by first reconstructing the message digest $z = h(x)$ using the public hash function h, and then checking that $ver_K(z, y) = $ true.

FIGURE 7.1
Signing a message digest

message	x	arbitrary length
$\downarrow$		
message digest	$z = h(x)$	160 bits
$\downarrow$		
signature	$y = sig_K(z)$	320 bits

7.2 Collision-free Hash Functions

We have to be careful that the use of a hash function h does not weaken the security of the signature scheme, for it is the message digest that is signed, not the message. It will be necessary for h to satisfy certain properties in order to prevent various forgeries.

The most obvious type of attack is for an opponent, Oscar, to start with a valid signed message (x, y), where $y = sig_K(h(x))$. (The pair (x, y) could be any message previously signed by Bob.) Then he computes $z = h(x)$ and attempts to find $x' \neq x$ such that $h(x') = h(x)$. If Oscar can do this, (x', y) would be a valid signed message, i.e., a *forgery*. In order to prevent this type of attack, we require that h satisfy the following collision-free property:

DEFINITION 7.1 *Let x be a message. A hash function h is weakly collision-free for x if it is computationally infeasible to find a message $x' \neq x$ such that $h(x') = h(x)$.*

Another possible attack is the following: Oscar first finds two messages $x \neq x'$ such that $h(x) = h(x')$. Oscar then gives x to Bob and persuades him to sign the message digest $h(x)$, obtaining y. Then (x', y) is a valid forgery.

This motivates a different collision-free property:

DEFINITION 7.2 *A hash function h is strongly collision-free if it is computationally infeasible to find messages x and x' such that $x' \neq x$ and $h(x') = h(x)$.*

Observe that a hash function h is strongly collision-free if and only if it in computationally infeasible to find a message x such that h is not weakly collision-free for x.

Here is a third variety of attack. As we mentioned in Section 6.2, it is often possible with certain signature schemes to forge signatures on random message digests z. Suppose Oscar computes a signature on such a random z, and then he finds a message x such that $z = h(x)$. If he can do this, then (x, y) is a valid forgery. To prevent this attack, we desire that h satisfy the same one-way property that was mentioned previously in the context of public-key cryptosystems and the **Lamport Signature Scheme**:

DEFINITION 7.3 *A hash function h is one-way if, given a message digest z, it is computationally infeasible to find a message x such that $h(x) = z$.*

We are now going to prove that the strongly collision-free property implies the one-way property. This is done by proving the contrapositive statement. More specifically, we will prove that an arbitrary inversion algorithm for a hash function can be used as an oracle in a Las Vegas probabilistic algorithm that finds collisions.

This reduction can be accomplished with a fairly weak assumption on the relative sizes of the domain and range of the hash function. We will assume for the time being that the hash function $h : X \rightarrow Z$, where X and Z are finite sets and $|X| \geq 2|Z|$. This is a reasonable assumption: If we think of an element of X as being encoded as a bitstring of length $\log_2 |X|$ and an element of Z as being encoded as a bitstring of length $\log_2 |Z|$, then the message digest $z = h(x)$ is at least one bit shorter than the message x. (Eventually, we will be interested in the situation where the message domain X is infinite, since we want to be able to deal with messages of arbitrary length. Our argument also applies in this situation.)

We are assuming that we have an inversion algorithm for h. That is, we have an algorithm $\mathbf{A}$ which accepts as input a message digest $z \in Z$, and finds an element $\mathbf{A}(z) \in X$ such that $h(\mathbf{A}(z)) = z$.

We prove the following theorem.

THEOREM 7.1
Suppose $h : X \rightarrow Z$ is a hash function where $|X|$ and $|Z|$ are finite and $|X| \geq 2|Z|$. Suppose $\mathbf{A}$ is an inversion algorithm for h. Then there exists a probabilistic Las Vegas algorithm which finds a collision for h with probability at least $1/2$.

PROOF Consider the algorithm $\mathbf{B}$ presented in Figure 7.2.   Clearly $\mathbf{B}$ is a probabilistic algorithm of the Las Vegas type, since it either finds a collision or returns no answer. Thus our main task is to compute the probability of success. For any $x \in X$, define $x \sim x_1$ if $h(x) = h(x_1)$. It is easy to see that $\sim$ is an equivalence relation. Define

$$[x] = \{x_1 \in X : x \sim x_1\}.$$

Each equivalence class $[x]$ consists of the inverse image of an element of Z, so the number of equivalence classes is at most $|Z|$. Denote the set of equivalence classes by C.

Now, suppose x is the element of X chosen in step 1. For this x, there are $|[x]|$ possible x_1's that could be returned in step 3. $|[x]| - 1$ of these x_1's are different

FIGURE 7.2
Using an inversion algorithm A to find collisions for a hash function h

---

1. choose a random $x \in X$
2. compute $z = h(x)$
3. compute $x_1 = \mathbf{A}(z)$
4. **if** $x_1 \neq x$ **then**

   x_1 and x collide under h (success)

   else

   QUIT (failure).

---

from x and thus lead to success in step 4. (Note that the algorithm $\mathbf{A}$ does not know the representative of the equivalence class $[x]$ that was chosen in step 1.) So, given a particular choice $x \in X$, the probability of success is $(|[x]| - 1)/|[x]|$.

The probability of success of the algorithm $\mathbf{B}$ is computed by averaging over all possible choices for x:

$$
\begin{aligned}
p(\text{success}) &= \frac{1}{|X|} \sum_{x \in X} \frac{|[x]| - 1}{|[x]|} \\
&= \frac{1}{|X|} \sum_{c \in C} \sum_{x \in c} \frac{|c| - 1}{|c|} \\
&= \frac{1}{|X|} \sum_{c \in C} (|c| - 1) \\
&= \frac{1}{|X|} \left(\sum_{c \in C} |c| - \sum_{c \in C} 1 \right) \\
&\geq \frac{|X| - |Z|}{|X|} \\
&\geq \frac{|X| - |X|/2}{|X|} \\
&= \frac{1}{2}.
\end{aligned}
$$

Hence we have constructed a Las Vegas algorithm with success probability at least $1/2$. ∎

Hence, it is sufficient that a hash function satisfy the strongly collision-free property, since it implies the other two properties. So in the remainder of this chapter we restrict our attention to strongly collision-free hash functions.

7.3   The Birthday Attack

In this section, we determine a necessary security condition for hash functions that depends only on the cardinality of the set Z (equivalently, on the size of the message digest). This necessary condition results from a simple method of finding collisions which is informally known as the *birthday attack*. This terminology arises from the so-called *birthday paradox*, which says that in a group of 23 random people, at least two will share a birthday with probability at least $1/2$. (Of course this is not a paradox, but it is probably counter-intuitive). The reason for the terminology "birthday attack" will become clear as we progress.

As before, let us suppose that $h : X \to Z$ is a hash function, X and Z are finite, and $|X| \geq 2|Z|$. Denote $|X| = m$ and $|Z| = n$. It is not hard to see that there are at least n collisions — the question is how to find them. A very naive approach is to choose k random distinct elements $x_1, \ldots, x_k \in X$, compute $z_i = h(x_i)$, $1 \leq i \leq k$, and then determine if a collision has taken place (by sorting the z_i's, for example).

This process is analogous to throwing k balls randomly into n bins and then checking to see if some bin contains at least two balls. (The k balls correspond to the k random x_i's, and the n bins correspond to the n possible elements of Z.)

We will compute a lower bound on the probability of finding a collision by this method. This lower bound will depend on k and n, but not on m. Since we are interested in a lower bound on the collision probability, we will make the assumption that $|h^{-1}(z)| \approx m/n$ for all $z \in Z$. (This is a reasonable assumption: if the inverse images are not approximately equal, then the probability of finding a collision will increase.)

Since the inverse images are all (roughly) the same size and the x_i's are chosen at random, the resulting z_i's can be thought of as random (not necessarily distinct) elements of Z. But it is a simple matter to compute the probability that k random elements $z_1, \ldots, z_k \in Z$ are distinct. Consider the z_i's in the order $z_1, \ldots, z_k$. The first choice z_1 is arbitrary; the probability that $z_2 \neq z_1$ is $1 - 1/n$; the probability that z_3 is distinct from z_1 and z_2 is $1 - 2/n$, etc.

Hence, we estimate the probability of no collisions to be

$$\left(1 - \frac{1}{n}\right)\left(1 - \frac{2}{n}\right)\cdots\left(1 - \frac{k-1}{n}\right) = \prod_{i=1}^{k-1}\left(1 - \frac{i}{n}\right).$$

If x is a small real number, then $1 - x \approx e^{-x}$. This estimate is derived by taking the first two terms of the series expansion

$$e^{-x} = 1 - x + \frac{x^2}{2!} - \frac{x^3}{3!} \cdots .$$

Then our estimated probability of no collisions is

$$\prod_{i=1}^{k-1} \left(1 - \frac{i}{n}\right) \approx \prod_{i=1}^{k-1} e^{\frac{-i}{n}}$$

$$= e^{\frac{-k(k-1)}{2n}} .$$

So we estimate the probability of at least one collision to be

$$1 - e^{\frac{-k(k-1)}{2n}} .$$

If we denote this probability by ϵ, then we can solve for k as a function of n and ϵ:

$$e^{\frac{-k(k-1)}{2n}} \approx 1 - \epsilon$$

$$\frac{-k(k-1)}{2n} \approx \ln(1 - \epsilon)$$

$$k^2 - k \approx 2n \ln \frac{1}{1 - \epsilon} .$$

If we ignore the term $-k$, then we estimate

$$k \approx \sqrt{2n \ln \frac{1}{1 - \epsilon}} .$$

If we take $\epsilon = .5$, then our estimate is

$$k \approx 1.17\sqrt{n}.$$

So this says that hashing just over $\sqrt{n}$ random elements of X yields a collision with a probability of 50%. Note that a different choice of ϵ leads to a different constant factor, but k will still be proportional to $\sqrt{n}$.

If X is the set of all human beings, Y is the set of 365 days in a non-leap year (i.e., excluding February 29), and $h(x)$ denotes the birthday of person x, then we are dealing with the birthday paradox. Taking $n = 365$ in our estimate, we get $k \approx 22.3$. Hence, as mentioned earlier, there will be at least one duplicated birthday among 23 random people with probability at least $1/2$.

This birthday attack imposes a lower bound on the sizes of message digests. A 40-bit message digest would be very insecure, since a collision could be found with probability $1/2$ with just over 2^{20} (about a million) random hashes. It is usually suggested that the minimum acceptable size of a message digest is 128 bits (the birthday attack will require over 2^{64} hashes in this case). The choice of a 160-bit message digest for use in the **DSS** was undoubtedly motivated by these considerations.

FIGURE 7.3
Chaum-van Heijst-Pfitzmann Hash Function

Suppose p is a large prime and $q = (p-1)/2$ is also prime. Let α and β be two primitive elements of $\mathbb{Z}_p$. The value $\log_\alpha \beta$ is not public, and we assume that it is computationally infeasible to compute its value. The hash function

$$h : \{0, \dots, q-1\} \times \{0, \dots, q-1\} \to \mathbb{Z}_p \backslash \{0\}$$

is defined as follows:

$$h(x_1, x_2) = \alpha^{x_1} \beta^{x_2} \bmod p.$$

7.4 A Discrete Log Hash Function

In this section, we describe a hash function, due to Chaum, van Heijst, and Pfitzmann, that will be secure provided a particular discrete logarithm cannot be computed. This hash function is not fast enough to be of practical use, but it is conceptually simple and provides a nice example of a hash function that can be proved secure under a reasonable computational assumption. The **Chaum-van Heijst-Pfitzmann Hash Function** is presented in Figure 7.3. We now prove a theorem concerning the security of this hash function.

THEOREM 7.2
Given one collision for the **Chaum-van Heijst-Pfitzmann Hash Function** *h, the discrete logarithm* $\log_\alpha \beta$ *can be computed efficiently.*

PROOF   Suppose we are given a collision

$$h(x_1, x_2) = h(x_3, x_4),$$

where $(x_1, x_2) \neq (x_3, x_4)$. So we have the following congruence:

$$\alpha^{x_1} \beta^{x_2} \equiv \alpha^{x_3} \beta^{x_4} \pmod{p},$$

or

$$\alpha^{x_1 - x_3} \equiv \beta^{x_4 - x_2} \pmod{p}.$$

Denote

$$d = \gcd(x_4 - x_2, p - 1).$$

Since $p - 1 = 2q$ and q is prime, it must be the case that $d \in \{1, 2, q, p-1\}$. Hence, we have four possibilities for d, which we will consider in turn.

First, suppose that $d = 1$. Then let

$$y = (x_4 - x_2)^{-1} \bmod (p - 1).$$

We have that

$$\beta \equiv \beta^{(x_4 - x_2)y} \pmod{p}$$
$$\equiv \alpha^{(x_1 - x_3)y} \pmod{p},$$

so we can compute the discrete logarithm $\log_\alpha \beta$ as follows:

$$\log_\alpha \beta = (x_1 - x_3)(x_4 - x_2)^{-1} \bmod (p - 1).$$

Next, suppose that $d = 2$. Since $p - 1 = 2q$ where q is odd, we must have $\gcd(x_4 - x_2, q) = 1$. Let

$$y = (x_4 - x_2)^{-1} \bmod q.$$

Now

$$(x_4 - x_2)y = kq + 1$$

for some integer k, so we have

$$\beta^{(x_4 - x_2)y} \equiv \beta^{kq+1} \pmod{p}$$
$$\equiv (-1)^k \beta \pmod{p}$$
$$\equiv \pm\beta \pmod{p},$$

since

$$\beta^q \equiv -1 \pmod{p}.$$

So we have

$$\beta^{(x_4 - x_2)y} \equiv \alpha^{(x_1 - x_3)y} \pmod{p}$$
$$\equiv \pm\beta \pmod{p}.$$

It follows that

$$\log_\alpha \beta = (x_1 - x_3)y \bmod (p - 1)$$

or

$$\log_\alpha \beta = (x_1 - x_3)y + q \bmod (p - 1).$$

We can easily test which of these two possibilities is the correct one. Hence, as in the case $d = 1$, we have calculated the discrete logarithm $\log_\alpha \beta$.

The next possibility is that $d = q$. But

$$0 \le x_2 \le q - 1$$

and
$$0 \le x_4 \le q - 1,$$
so
$$-(q - 1) \le x_4 - x_2 \le q - 1.$$

So it is impossible that $\gcd(x_4 - x_2, p - 1) = q$; in other words, this case does not arise.

The final possibility is that $d = p - 1$. This happens only if $x_2 = x_4$. But then we have
$$\alpha^{x_1} \beta^{x_2} \equiv \alpha^{x_3} \beta^{x_2} \pmod{p},$$
so
$$\alpha^{x_1} \equiv \alpha^{x_3} \pmod{p},$$

and $x_1 = x_3$. Thus $(x_1, x_2) = (x_3, x_4)$, a contradiction. So this case is not possible, either.

Since we have considered all possible values for d, we conclude that the hash function h is strongly collision-free provided that it is infeasible to compute the discrete logarithm $\log_\alpha \beta$ in $\mathbb{Z}_p$.   ∎

We illustrate the result of the above theorem with an example.

Example 7.1
Suppose $p = 12347$ (so $q = 6173$), $\alpha = 2$ and $\beta = 8461$. Suppose we are given the collision
$$\alpha^{5692} \beta^{144} \equiv \alpha^{212} \beta^{4214} \pmod{12347}.$$

Thus $x_1 = 5692$, $x_2 = 144$, $x_3 = 212$ and $x_4 = 4214$. Now, $\gcd(x_4 - x_2, p - 1) = 2$, so we begin by computing
$$\begin{aligned} y &= (x_4 - x_2)^{-1} \bmod q \\ &= (4214 - 144)^{-1} \bmod 6173 \\ &= 4312. \end{aligned}$$

Next, we compute
$$\begin{aligned} y' &= (x_1 - x_3) y \bmod (p - 1) \\ &= (5692 - 212) 4312 \bmod 12346 \\ &= 11862. \end{aligned}$$

Now it is the case that $\log_\alpha \beta \in \{y', y' + q \bmod (p - 1)\}$. Since
$$\alpha^{y'} \bmod p = 2^{11862} \bmod 12346 = 9998,$$

we conclude that

$$\log_\alpha \beta = y' + q \bmod (p-1)$$
$$= 11862 + 6173 \bmod 12346$$
$$= 5689.$$

As a check, we can verify that

$$2^{5689} \equiv 8461 \pmod{12347}.$$

Hence, we have determined $\log_\alpha \beta$.   ▯

7.5 Extending Hash Functions

So far, we have considered hash functions with a finite domain. We now study how a strongly collision-free hash function with a finite domain can be extended to a strongly collision-free hash function with an infinite domain. This will enable us to sign messages of arbitrary length.

Suppose $h : (\mathbb{Z}_2)^m \to (\mathbb{Z}_2)^t$ is a strongly collision-free hash function, where $m \geq t+1$. We will use h to construct a strongly collision-free hash function $h^* : X \to (\mathbb{Z}_2)^t$, where

$$X = \bigcup_{i=m}^{\infty} (\mathbb{Z}_2)^i.$$

We first consider the situation where $m \geq t+2$.

We will think of elements of X as bit-strings. $|x|$ denotes the length of x (i.e., the number of bits in x), and $x \parallel y$ denotes the concatenation of the bit-strings x and y. Suppose $|x| = n > m$. We can express x as the concatenation

$$x = x_1 \parallel x_2 \parallel \ldots \parallel x_k,$$

where

$$|x_1| = |x_2| = \ldots = |x_{k-1}| = m - t - 1$$

and

$$|x_k| = m - t - 1 - d,$$

where $0 \leq d \leq m - t - 2$. Hence, we have that

$$k = \left\lceil \frac{n}{m - t - 1} \right\rceil.$$

We define $h^*(x)$ by the algorithm presented in Figure 7.4.

FIGURE 7.4
Extending a hash function h to h^* ($m \geq t + 2$)

---

1. **for** $i = 1$ **to** $k - 1$ **do**
      $y_i = x_i$
2. $y_k = x_k \parallel 0^d$
3. let y_{k+1} be the binary representation of d
4. $g_1 = h(0^{t+1} \parallel y_1)$
5. **for** $i = 1$ **to** k **do**
      $g_{i+1} = h(g_i \parallel 1 \parallel y_{i+1})$
6. $h^*(x) = g_{k+1}$.

Denote
$$y(x) = y_1 \parallel y_2 \parallel \ldots \parallel y_{k+1}.$$

Observe that y_k is formed from x_k by padding on the right with d zeroes, so that all the blocks y_i ($1 \leq i \leq k$) are of length $m - t - 1$. Also, in step 3, y_{k+1} should be padded on the left with zeroes so that $|y_{k+1}| = m - t - 1$.

In order to hash x, we first construct $y(x)$, and then "process" the blocks $y_1, y_2, \ldots, y_{k+1}$ in a particular fashion. It is important that $y(x) \neq y(x')$ whenever $x \neq x'$. In fact, y_{k+1} is defined in such a way that the mapping $x \mapsto y(x)$ will be an injection.

The following theorem proves that h^* is secure provided that h is secure.

THEOREM 7.3
Suppose $h : (\mathbb{Z}_2)^m \rightarrow (\mathbb{Z}_2)^t$ is a strongly collision-free hash function, where $m \geq t + 2$. Then the function $h^* : \bigcup_{i=m}^{\infty} (\mathbb{Z}_2)^i \rightarrow (\mathbb{Z}_2)^t$, as constructed in Figure 7.4, is a strongly collision-free hash function.

PROOF   Suppose that we can find $x \neq x'$ such that $h^*(x) = h^*(x')$. Given such a pair, we will show how we can find a collision for h in polynomial time. Since h is assumed to be strongly collision-free, we will obtain a contradiction, and thus h^* will be proved to be strongly collision-free.

Denote
$$y(x) = y_1 \parallel y_2 \parallel \ldots \parallel y_{k+1}$$

and
$$y(x') = y_1' \parallel y_2' \parallel \ldots \parallel y_{\ell+1}',$$

where x and x' are padded with d and d' 0's, respectively, in step 2. Denote the values computed in steps 4 and 5 by $g_1, \ldots, g_{k+1}$ and $g_1', \ldots, g_{\ell+1}'$, respectively.

We identify two cases, depending on whether or not $|x| \equiv |x'| \pmod{m-t-1}$.

case 1: $|x| \not\equiv |x'| \pmod{m-t-1}$.

Here $d \neq d'$ and $y_{k+1} \neq y'_{\ell+1}$. We have

$$
\begin{aligned}
h(g_k \parallel 1 \parallel y_{k+1}) &= g_{k+1} \\
&= h^*(x) \\
&= h^*(x') \\
&= g'_{\ell+1} \\
&= h(g'_\ell \parallel 1 \parallel y'_{\ell+1}),
\end{aligned}
$$

which is a collision for h since $y_{k+1} \neq y'_{\ell+1}$.

case 2: $|x| \equiv |x'| \pmod{m-t-1}$.

It is convenient to split this into two subcases:

case 2a: $|x| = |x'|$.

Here we have $k = \ell$ and $y_{k+1} = y'_{k+1}$. We begin as in case 1:

$$
\begin{aligned}
h(g_k \parallel 1 \parallel y_{k+1}) &= g_{k+1} \\
&= h^*(x) \\
&= h^*(x') \\
&= g'_{k+1} \\
&= h(g'_k \parallel 1 \parallel y'_{k+1}).
\end{aligned}
$$

If $g_k \neq g'_k$, then we find a collision for h, so assume $g_k = g'_k$. Then we have

$$
\begin{aligned}
h(g_{k-1} \parallel 1 \parallel y_k) &= g_k \\
&= g'_k \\
&= h(g'_{k-1} \parallel 1 \parallel y'_k).
\end{aligned}
$$

Either we find a collision for h, or $g_{k-1} = g'_{k-1}$ and $y_k = y'_k$. Assuming we do not find a collision, we continue working backwards, until finally we obtain

$$
\begin{aligned}
h(0^{t+1} \parallel y_1) &= g_1 \\
&= g'_1 \\
&= h(0^{t+1} \parallel y'_1).
\end{aligned}
$$

If $y_1 \neq y'_1$, then we find a collision for h, so we assume $y_1 = y'_1$. But then $y_i = y'_i$ for $1 \leq i \leq k+1$, so $y(x) = y(x')$. But this implies $x = x'$ since

FIGURE 7.5
Extending a hash function h to h^* ($m = t + 1$)

---

1. let $y = y_1 y_2 \ldots y_k = 11 \parallel f(x_1) \parallel f(x_2) \parallel \ldots \parallel f(x_n)$
2. $g_1 = h(0^t \parallel y_1)$
3. **for** $i = 1$ **to** $k - 1$ **do**
   $$g_{i+1} = h(g_i \parallel y_{i+1})$$
4. $h^*(x) = g_k.$

---

the mapping $x \mapsto y(x)$ is an injection. Since we assumed $x \neq x'$, we have a contradiction.

case 2b: $|x| \neq |x'|$.

Without loss of generality, assume $|x'| > |x|$, so $\ell > k$. This case proceeds in a similar fashion as case 2a. Assuming we find no collisions for h, we eventually reach the situation where

$$h(0^{t+1} \parallel y_1) = g_1$$
$$= g'_{\ell-k+1}$$
$$= h(g'_{\ell-k} \parallel 1 \parallel y'_{\ell-k+1}).$$

But the $(t + 1)$st bit of $0^{t+1} \parallel y_1$ is a 0 and the $(t + 1)$st bit of $g'_{\ell-k} \parallel 1 \parallel y'_{\ell-k+1}$ is a 1. So we find a collision for h.

Since we have considered all possible cases, we have the desired conclusion. ∎

The construction of Figure 7.4 can be used only when $m \geq t + 2$. Let's now look at the situation where $m = t + 1$. We need to use a different construction for h^*. As before, suppose $|x| = n > m$. We first encode x in a special way. This will be done using the function f defined as follows:

$$f(0) = 0$$
$$f(1) = 01.$$

The algorithm to construct $h^*(x)$ is presented in Figure 7.5.

The encoding $x \mapsto y = y(x)$, defined in step 1, satisfies two important properties:

1. If $x \neq x'$, then $y(x) \neq y(x')$ (i.e., $x \mapsto y(x)$ is an injection).

2. There do not exist two strings $x \neq x'$ and a string z such that $y(x) = z \parallel y(x')$. (In other words, no encoding is a *postfix* of another encoding. This is easily seen because each string $y(x)$ begins with 11, and there do not exist two consecutive 1's in the remainder of the string.)

THEOREM 7.4
Suppose $h : (\mathbb{Z}_2)^{t+1} \to (\mathbb{Z}_2)^t$ is a strongly collision-free hash function. Then the function $h^* : \bigcup_{i=t+1}^{\infty} (\mathbb{Z}_2)^i \to (\mathbb{Z}_2)^t$, as constructed in Figure 7.5, is a strongly collision-free hash function.

PROOF Suppose that we can find $x \neq x'$ such that $h^*(x) = h^*(x')$. Denote

$$y(x) = y_1 y_2 \dots y_k$$

and

$$y(x') = y_1' y_2' \dots y_\ell'.$$

We consider two cases.

case 1: $k = \ell$.

As in Theorem 7.3, either we find a collision for h, or we obtain $y = y'$. But this implies $x = x'$, a contradiction.

case 2: $k \neq \ell$.

Without loss of generality, assume $\ell > k$. This case proceeds in a similar fashion. Assuming we find no collisions for h, we have the following sequence of equalities:

$$y_k = y_\ell'$$

$$y_{k-1} = y_{\ell-1}'$$

$$\vdots \quad \vdots$$

$$y_1 = y_{\ell-k+1}'.$$

But this contradicts the "postfix-free" property stated above.

We conclude that h^* is collision-free. ∎

We summarize the two constructions of in this section, and the number of applications of h needed to compute h^*, in the following theorem.

THEOREM 7.5
Suppose $h : (\mathbb{Z}_2)^m \to (\mathbb{Z}_2)^t$ is a strongly collision-free hash function, where $m \geq t + 1$. Then there exists a strongly collision-free hash function

$$h^* : \bigcup_{i=m}^{\infty} (\mathbb{Z}_2)^i \to (\mathbb{Z}_2)^t.$$

The number of times h is computed in the evaluation of h* is at most

$$1 + \left\lceil \frac{n}{m-t-1} \right\rceil \quad \text{if } m \geq t+2$$
$$2n + 2 \quad \text{if } m = t+1,$$

where $|x| = n$.

7.6 Hash Functions from Cryptosystems

So far, the methods we have described lead to hash functions that are probably too slow to be useful in practice. Another approach is to use an existing private-key cryptosystem to construct a hash function. Let us suppose that $(\mathcal{P}, \mathcal{C}, \mathcal{K}, \mathcal{E}, \mathcal{D})$ is a computationally secure cryptosystem. For convenience, let us assume also that $\mathcal{P} = \mathcal{C} = \mathcal{K} = (\mathbb{Z}_2)^n$. Here we should have $n \geq 128$, say, in order to prevent birthday attacks. This precludes using **DES** (as does the fact that the key length of **DES** is different from the plaintext length).

Suppose we are given a bitstring

$$x = x_1 \parallel x_2 \parallel \ldots \parallel x_k,$$

where $x_i \in (\mathbb{Z}_2)^n$, $1 \leq i \leq k$. (If the number of bits in x is not a multiple of n, then it will be necessary to pad x in some way, such as was done in Section 7.5. For simplicity, we will ignore this now.)

The basic idea is to begin with a fixed "initial value" $g_0 = \text{IV}$, and then construct $g_1, \ldots, g_k$ in order by a rule of the form

$$g_i = f(x_i, g_{i-1}),$$

where f is a function that incorporates the encryption function of our cryptosystem. Finally, define the message digest $h(x) = g_k$.

Several hash functions of this type have been proposed, and many of them have been shown to be insecure (independent of whether or not the underlying cryptosystem is secure). However, four variations of this theme that appear to be secure are as follows:

$$g_i = e_{g_{i-1}}(x_i) \oplus x_i$$
$$g_i = e_{g_{i-1}}(x_i) \oplus x_i \oplus g_{i-1}$$
$$g_i = e_{g_{i-1}}(x_i \oplus g_{i-1}) \oplus x_i$$
$$g_i = e_{g_{i-1}}(x_i \oplus g_{i-1}) \oplus x_i \oplus g_{i-1}.$$

FIGURE 7.6
Constructing M in MD4

---

1. $d = (447 - |x|) \bmod 512$
2. let ℓ denote the binary representation of $|x| \bmod 2^{64}$, $|\ell| = 64$
3. $M = x \parallel 1 \parallel 0^d \parallel \ell$

7.7 The MD4 Hash Function

The **MD4 Hash Function** was proposed in 1990 by Rivest, and a strengthened version, called **MD5**, was presented in 1991. The **Secure Hash Standard** (or **SHS**) is more complicated, but it is based on the same underlying methods. It was published in the Federal Register on January 31, 1992, and adopted as a standard on May 11, 1993. (A proposed revision was put forward on July 11, 1994, to correct a "technical flaw" in the **SHS**.) All of the above hash functions are very fast, so they are practical for signing very long messages.

In this section, we will describe **MD4** in detail, and discuss some of the modifications that are employed in **MD5** and the **SHS**.

Given a bitstring x, we will first produce an array

$$M = M[0]M[1]\ldots M[N-1],$$

where each $M[i]$ is a bitstring of length 32 and $N \equiv 0 \bmod 16$. We will call each $M[i]$ a *word*. M is constructed from x using the algorithm presented in Figure 7.6.

In the construction of M, we append a single 1 to x, then we concatenate enough 0's so that the length becomes congruent to 448 modulo 512, and finally we concatenate 64 bits that contain the binary representation of the (original) length of x (reduced modulo 2^{64}, if necessary). The resulting string M has length divisible by 512. So when we break M up into 32-bit words, the resulting number of words, denoted by N, will be divisible by 16.

Now we proceed to construct a 128-bit message digest. A high-level description of the algorithm is presented in Figure 7.7. The message digest is constructed as the concatenation of the four words A, B, C and D, which we refer to as *registers*. The four registers are initialized in step 1. Now we process the array M 16 words at a time. In each iteration of the loop in step 2, we first take the "next" 16 words of M and store them in an array X (step 3). The values of the four registers are then stored (step 4). Then we perform three "rounds" of hashing. Each round consists of one operation on each of the 16 words in X (we will describe

FIGURE 7.7
The MD4 hash function

1. $A = 67452301$ (hex)
     $B = efcdab89$ (hex)
     $C = 98badcfe$ (hex)
     $D = 10325476$ (hex)
2.   **for** $i = 0$ **to** $N/16 - 1$ **do**
3.   **for** $j = 0$ **to** 15 **do**
             $X[j] = M[16i + j]$
4. $AA = A$
         $BB = B$
         $CC = C$
         $DD = D$
5.   **Round1**
6.   **Round2**
7.   **Round3**
8. $A = A + AA$
         $B = B + BB$
         $C = C + CC$
         $D = D + DD$

these operations in more detail shortly). The operations done in the three rounds produce new values in the four registers. Finally, the four registers are updated in step 8 by adding back the values that were stored in step 4. This addition is defined to be addition of positive integers, reduced modulo 2^{32}.

The three rounds in **MD4** are different (unlike **DES**, say, where the 16 rounds are identical). We first describe several different operations that are employed in these three rounds. In the following description, X and Y denote input words, and each operation produces a word as output. Here are the operations employed:

$X \wedge Y$	bitwise "and" of X and Y
$X \vee Y$	bitwise "or" of X and Y
$X \oplus Y$	bitwise "xor" of X and Y
$\neg X$	bitwise complement of X
$X + Y$	integer addition modulo 2^{32}
$X \lll s$	circular left shift of X by s positions ($0 \leq s \leq 31$)

Note that all of these operations are very fast, and the only arithmetic operation that is used is addition modulo 2^{32}. If **MD4** is actually implemented, it will be necessary to take into account the underlying architecture of the computer it is run on in order to perform addition correctly. Suppose $a_1 a_2 a_3 a_4$ are the four bytes in a word. We think of each a_i as being an integer in the range $0, \ldots, 255$, represented in binary. In a *big-endian* architecture (such as a Sun SPARCstation), this word represents the integer

$$a_1 2^{24} + a_2 2^{16} + a_3 2^8 + a_4.$$

In a *little-endian* architecture (such as the Intel 80xxx line), this word represents the integer

$$a_4 2^{24} + a_3 2^{16} + a_2 2^8 + a_1.$$

MD4 assumes a little-endian architecture. It is important that the message digest is independent of the underlying architecture. So if we wish to run **MD4** on a big-endian computer, it will be necessary to perform the addition operation $X + Y$ as follows:

1. Interchange x_1 and x_4; x_2 and x_3; y_1 and y_4; and y_2 and y_3.
2. Compute $Z = X + Y \bmod 2^{32}$
3. Interchange z_1 and z_4; and z_2 and z_3.

Rounds 1, 2, and 3 of **MD4** respectively use three functions f, g and h. Each of f, g and h is a bitwise boolean function that takes three words as input and produces a word as output. They are defined as follows:

$$f(X, Y, Z) = (X \wedge Y) \vee ((\neg X) \wedge Z)$$

$$g(X, Y, Z) = (X \wedge Y) \vee (X \wedge Z) \vee (Y \wedge Z)$$

$$h(X, Y, Z) = X \oplus Y \oplus Z.$$

The complete description of Rounds 1, 2 and 3 of **MD4** are presented in Figures 7.8–7.10.

MD4 was designed to be very fast, and indeed, software implementations on Sun SPARCstations attain speeds of 1.4 Mbytes/sec. On the other hand, it is difficult to say something concrete about the security of a hash function such as **MD4** since it is not "based" on a well-studied problem such as factoring or the

FIGURE 7.8
Round 1 of MD4

1. $A = (A + f(B, C, D) + X[0]) \lll 3$
2. $D = (D + f(A, B, C) + X[1]) \lll 7$
3. $C = (C + f(D, A, B) + X[2]) \lll 11$
4. $B = (B + f(C, D, A) + X[3]) \lll 19$
5. $A = (A + f(B, C, D) + X[4]) \lll 3$
6. $D = (D + f(A, B, C) + X[5]) \lll 7$
7. $C = (C + f(D, A, B) + X[6]) \lll 11$
8. $B = (B + f(C, D, A) + X[7]) \lll 19$
9. $A = (A + f(B, C, D) + X[8]) \lll 3$
10. $D = (D + f(A, B, C) + X[9]) \lll 7$
11. $C = (C + f(D, A, B) + X[10]) \lll 11$
12. $B = (B + f(C, D, A) + X[11]) \lll 19$
13. $A = (A + f(B, C, D) + X[12]) \lll 3$
14. $D = (D + f(A, B, C) + X[13]) \lll 7$
15. $C = (C + f(D, A, B) + X[14]) \lll 11$
16. $B = (B + f(C, D, A) + X[15]) \lll 19$

Discrete Log problem. So, as is the case with **DES**, confidence in the security of the system can only be attained over time, as the system is studied and (one hopes) not found to be insecure.

Although **MD4** has not been broken, weakened versions that omit either the first or the third round can be broken without much difficulty. That is, it is easy to find collisions for these two-round versions of **MD4**. A strengthened version of **MD4**, called **MD5**, was proposed in 1991. **MD5** uses four rounds instead of three, and runs about 30% slower than **MD4** (about .9 Mbytes/sec on a SPARCstation).

The **Secure Hash Standard** is yet more complicated, and slower (about .2 Mbytes/sec on a SPARCstation). We will not give a complete description, but we will indicate a few of the modifications employed in the **SHS**.

1. **SHS** is designed to run on a big-endian architecture, rather than a little-endian architecture.

2. **SHS** produces a 5-register (160-bit) message digest.

FIGURE 7.9
Round 2 of MD4

1.	$A = (A + g(B, C, D) + X[0] + 5A827999) \lll 3$
2.	$D = (D + g(A, B, C) + X[4] + 5A827999) \lll 5$
3.	$C = (C + g(D, A, B) + X[8] + 5A827999) \lll 9$
4.	$B = (B + g(C, D, A) + X[12] + 5A827999) \lll 13$
5.	$A = (A + g(B, C, D) + X[1] + 5A827999) \lll 3$
6.	$D = (D + g(A, B, C) + X[5] + 5A827999) \lll 5$
7.	$C = (C + g(D, A, B) + X[9] + 5A827999) \lll 9$
8.	$B = (B + g(C, D, A) + X[13] + 5A827999) \lll 13$
9.	$A = (A + g(B, C, D) + X[2] + 5A827999) \lll 3$
10.	$D = (D + g(A, B, C) + X[6] + 5A827999) \lll 5$
11.	$C = (C + g(D, A, B) + X[10] + 5A827999) \lll 9$
12.	$B = (B + g(C, D, A) + X[14] + 5A827999) \lll 13$
13.	$A = (A + g(B, C, D) + X[3] + 5A827999) \lll 3$
14.	$D = (D + g(A, B, C) + X[7] + 5A827999) \lll 5$
15.	$C = (C + g(D, A, B) + X[11] + 5A827999) \lll 9$
16.	$B = (B + g(C, D, A) + X[15] + 5A827999) \lll 13$

3. **SHS** processes the message 16 words at a time, as does **MD4**. However, the 16 words are first "expanded" into 80 words. Then a sequence of 80 operations is performed, one on each word.

The following "expansion function" is used. Given the 16 words $X[0], \ldots,$ $X[15]$, we compute 64 more words by the recurrence relation

$$X[j] = X[j - 3] \oplus X[j - 8] \oplus X[j - 14] \oplus X[j - 16], 16 \le j \le 79. \quad (7.1)$$

The result of Equation 7.1 is that each of the words $X[16], \ldots, X[79]$ is formed as the exclusive-or of a predetermined subset of the words $X[0], \ldots, X[15]$.
For example, we have

$$X[16] = X[0] \oplus X[2] \oplus X[8] \oplus X[13]$$

$$X[17] = X[1] \oplus X[3] \oplus X[9] \oplus X[14]$$

$$X[18] = X[2] \oplus X[4] \oplus X[10] \oplus X[15]$$

$$X[19] = X[0] \oplus X[2] \oplus X[3] \oplus X[5] \oplus X[8] \oplus X[11] \oplus X[13]$$

FIGURE 7.10
Round 3 of MD4

1. $A = (A + h(B, C, D) + X[0] + 6ED9EBA1) \lll 3$
2. $D = (D + h(A, B, C) + X[8] + 6ED9EBA1) \lll 9$
3. $C = (C + h(D, A, B) + X[4] + 6ED9EBA1) \lll 11$
4. $B = (B + h(C, D, A) + X[12] + 6ED9EBA1) \lll 15$
5. $A = (A + h(B, C, D) + X[2] + 6ED9EBA1) \lll 3$
6. $D = (D + h(A, B, C) + X[10] + 6ED9EBA1) \lll 9$
7. $C = (C + h(D, A, B) + X[6] + 6ED9EBA1) \lll 11$
8. $B = (B + h(C, D, A) + X[14] + 6ED9EBA1) \lll 15$
9. $A = (A + h(B, C, D) + X[1] + 6ED9EBA1) \lll 3$
10. $D = (D + h(A, B, C) + X[9] + 6ED9EBA1) \lll 9$
11. $C = (C + h(D, A, B) + X[5] + 6ED9EBA1) \lll 11$
12. $B = (B + h(C, D, A) + X[13] + 6ED9EBA1) \lll 15$
13. $A = (A + h(B, C, D) + X[3] + 6ED9EBA1) \lll 3$
14. $D = (D + h(A, B, C) + X[11] + 6ED9EBA1) \lll 9$
15. $C = (C + h(D, A, B) + X[7] + 6ED9EBA1) \lll 11$
16. $B = (B + h(C, D, A) + X[15] + 6ED9EBA1) \lll 15$

$$\vdots$$

$$X[79] = X[1] \oplus X[4] \oplus X[5] \oplus X[8] \oplus X[9] \oplus X[12] \oplus X[13].$$

The proposed revision of the **SHS** concerns the expansion function. It is proposed that Equation 7.1 be replaced by the following:

$$X[j] = (X[j-3] \oplus X[j-8] \oplus X[j-14] \oplus X[j-16]) \lll 1, 16 \le j \le 79. \quad (7.2)$$

As before, the operation "$\lll 1$" means a circular left shift of one position.

7.8 Timestamping

One difficulty with signature schemes is that a signing algorithm may be compromised. For example, suppose that Oscar is able to determine Bob's secret

FIGURE 7.11
Timestamping a signature on a message x

1.	Bob computes $z = h(x)$
2.	Bob computes $z' = h(z \parallel pub)$
3.	Bob computes $y = sig_K(z')$
4.	Bob publishes (z, pub, y) in the next day's newspaper.

exponent a in the **DSS**. Then, of course, Oscar can forge Bob's signature on any message he likes. But another (perhaps even more serious) problem is that the compromise of a signing algorithm calls in to question the authenticity of all messages signed by Bob, including those he signed before Oscar stole the signing algorithm.

Here is yet another undesirable situation that could arise: Suppose Bob signs a message and later wishes to disavow it. Bob might publish his signing algorithm and then claim that his signature on the message in question is a forgery.

The reason these types of events can occur is that there is no way to determine when a message was signed. This suggests that we consider ways of *timestamping* a (signed) message. A timestamp should provide proof that a message was signed at a particular time. Then, if Bob's signing algorithm is compromised, it would not invalidate any signatures he made previously. This is similar conceptually to the way credit cards work: if someone loses a credit card and notifies the bank that isssued it, it becomes invalid. But purchases made prior to the loss of the card are not affected.

In this section, we will describe a few methods of timestamping. First, we observe that Bob can produce a convincing timestamp on his own. First, Bob obtains some "current" publicly available information which could not have been predicted before it happened. For example, such information might consist of all the major league baseball scores from the previous day, or the values of all the stocks listed on the New York Stock Exchange. Denote this information by pub.

Now, suppose Bob wants to timestamp his signature on a message x. We assume that h is a publicly known hash function. Bob will proceed according to the algorithm presented in Figure 7.11. Here is how the scheme works: The presence of the information pub means that Bob could not have produced y before the date in question. And the fact that y is published in the next day's newspaper proves that Bob did not compute y after the date in question. So Bob's signature y is bounded within a period of one day. Also observe that Bob does not reveal the message x in this scheme since only z is published. If necessary, Bob can prove that x was the message he signed and timestamped simply by revealing it.

FIGURE 7.12
Timestamping $(z_n, y_n, \mathrm{ID}_n)$

1.   The TSS computes $L_n = (t_{n-1}, \mathrm{ID}_{n-1}, z_{n-1}, y_{n-1}, h(L_{n-1}))$
2.   The TSS computes $C_n = (n, t_n, z_n, y_n, \mathrm{ID}_n, L_n)$
3.   The TSS computes $s_n = sig_{\mathrm{TSS}}(h(C_n))$
4.   The TSS sends $(C_n, s_n, \mathrm{ID}_{n+1})$ to ID_n.

It is also straightforward to produce timestamps if there is a trusted times-tamping service available (i.e., an electronic notary public). Bob can compute $z = h(x)$ and $y = sig_K(z)$ and then send (z, y) to the timestamping service, or TSS. The TSS will then append the date D and sign the triple (z, y, D). This works perfectly well provided that the signing algorithm of the TSS remains se-cure and provided that the TSS cannot be bribed to backdate timestamps. (Note also that this method establishes only that Bob signed a message before a certain time. If Bob also wanted to establish that he signed it after a certain date, he could incorporate some public information pub as in the previous method.)

If it is undesirable to trust the TSS unconditionally, the security can be in-creased by sequentially linking the messages that are timestamped. In such a scheme, Bob would send an ordered triple $(z, y, \mathrm{ID}(\mathrm{Bob}))$ to the TSS. Here z is the message digest of the message x; y is Bob's signature on z; and $\mathrm{ID}(\mathrm{Bob})$ is Bob's identifying information. The TSS will be timestamping a sequence of triples of this form. Denote by $(z_n, y_n, \mathrm{ID}_n)$ the nth triple to be timestamped by the TSS, and let t_n denote the time at which the nth request is made.

The TSS will timestamp the nth triple using the algorithm in Figure 7.12. The quantity L_n is "linking information" that ties the nth request to the previous one. (L_0 will be taken to be some predetermined dummy information to get the process started.)

Now, if challenged, Bob can reveal his message x_n, and then y_n can be veri-fied. Next, the signature s_n of the TSS can be verified. If desired, then ID_{n-1} or ID_{n+1} can be requested to produce their timestamps, $(C_{n-1}, s_{n-1}, \mathrm{ID}_n)$ and $(C_{n+1}, s_{n+1}, \mathrm{ID}_{n+2})$, respectively. The signatures of the TSS can be checked in these timestamps. Of course, this process can be continued as far as desired, backwards and/or forwards.

7.9 Notes and References

The discrete log hash function described in Section 7.4 is due to Chaum, van Heijst, and Pfitzmann [CVHP92]. A hash function that can be proved secure provided that a composite integer n cannot be factored is given by Gibson [GIB91] (see Exercise 7.4 for a description of this scheme).

The material on extending hash functions in Section 7.5 is based on Damgård [DA90]. Similar methods were discovered by Merkle [ME90].

For infomation concerning the construction of hash functions from private-key cryptosystems, see Preneel, Govaerts, and Vandewalle [PGV94].

The **MD4** hashing algorithm was presented in Rivest [RI91], and the **Secure Hash Standard** is described in [NBS93]. An attack against two of the three rounds of **MD4** is given by den Boer and Bossalaers [DBB92]. Other recently proposed hash functions include N-**hash** [MOI90] and **Snefru** [ME90A].

Timestamping is discussed in Haber and Stornetta [HS91] and Bayer, Haber, and Stornetta [BHS93].

A thorough survey of hashing techniques can be found in Preneel, Govaerts, and Vandewalle [PGV93].

Exercises

7.1 Suppose $h : X \to Y$ is a hash function. For any $y \in Y$, let

$$h^{-1}(y) = \{x : h(x) = y\}$$

and denote $s_y = |h^{-1}(y)|$. Define

$$N = |\{\{x_1, x_2\} : h(x_1) = h(x_2)\}|.$$

Note that N counts the number of unordered pairs in X that collide under h. Answer the following:

(a) Prove that

$$\sum_{y \in Y} s_y = |X|,$$

so the mean of the s_y's is

$$\bar{s} = \frac{|X|}{|Y|}.$$

(b) Prove that

$$N = \sum_{y \in Y} \binom{s_y}{2} = \frac{1}{2} \sum_{y \in Y} s_y{}^2 - \frac{|X|}{2}.$$

(c) Prove that

$$\sum_{y \in Y} (s_y - \bar{s})^2 = 2N + |X| - \frac{|X|^2}{|Y|}.$$

FIGURE 7.13
Hashing $4m$ bits to m bits

---

1. write $x \in (\mathbb{Z}_2)^{4m}$ as $x = x_1 \| x_2$, where $x_1, x_2 \in (\mathbb{Z}_2)^{2m}$
2. define $h_2(x) = h_1(h_1(x_1) \| h_1(x_2))$.

(d) Using the result proved in part (c), prove that

$$N \geq \frac{1}{2} \left(\frac{|X|^2}{|Y|} - |X| \right).$$

Further, show that equality is attained if and only if

$$s_y = \frac{|X|}{|Y|}$$

for every $y \in Y$.

7.2 As in Exercise 7.1, suppose $h : X \to Y$ is a hash function, and let

$$h^{-1}(y) = \{x : h(x) = y\}$$

for any $y \in Y$. Let ϵ denote the probability that $h(x_1) = h(x_2)$, where x_1 and x_2 are random (not necessarily distinct) elements of X. Prove that

$$\epsilon \geq \frac{1}{|Y|},$$

with equality if and only if

$$|h^{-1}(y)| = \frac{|X|}{|Y|}$$

for every $y \in Y$.

7.3 Suppose $p = 15083$, $\alpha = 154$ and $\beta = 2307$ in the **Chaum-van Heijst-Pfitzmann Hash Function**. Given the collision

$$\alpha^{7431} \beta^{5564} \equiv \alpha^{1459} \beta^{954} \pmod{p},$$

compute $\log_\alpha \beta$.

7.4 Suppose $n = pq$, where p and q are two (secret) distinct large primes such that $p = 2p_1 + 1$ and $q = 2q_1 + 1$, where p_1 and q_1 are prime. Suppose that α is an element of order $2p_1 q_1$ in $\mathbb{Z}_n^*$ (this is the largest order of any element in $\mathbb{Z}_n^*$). Define a hash function $h : \{1, \ldots, n^2\} \to \mathbb{Z}_n^*$ by the rule $h(x) = \alpha^x \bmod n$.

Now, suppose that $n = 603241$ and $\alpha = 11$ are used to define a hash function h of this type. Suppose that we are given three collisions for h: $h(1294755) = h(80115359) = h(52738737)$. Use this information to factor n.

7.5 Suppose $h_1 : (\mathbb{Z}_2)^{2m} \to (\mathbb{Z}_2)^m$ is a strongly collision-free hash function.

    (a) Define $h_2 : (\mathbb{Z}_2)^{4m} \to (\mathbb{Z}_2)^m$ as in Figure 7.13. Prove that h_2 is strongly collision-free.

    (b) For an integer $i \geq 2$, define a hash function $h_i : (\mathbb{Z}_2)^{2^i m} \to (\mathbb{Z}_2)^m$ recursively from h_{i-1}, as indicated in Figure 7.14. Prove that h_i is strongly collision-free.

7.6 Using the (original) expansion function of the **SHS**, Equation 7.1, express each of $X[16], \ldots, X[79]$ in terms of $X[0], \ldots, X[15]$. Now, for each pair $X[i], X[j]$,

FIGURE 7.14
Hashing $2^i m$ bits to m bits

---

1. write $x \in (\mathbb{Z}_2)^{2^i m}$ as $x = x_1 \parallel x_2$, where $x_1, x_2 \in (\mathbb{Z}_2)^{2^{i-1} m}$
2. define $h_i(x) = h_1(h_{i-1}(x_1) \parallel h_{i-1}(x_2))$.

---

where $1 \le i < j \le 15$, use a computer program to determine λ_{ij}, which denotes the number of $X[k]$'s ($16 \le k \le 79$) such that $X[i]$ and $X[j]$ both occur in the expression for $X[k]$. What is the range of values λ_{ij}?

8

Key Distribution and Key Agreement

8.1 Introduction

We have observed that public-key systems have the advantage over private-key systems that a secure channel is not needed to exchange a secret key. But, unfortunately, most public-key systems are much slower than private-key systems such as **DES**, for example. So, in practice, private-key systems are usually used to encrypt "long" messages. But then we come back to the problem of exchanging secret keys.

In this chapter, we discuss several approaches to the problem of establishing secret keys. We will distinguish between key distribution and key agreement. *Key distribution* is defined to be a mechanism whereby one party chooses a secret key and then transmits it to another party or parties. *Key agreement* denotes a protocol whereby two (or more) parties jointly establish a secret key by communicating over a public channel. In a key agreement scheme, the value of the key is determined as a function of inputs provided by both parties.

As our setting, we have an insecure network of n users. In some of our schemes, we will have a *trusted authority* (denoted by TA) that is reponsible for such things as verifying the identities of users, choosing and transmitting keys to users, etc.

Since the network is insecure, we need to protect against potential opponents. Our opponent, Oscar, might be a *passive adversary*, which means that his actions are restricted to eavesdropping on messages that are transmitted over the channel. On the other hand, we might want to guard against the possibility that Oscar is an *active adversary*. An active adversary can do various types of nasty things such as the following:

1. alter messages that he observes being transmitted over the network

2. save messages for reuse at a later time

3. attempt to masquerade as various users in the network.

The objective of an active adversary might be one of the following:

1. to fool U and V into accepting an "invalid" key as valid (an invalid key could be an old key that has expired, or a key chosen by the adversary, to mention two possibilities)

2. to make U or V believe that they have exchanged a key with other when they have not.

The objective of a key distribution or key agreement protocol is that, at the end of the protocol, the two parties involved both have possession of the same key K, and the value of K is not known to any other party (except possibly the TA). Certainly it is much more difficult to design a protocol providing this type of security in the presence of an active adversary as opposed to a passive one.

We first consider the idea of *key predistribution* in Section 8.2. For every pair of users $\{U, V\}$, the TA chooses a random key $K_{U,V} = K_{V,U}$ and transmits it "off-band" to U and V over a secure channel. (That is, the transmission of keys does not take place over the network, since the network is not secure.) This approach is unconditionally secure, but it requires a secure channel between the TA and every user in the network. But, of possibly even more significance is the fact that each user must store $n - 1$ keys, and the TA needs to transmit a total of $\binom{n}{2}$ keys securely (this is sometimes called the "n^2 problem"). Even for relatively small networks, this can become prohibitively expensive, and thus it is not really a practical solution.

In Section 8.2.1, we discuss an interesting unconditionally secure key predistribution scheme, due to Blom, that allows a reduction in the amount of secret information to be stored by the users in the network. We also present in Section 8.2.2 a computationally secure key predistribution scheme based on the discrete logarithm problem.

A more practical approach can be described as *on-line key distribution by TA*. In such a scheme, the TA acts as a *key server*. The TA shares a secret key K_U with every user U in the network. When U wishes to communicate with V, she requests a *session key* from the TA. The TA generates a session key K and sends it in encrypted form for U and V to decrypt. The well-known **Kerberos** system, which we describe in Section 8.3, is based on this approach.

If it is impractical or undesirable to have an on-line TA, then a common approach is to use a *key agreement protocol*. In a key agreement protocol, U and V jointly choose a key by communicating over a public channel. This remarkable idea is due to Diffie and Hellman, and (independently) to Merkle. We describe a few of the more popular key agreement protocols. A variation of the original protocol of Diffie and Hellman, modified to protect against an active adversary, is presented in Section 8.4.1. Two other interesting protocols are also discussed: the **MTI** scheme is presented in Section 8.4.2 and the **Girault** scheme is covered in Section 8.4.3.

8.2   Key Predistribution

In the basic method, the TA generates $\binom{n}{2}$ keys, and gives each key to a unique pair of users in a network of n users. As mentioned above, we require a secure channel between the TA and each user to transmit these keys. This is a significant improvement over each pair of users independently exchanging keys over a secure channel, since the number of secure channels required has been reduced from $\binom{n}{2}$ to n. But if n is large, this solution is not very practical, both in terms of the amount of information to be transmitted securely, and in the amount of information that each user must store securely (namely, the secret keys of the other other $n-1$ users).

Thus, it is of interest to try to reduce the amount of information that needs to be transmitted and stored, while still allowing each pair of users U and V to be able to (independently) compute a secret key $K_{U,V}$. An elegant scheme to accomplish this, called the **Blom Key Predistribution Scheme**, is discussed in the next subsection.

8.2.1   Blom's Scheme

As above, we suppose that we have a network of n users. For convenience, we suppose that keys are chosen from a finite field $\mathbb{Z}_p$, where $p \geq n$ is prime. Let k be an integer, $1 \leq k \leq n-2$. The value k is the largest size coalition against which the scheme will remain secure. In the **Blom Scheme**, the TA will transmit $k+1$ elements of $\mathbb{Z}_p$ to each user over a secure channel (as opposed to $n-1$ in the basic key predistribution scheme). Each pair of users, U and V, will be able to compute a key $K_{U,V} = K_{V,U}$, as before. The security condition is as follows: any set of at most k users disjoint from $\{U, V\}$ must be unable to determine any information about $K_{U,V}$ (note that we are speaking here about unconditional security).

We first present the special case of Blom's scheme where $k = 1$. Here, the TA will transmit two elements of $\mathbb{Z}_p$ to each user over a secure channel, and any individual user W will be unable to determine any information about $K_{U,V}$ if $W \neq U, V$. Blom's scheme is presented in Figure 8.1. We illustrate the **Blom Scheme** with $k = 1$ in the following example.

Example 8.1
Suppose the three users are U, V and W, $p = 17$, and their public elements are $r_U = 12$, $r_V = 7$ and $r_W = 1$. Suppose that the TA chooses $a = 8$, $b = 7$ and $c = 2$, so the polynomial f is

$$f(x, y) = 8 + 7(x + y) + 2xy.$$

The g polynomials are as follows:

$$g_U(x) = 7 + 14x$$

FIGURE 8.1
Blom Key Distribution Scheme ($k = 1$)

---

1. A prime number p is made public, and for each user U, an element $r_U \in \mathbb{Z}_p$ is made public. The elements r_U must be distinct.

2. The TA chooses three random elements $a, b, c \in \mathbb{Z}_p$ (not necessarily distinct), and forms the polynomial

$$f(x, y) = a + b(x + y) + cxy \bmod p.$$

3. For each user U, the TA computes the polynomial

$$g_U(x) = f(x, r_U) \bmod p$$

and transmits $g_U(x)$ to U over a secure channel. Note that $g_U(x)$ is a linear polynomial in x, so it can be written as

$$g_U(x) = a_U + b_U x,$$

where

$$a_U = a + br_U \bmod p$$

and

$$b_U = b + cr_U \bmod p.$$

4. If U and V want to communicate, then they use the common key

$$K_{U,V} = K_{V,U} = f(r_U, r_V) = a + b(r_U + r_V) + cr_U r_V \bmod p,$$

where U computes $K_{U,V}$ as

$$f(r_U, r_V) = g_U(r_V)$$

and V computes $K_{U,V}$ as

$$f(r_U, r_V) = g_V(r_U).$$

$$g_V(x) = 6 + 4x$$
$$g_W(x) = 15 + 9x.$$

The three keys are thus

$$K_{U,V} = 3$$
$$K_{U,W} = 4$$
$$K_{V,W} = 10.$$

U would compute $K_{U,V}$ as

$$g_U(r_V) = 7 + 14 \times 7 \bmod 17 = 3$$

V would compute $K_{U,V}$ as

$$g_V(r_U) = 6 + 4 \times 12 \bmod 17 = 3.$$

We leave the computation of the other keys as an exercise for the reader.   ⬚

We now prove that no one user can determine any information about the key of two other users.

THEOREM 8.1
*The **Blom Scheme** with $k = 1$ is unconditionally secure against any individual user.*

PROOF   Let's suppose that user W wants to try to compute the key

$$K_{U,V} = a + b(r_U + r_V) + c r_U r_V \bmod p.$$

The values r_U and r_V are public, but a, b and c are unknown. W does know the values

$$a_W = a + b r_W \bmod p$$

and

$$b_W = b + c r_W \bmod p$$

since these are the coefficients of the polynomial $g_W(x)$ that was sent to W by the TA.
   What we will do is show that the information known by W is consistent with any possible value $\ell \in \mathbb{Z}_p$ of the key $K_{U,V}$. Hence, W cannot rule out any values for $K_{U,V}$. Consider the following matrix equation (in $\mathbb{Z}_p$):

$$\begin{pmatrix} 1 & r_U + r_V & r_U r_V \\ 1 & r_W & 0 \\ 0 & 1 & r_W \end{pmatrix} \begin{pmatrix} a \\ b \\ c \end{pmatrix} = \begin{pmatrix} \ell \\ a_W \\ b_W \end{pmatrix}.$$

The first equation represents the hypothesis that $K_{U,V} = \ell$; the second and third equations contain the information that W knows about a, b and c from $g_W(x)$.

The determinant of the coefficient matrix is

$$r_W^2 + r_U r_V - (r_U + r_V)r_W = (r_W - r_U)(r_W - r_V),$$

where all arithmetic is done in $\mathbb{Z}_p$. Since $r_W \neq r_U$ and $r_W \neq r_V$, it follows that the coefficient matrix has non-zero determinant, and hence the matrix equation has a unique solution for a, b, c. In other words, any possible value ℓ of $K_{U,V}$ is consistent with the information known to W. $\blacksquare$

On the other hand, a coalition of two users, say $\{W, X\}$, will be able to determine any key $K_{U,V}$ where $\{W, X\} \cap \{U, V\} = \emptyset$. W and X together know that

$$a_W = a + br_W$$
$$b_W = b + cr_W$$
$$a_X = a + br_X$$
$$b_X = b + cr_X.$$

Thus they have four equations in three unknowns, and they can easily compute a unique solution for a, b and c. Once they know a, b and c, they can form the polynomial $f(x, y)$ and compute any key they wish.

It is straightforward to generalize the scheme to remain secure against coalitions of size k. The only thing that changes is step 2. The TA will use a polynomial $f(x, y)$ having the form

$$f(x, y) = \sum_{i=0}^{k} \sum_{j=0}^{k} a_{i,j} x^i y^j \bmod p,$$

where $a_{i,j} \in \mathbb{Z}_p$ ($0 \leq i \leq k, 0 \leq j \leq k$), and $a_{i,j} = a_{j,i}$ for all i, j. The remainder of the protocol is unchanged.

8.2.2 Diffie-Hellman Key Predistribution

In this section, we describe a key predistribution scheme that is a modification of the well-known Diffie-Hellman key exchange protocol that we will discuss a bit later, in Section 8.4. We call this the **Diffie-Hellman Key Predistribution Scheme**. The scheme is computationally secure provided a problem related to the **Discrete Logarithm** problem is intractible.

We will describe the scheme over $\mathbb{Z}_p$, where p is prime, though it can be implemented in any finite group in which the **Discrete Logarithm** problem is intractible. We will assume that α is a primitive element of $\mathbb{Z}_p$, and that the values p and α are publicly known to everyone in the network.

In this scheme, ID(U) will denote certain identification information for each user U in the network, e.g., his or her name, e-mail address, telephone number, or other relevant information. Also, each user U has a secret exponent a_U (where $0 \leq a_U \leq p - 2$), and a corresponding public value

$$b_U = \alpha^{a_U} \bmod p.$$

The TA will have a signature scheme with a (public) verification algorithm ver_{TA} and a secret signing algorithm sig_{TA}. Finally, we will implicitly assume that all information is hashed, using a public hash function, before it is signed. To make the procedures easier to read, we will not include the necessary hashing in the description of the protocols.

Certain information pertaining to a user U will be authenticated by means of a *certificate* which is issued and signed by the TA. Each user U will have a certificate

$$C(U) = (ID(U), b_U, sig_{TA}(ID(U), b_U)),$$

where b_U is formed as described above (note that the TA does not need to know the value of a_U). A certificate for a user U will be issued when U joins the network. Certificates can be stored in a public database, or each user can store his or her own certificate. The signature of the TA on a certificate allows anyone in the network to verify the information it contains.

It is very easy for U and V to compute the common key

$$K_{U,V} = \alpha^{a_U a_V} \bmod p,$$

as shown in Figure 8.2.

We illustrate the algorithm with a small example.

Example 8.2
Suppose $p = 25307$ and $\alpha = 2$ are publicly known (p is prime and α is a primitive root modulo p). Suppose U chooses $a_U = 3578$. Then she computes

$$b_U = \alpha^{a_U} \bmod p$$
$$= 2^{3578} \bmod 25307$$
$$= 6113,$$

which is placed on her certificate. Suppose V chooses $a_V = 19956$. Then he computes

$$b_V = \alpha^{a_V} \bmod p$$
$$= 2^{19956} \bmod 25307$$
$$= 7984,$$

FIGURE 8.2
Diffie-Hellman Key Predistribution

---

1. A prime p and a primitive element $\alpha \in \mathbb{Z}_p^*$ are made public.
2. V computes

$$K_{U,V} = \alpha^{a_U a_V} \bmod p = b_U^{\,a_V} \bmod p,$$

using the public value b_U from U's certificate, together with his own secret value a_V.

3. U computes

$$K_{U,V} = \alpha^{a_U a_V} \bmod p = b_V^{\,a_U} \bmod p,$$

using the public value b_V from V's certificate, together with her own secret value a_U.

---

which is placed on his certificate.

Now U can compute the key

$$\begin{aligned}
K_{U,V} &= b_V^{\,a_U} \bmod p \\
&= 7984^{3578} \bmod 25307 \\
&= 3694,
\end{aligned}$$

and V can compute the same key

$$\begin{aligned}
K_{U,V} &= b_U^{\,a_V} \bmod p \\
&= 6113^{19956} \bmod 25307 \\
&= 3694.
\end{aligned}$$

$\square$

Let us think about the security of this scheme in the presence of a passive or active adversary. The signature of the TA on users' certificates effectively prevents W from altering any information on someone else's certificate. Hence we need only worry about passive attacks. So the pertinent question is: Can a user W compute $K_{U,V}$ if $W \neq U, V$? In other words, given $\alpha^{a_U} \bmod p$ and $\alpha^{a_V} \bmod p$ (but not a_U nor a_V), is it feasible to compute $\alpha^{a_U a_V} \bmod p$? This problem is called the **Diffie-Hellman** problem, and it is formally defined (using

FIGURE 8.3
The Diffie-Hellman problem

Problem Instance $I = (p, \alpha, \beta, \gamma)$, where p is prime, $\alpha \in \mathbb{Z}_p^*$ is a primitive element, and $\beta, \gamma \in \mathbb{Z}_p^*$.

Objective   Compute $\beta^{\log_\alpha \gamma} \bmod p \, (= \gamma^{\log_\alpha \beta} \bmod p)$.

---

an equivalent but slightly different presentation) in Figure 8.3. It is clear that **Diffie-Hellman Key Predistribution** is secure against a passive adversary if and only if the **Diffie-Hellman** problem is intractible.

If W could determine a_U from b_U, or if he could determine a_V from b_V, then he could compute $K_{U,V}$ exactly as U (or V) does. But both these computations are instances of the **Discrete Log** problem. So, provided that the **Discrete Log** problem in $\mathbb{Z}_p$ is intractible, **Diffie-Hellman Key Predistribution** is secure against this particular type of attack. However, it is an unproven conjecture that any algorithm that solves the **Diffie-Hellman** problem could also be used to solve the **Discrete Log** problem. (This is very similar to the situation with **RSA**, where it is conjectured, but not proved, that breaking **RSA** is polynomially equivalent to factoring.)

By the remarks made above, the **Diffie-Hellman** problem is no more difficult than the **Discrete Log** problem. Although we cannot say precisely how difficult this problem is, we can relate its security to that of another cryptosystem we have already studied, namely the **ElGamal Cryptosystem**.

THEOREM 8.2
Breaking the **ElGamal Cryptosystem** *is equivalent to solving the* **Diffie-Hellman** *problem.*

PROOF   First we recall how **ElGamal** encryption and decryption work. The key is $K = (p, \alpha, a, \beta)$, where $\beta = \alpha^a \bmod p$ (a is secret and p, α, and β are public). For a (secret) random number $k \in \mathbb{Z}_{p-1}$,

$$e_K(x, k) = (y_1, y_2),$$

where

$$y_1 = \alpha^k \bmod p$$

and

$$y_2 = x\beta^k \bmod p.$$

For $y_1, y_2 \in \mathbb{Z}_p^*$,

$$d_K(y_1, y_2) = y_2(y_1{}^a)^{-1} \bmod p.$$

Suppose we have an algorithm **A** to solve the **Diffie-Hellman** problem, and we are given an **ElGamal** encryption (y_1, y_2). We will apply the algorithm **A** with inputs p, α, y_1, and β. Then, we obtain the value

$$\mathbf{A}(p, \alpha, y_1, \beta) = \mathbf{A}(p, \alpha, \alpha^k, \alpha^a)$$
$$= \alpha^{ka} \bmod p$$
$$= \beta^k \bmod p.$$

Then, the decryption of (y_1, y_2) can easily be computed as

$$x = y_2(\beta^k)^{-1} \bmod p.$$

Conversely, suppose we have an algorithm **B** that performs **ElGamal** decryption. That is, **B** takes as inputs p, α, β, y_1, and y_2, and computes the quantity

$$x = y_2(y_1^{\log_\alpha \beta})^{-1} \bmod p.$$

Now, given inputs p, α, β, and γ for the **Diffie-Hellman** problem, it is easy to see that

$$\mathbf{B}(p, \alpha, \beta, \gamma, 1)^{-1} = 1((\gamma^{\log_\alpha \beta})^{-1})^{-1} \bmod p$$
$$= \gamma^{\log_\alpha \beta} \bmod p,$$

as desired. $\blacksquare$

8.3 Kerberos

In the key predistribution methods we discussed in the previous section, each pair of users can compute one fixed key. If the same key is used for a long period of time, there is a danger that it might be compromised. Thus it is often preferable to use an on-line method in which a new session key is produced every time a pair of users want to communicate (this property is called *key freshness*).

If on-line key distribution is used, there is no need for any network user to store keys to communicate with other users (each user will share a key with the TA, however). Session keys will be transmitted on request by the TA. It is the responsibility of the TA to ensure key freshness.

Kerberos is a popular key serving system based on private-key cryptography. In this section, we give an overview of the protocol for issuing session keys in **Kerberos**. Each user U shares a secret **DES** key K_U with the TA. In the most recent version of **Kerberos** (version V), all messages to be transmitted are encrypted using cipher block chaining (CBC) mode, as described in Section 3.4.1.

FIGURE 8.4
Transmission of a session key using Kerberos

---

1. U asks the TA for a session key to communicate with V.

2. The TA chooses a random session key K, a timestamp T, and a lifetime L.

3. The TA computes

$$m_1 = e_{K_U}(K, \text{ID}(V), T, L)$$

and

$$m_2 = e_{K_V}(K, \text{ID}(U), T, L)$$

and sends m_1 and m_2 to U.

4. U uses the decryption function d_{K_U} to compute K, T, L, and $\text{ID}(V)$ from m_1. She then computes

$$m_3 = e_K(\text{ID}(U), T)$$

and sends m_3 to V along with the message m_2 she received from the TA.

5. V uses the decryption function d_{K_V} to compute K, T, L and $\text{ID}(U)$ from m_2. He then uses d_K to compute T and $\text{ID}(U)$ from m_3. He checks that the two values of T and the two values of $\text{ID}(U)$ are the same. If so, then V computes

$$m_4 = e_K(T + 1)$$

and sends it to U.

6. U decrypts m_4 using d_K and verifies that the result is $T + 1$.

As in Section 8.2.2, $\text{ID}(U)$ will denote public identification information for user U. When a request for a session key is sent to the TA, the TA will generate a new random session key K. Also, the TA will record the time at which the request is made as a *timestamp*, T, and specify the *lifetime*, L, during which K will be valid. That is, the session key K is to be regarded as a valid key from time T to time $T + L$. All this information is encrypted and transmitted to U and (eventually) to V. Before going into more details, we will present the protocol in Figure 8.4.

The information transmitted in the protocol is illustrated in the following diagram:

$$\text{TA} \quad \xrightarrow{\begin{array}{c} e_{K_U}(K, \text{ID(V)}, T, L) \\ e_{K_V}(K, \text{ID(U)}, T, L) \end{array}} \quad \text{U} \quad \xrightarrow{\begin{array}{c} e_K(\text{ID(U)}, T) \\ e_{K_V}(K, \text{ID(U)}, T, L) \end{array}} \quad \text{V}$$

$$\xleftarrow{\quad e_K(T+1) \quad}$$

We will now explain what is going on in the various steps of the protocol. Although we have no formal proof that **Kerberos** is "secure" against an active adversary, we can at least give some informal motivation of the features of the protocol.

As mentioned above, the TA generates K, T, and L in step 2. In step 3, this information, along with ID(V), is encrypted using the key K_U shared by U and the TA to form m_1. Also, K, T, L, and ID(U) are encrypted using the key K_V shared by V and the TA to form m_2. Both these encrypted messages are sent to U.

U can use her key to decrypt m_1, and thus obtain K, T, and L. She will verify that the current time is in the interval from T to $T + L$. She can also check that the session key K has been issued for her desired communicant V by verifying the information ID(V) decrypted from m_1.

Next, U will relay m_2 to V. As well, U will use the new session key K to encrypt T and ID(U) and send the resulting message m_3 to V.

When V receives m_2 and m_3 from U, he decrypts m_2 to obtain T, K, L and ID(U). Then he uses the new session key K to decrypt m_3 and he verifies that T and ID(U), as decrypted from m_2 and m_3, are the same. This ensures V that the session key encrypted within m_2 is the same key that was used to encrypt m_3. Then V uses K to encrypt $T + 1$, and sends the result back to U as message m_4.

When U receives m_4, she decrypts it using K and verifies that the result is $T + 1$. This ensures U that the session key K has been successfully transmitted to V, since K was needed in order to produce the message m_4.

It is important to note the different functions of the messages transmitted in this protocol. The messages m_1 and m_2 are used to provide secrecy in the transmission of the session key K. On the other hand, m_3 and m_4 are used to provide *key confirmation*, that is, to enable U and V to convince each other that they possess the same session key K. In most key distribution schemes, (session) key confirmation can be included as a feature if it is not already present. Usually this is done in a similar fashion as it is done in **Kerberos**, namely by using the new session key K to encrypt known quantities. In **Kerberos**, U uses K to encrypt ID(U) and T, which are already encrypted in m_2. Similarly, V uses K to encrypt $T + 1$.

The purpose of the timestamp T and lifetime L is to prevent an active adversary from storing "old" messages for retransmission at a later time (this is called a *replay attack*). This method works because keys are not accepted as valid once they have expired.

FIGURE 8.5
Diffie-Hellman Key Exchange

1. U chooses a_U at random, $0 \le a_U \le p - 2$.
2. U computes $\alpha^{a_U} \bmod p$ and sends it to V.
3. V chooses a_V at random, $0 \le a_V \le p - 2$.
4. V computes $\alpha^{a_V} \bmod p$ and sends it to U.
5. U computes
$$K = (\alpha^{a_V})^{a_U} \bmod p$$
and V computes
$$K = (\alpha^{a_U})^{a_V} \bmod p.$$

One of the drawbacks of **Kerberos** is that all the users in the network should have synchronized clocks, since the current time is used to determine if a given session key K is valid. In practice, it is very difficult to provide perfect synchronization, so some amount of variation in times must be allowed.

8.4   Diffie-Hellman Key Exchange

If we do not want to use an on-line key server, then we are forced to use a key agreement protocol to exchange secret keys. The first and best known key agreement protocol is **Diffie-Hellman Key Exchange**. We will assume that p is prime, α is a primitive element of $\mathbb{Z}_p$, and that the values p and α are publicly known. (Alternatively, they could be chosen by U and communicated to V in the first step of the protocol.) **Diffie-Hellman Key Exchange** is presented in Figure 8.5.

At the end of the protocol, U and V have computed the same key

$$K = \alpha^{a_U a_V} \bmod p.$$

This protocol is very similar to **Diffie-Hellman Key Predistribution** described earlier. The difference is that the exponents a_U and a_V of users U and V (respectively) are chosen anew each time the protocol is run, instead of being fixed. Also, in this protocol, both U and V are assured of key freshness, since the session key depends on both random exponents a_U and a_V.

8.4.1 The Station-to-station Protocol

Diffie-Hellman Key Exchange is supposed to look like this:

$$U \quad \xrightarrow{\alpha^{a_U}} \quad V$$
$$\xleftarrow{\alpha^{a_V}}$$

Unfortunately, the protocol is vulnerable to an active adversary who uses an *intruder-in-the-middle* attack. There is an episode of *The Lucy Show* in which Vivian Vance is having dinner in a restaurant with a date, and Lucille Ball is hiding under the table. Vivian and her date decide to hold hands under the table. Lucy, trying to avoid detection, holds hands with each of them and they think they are holding hands with each other.

An intruder-in-the-middle attack on the **Diffie-Hellman Key Exchange** protocol works in the same way. W will intercept messages between U and V and substitute his own messages, as indicated in the following diagram:

$$U \quad \xrightarrow{\alpha^{a_U}} \quad W \quad \xrightarrow{\alpha^{a'_U}} \quad V$$
$$\xleftarrow{\alpha^{a'_V}} \qquad \xleftarrow{\alpha^{a_V}}$$

At the end of the protocol, U has actually established the secret key $\alpha^{a_U a'_V}$ with W, and V has established a secret key $\alpha^{a'_U a_V}$ with W. When U tries to encrypt a message to send to V, W will be able to decrypt it but V will not. (A similar situation holds if V sends a message to U.)

Clearly, it is essential for U and V to make sure that they are exchanging messages with each other and not with W. Before exchanging keys, U and V might carry out a separate protocol to establish each other's identity, for example by using one of the identification schemes that we will describe in Chapter 9. But this offers no protection against an intruder-in-the-middle attack if W simply remains inactive until after U and V have proved their identities to each other. Hence, the key agreement protocol should itself authenticate the participants' identities at the same time as the key is being established. Such a protocol will be called *authenticated key agreement*.

We will describe an authenticated key agreement protocol which is a modification of **Diffie-Hellman Key Exchange**. The protocol assumes a publicly known prime p and a primitive element α, and it makes use of certificates. Each user U will have a signature scheme with verification algorithm ver_U and signing algorithm sig_U. The TA also has a signature scheme with public verification algorithm ver_{TA}. Each user U has a certificate

$$\mathbf{C}(U) = (ID(U), ver_U, sig_{TA}(ID(U), ver_U)),$$

where $ID(U)$ is identification information for U.

FIGURE 8.6
Simplified Station-to-station Protocol

---

1. U chooses a random number a_U, $0 \leq a_U \leq p - 2$.
2. U computes
$$\alpha^{a_U} \bmod p$$
   and sends it to V.
3. V chooses a random number a_V, $0 \leq a_V \leq p - 2$.
4. V computes
$$\alpha^{a_V} \bmod p.$$
   Then he computes
$$K = (\alpha^{a_U})^{a_V} \bmod p$$
   and
$$y_V = sig_V(\alpha^{a_V}, \alpha^{a_U}).$$
4. V sends $(\mathbf{C}(V), \alpha^{a_V}, y_V)$ to U.
5. U computes
$$K = (\alpha^{a_V})^{a_U} \bmod p.$$
   She verifies y_V using ver_V and she verifies $\mathbf{C}(V)$ using ver_{TA}.
6. U computes
$$y_U = sig_U(\alpha^{a_U}, \alpha^{a_V})$$
   and she sends $(\mathbf{C}(U), y_U)$ to V.
7. V verifies y_U using ver_U and he verifies $\mathbf{C}(U)$ using ver_{TA}.

The authenticated key agreement known as the **Station-to-station Protocol** (or **STS** for short) is due to Diffie, Van Oorschot, and Wiener. The protocol we present in Figure 8.6 is a slight simplification; it can be used in such a way that it is conformant with the ISO 9798-3 protocols.

The information exchanged in the simplified **STS protocol** (excluding certificates) is illustrated as follows:

$$
\begin{array}{ccc}
 & \xrightarrow{\quad \alpha^{a_U} \quad} & \\
U & \xleftarrow{\quad \alpha^{a_V}, sig_V(\alpha^{a_V}, \alpha^{a_U}) \quad} & V \\
 & \xrightarrow{\quad sig_U(\alpha^{a_U}, \alpha^{a_V}) \quad} &
\end{array}
$$

Let's see how this protects against an intruder-in-the-middle attack. As before, W will intercept α^{a_U} and replace it with $\alpha^{a'_U}$. W then receives $\alpha^{a_V}, sig_V(\alpha^{a_V}, \alpha^{a'_U})$ from V. He would like to replace α^{a_V} with $\alpha^{a'_V}$, as before. However, this means that he must also replace $sig_V(\alpha^{a_V}, \alpha^{a'_U})$ by $sig_V(\alpha^{a'_V}, \alpha^{a_U})$. Unfortunately for W, he cannot compute V's signature on $(\alpha^{a'_V}, \alpha^{a_U})$ since he doesn't know V's signing algorithm sig_V. Similarly, W is unable to replace $sig_U(\alpha^{a_U}, \alpha^{a'_V})$ by $sig_U(\alpha^{a'_U}, \alpha^{a_V})$ because he does not know U's signing algorithm.

This is illustrated in the following diagram:

$$
\begin{array}{ccccc}
& \xrightarrow{\qquad \alpha^{a_U} \qquad} & & \xrightarrow{\qquad \alpha^{a'_U} \qquad} & \\
U & \xleftarrow{\alpha^{a'_V}, sig_V(\alpha^{a'_V}, \alpha^{a_U}) = ?} & W & \xleftarrow{\alpha^{a_V}, sig_V(\alpha^{a_V}, \alpha^{a'_U})} & V \\
& \xrightarrow{\qquad sig_U(\alpha^{a_U}, \alpha^{a'_V}) \qquad} & & \xrightarrow{sig_U(\alpha^{a'_U}, \alpha^{a_V}) = ?} &
\end{array}
$$

It is the use of signatures that thwarts the intruder-in-the-middle attack.

The protocol, as described in Figure 8.6, does not provide key confirmation. However, it is easy to modify so that it does, by defining

$$ y_V = e_K(sig_V(\alpha^{a_V}, \alpha^{a_U})) $$

in step 4 and defining

$$ y_U = e_K(sig_U(\alpha^{a_U}, \alpha^{a_V})) $$

in step 6. (As in **Kerberos**, we obtain key confirmation by encrypting a known quantity using the new session key.) The resulting protocol is known as the **Station-to-station Protocol**. We leave the remaining details for the interested reader to fill in.

8.4.2 MTI Key Agreement Protocols

Matsumoto, Takashima, and Imai have constructed several interesting key agreement protocols by modifying **Diffie-Hellman Key Exchange**. These protocols, which we call **MTI** protocols, do not require that U and V compute any signatures. They are *two-pass protocols* since there are only two separate transmissions of information performed (one from U to V and one from V to U). In contrast, the **STS** protocol is a three-pass protocol.

We present one of the **MTI** protocols. The setting for this protocol is the same as for **Diffie-Hellman Key Predistribution**. We assume a publicly known prime p and a primitive element α. Each user U has an ID string, ID(U), a secret exponent a_U ($0 \le a_U \le p - 2$), and a corresponding public value

$$ b_U = \alpha^{a_U} \bmod p. $$

The TA has a signature scheme with a (public) verification algorithm ver_{TA} and a secret signing algorithm sig_{TA}.

FIGURE 8.7
Matsumoto-Takashima-Imai Key Agreement Protocol

---

1.   U chooses r_U at random, $0 \leq r_U \leq p - 2$, and computes

$$s_U = \alpha^{r_U} \bmod p.$$

2.   U sends $(\mathbf{C}(U), s_U)$ to V.

3.   V chooses r_V at random, $0 \leq r_V \leq p - 2$, and computes

$$s_V = \alpha^{r_V} \bmod p.$$

4.   V sends $(\mathbf{C}(V), s_V)$ to U.

5.   U computes
$$K = s_V{}^{a_U} b_V{}^{r_U} \bmod p,$$

where she obtains the value b_V from $\mathbf{C}(V)$; and V computes

$$K = s_U{}^{a_V} b_U{}^{r_V} \bmod p,$$

where he obtains the value b_U from $\mathbf{C}(U)$.

Each user U will have a certificate

$$\mathbf{C}(U) = (\mathrm{ID}(U), b_U, sig_{\mathrm{TA}}(\mathrm{ID}(U), b_U)),$$

where b_U is formed as described above.

We present the MTI key agreement protocol in Figure 8.7. At the end of the protocol, U and V have both computed the same key

$$K = \alpha^{r_U a_V + r_V a_U} \bmod p.$$

We give an example to illustrate this protocol.

Example 8.3
Suppose $p = 27803$ and $\alpha = 5$ are publicly known. Assume U chooses $a_U = 21131$; then she will compute

$$b_U = 5^{21131} \bmod 27803 = 21420$$

which is placed on her certificate. As well, assume V chooses $a_V = 17555$. Then he will compute

$$b_V = 5^{17555} \bmod 27803 = 17100$$

which is placed on his certificate.

Now suppose that U chooses $r_U = 169$; then she will send the value

$$s_U = 5^{169} \bmod 27803 = 6268$$

to V. Suppose that V chooses $r_V = 23456$; then he will send the value

$$s_V = 5^{23456} \bmod 27803 = 26759$$

to U.

Now U can compute the key

$$K_{U,V} = s_V{}^{a_U} b_V{}^{r_U} \bmod p$$
$$= 26759^{21131} 17100^{169} \bmod 27803$$
$$= 21600,$$

and V can compute the key

$$K_{U,V} = s_U{}^{a_V} b_U{}^{r_V} \bmod p$$
$$= 6268^{17555} 21420^{23456} \bmod 27803$$
$$= 21600,$$

Thus U and V have computed the same key. □

The information transmitted during the protocol is depicted as follows:

$$\mathbf{C}(U), \alpha^{r_U} \bmod p$$
$$U \xrightarrow{\hspace{3cm}} V$$
$$\mathbf{C}(V), \alpha^{r_V} \bmod p$$
$$\xleftarrow{\hspace{3cm}}$$

Let's look at the security of the scheme. It is not too difficult to show that the security of the **MTI** protocol against a passive adversary is exactly the same as the **Diffie-Hellman** problem — see the exercises. As with many protocols, proving security in the presence of an active adversary is problematic. We will not attempt to prove anything in this regard, and we limit ourselves to some informal arguments.

Here is one threat we might consider: Without the use of signatures during the protocol, it might appear that there is no protection against an intruder-in-the-middle attack. Indeed, it is possible that W might alter the values that U and V send each other. We depict one typical scenario that might arise, as follows:

$$\mathbf{C}(U), \alpha^{r_U} \qquad\qquad \mathbf{C}(U), \alpha^{r'_U}$$
$$U \xrightarrow{\hspace{2.5cm}} W \xrightarrow{\hspace{2.5cm}} V$$
$$\mathbf{C}(V), \alpha^{r'_V} \qquad\qquad \mathbf{C}(V), \alpha^{r_V}$$
$$\xleftarrow{\hspace{2.5cm}} \qquad\qquad \xleftarrow{\hspace{2.5cm}}$$

In this situation, U and V will compute different keys: U will compute

$$K = \alpha^{r_U a_V + r'_V a_U} \bmod p.$$

while V will compute

$$K = \alpha^{r'_U a_V + r_V a_U} \bmod p.$$

However, neither of the key computations of U or V can be carried out by W, since they require knowledge of the secret exponents a_U and a_V, respectively. So even though U and V have computed different keys (which will of course be useless to them), neither of these keys can be computed by W (assuming the intractibility of the **Discrete Log** problem). In other words, both U and V are assured that the other is the only user in the network that could compute the key that they have computed. This property is sometimes called *implicit key authentication*.

8.4.3 Key Agreement Using Self-certifying Keys

In this section, we describe a method of key agreement, due to Girault, that does not require certificates. The value of a public key and the identity of its owner implicitly authenticate each other.

The **Girault Scheme** combines features of **RSA** and discrete logarithms. Suppose $n = pq$, where $p = 2p_1 + 1$, $q = 2q_1 + 1$, and p, q, p_1, and q_1 are all large primes. The multiplicative group $\mathbb{Z}_n^*$ is isomorphic to $\mathbb{Z}_p^* \times \mathbb{Z}_q^*$. The maximum order of any element in $\mathbb{Z}_n^*$ is therefore the least common multiple of $p - 1$ and $q - 1$, or $2p_1q_1$. Let α be an element of order $2p_1q_1$. Then the cyclic subgroup of $\mathbb{Z}_n^*$ generated by α is a suitable setting for the **Discrete Logarithm** problem.

In the **Girault Scheme**, the factorization of n is known only to the TA. The values n and α are public, but p, q, p_1, and q_1 are all secret. The TA chooses a public **RSA** encryption exponent, which we will denote by e. The corresponding decryption exponent, d, is secret (recall that $d = e^{-1} \bmod \phi(n)$).

Each user U has an ID string ID(U), as in previous schemes. A user U obtains a *self-certifying public key*, p_U, from the TA as indicated in Figure 8.8. Observe that U needs the help of the TA to produce p_U. Note also that

$$b_U = p_U{}^e + \text{ID(U)} \bmod n$$

can be computed from p_U and ID(U) using publicly available information.

The **Girault Key Agreement Protocol** is presented in Figure 8.9. The information transmitted during the protocol is depicted as follows:

$$U \quad \frac{\text{ID(U)}, p_U, \alpha^{r_U} \bmod n}{\text{ID(V)}, p_V, \alpha^{r_V} \bmod n} \quad V$$

FIGURE 8.8
Obtaining a self-certifying public key from the TA

---

1. U chooses a secret exponent a_U, and computes

$$b_U = \alpha^{a_U} \bmod n.$$

2. U gives a_U and b_U to the TA.
3. The TA computes

$$p_U = (b_U - \text{ID}(U))^d \bmod n.$$

4. The TA gives p_U to U.

At the end of the protocol, U and V each have computed the key

$$K = \alpha^{r_U a_V + r_V a_U} \bmod n.$$

Here is an example of key exchange using the **Girault Scheme**.

Example 8.4
Suppose $p = 839$ and $q = 863$. Then $n = 724057$ and $\phi(n) = 722356$. The element $\alpha = 5$ has order $2p_1 q_1 = \phi(n)/2$. Suppose the TA chooses $d = 125777$ as the **RSA** decryption exponent; then $e = 84453$.

Suppose U has $\text{ID}(U) = 500021$ and $a_U = 111899$. Then $b_U = 488889$ and $p_U = 650704$. Suppose also that V has $\text{ID}(V) = 500022$ and $a_V = 123456$. Then $b_V = 111692$ and $p_V = 683556$.

Now, U and V want to exchange a key. Suppose U chooses $r_U = 56381$, which means that $s_U = 171007$. Further, suppose V chooses $r_V = 356935$, which means that $s_V = 320688$.

Then both U and V will compute the same key $K = 42869$. ▯

Let's consider how the self-certifying keys guard against one specific type of attack. Since the values b_U, p_U, and $\text{ID}(U)$ are not signed by the TA, there is no way for anyone else to verify their authenticity directly. Suppose this information is forged by W (i.e., it is not produced in cooperation with the TA), who wants to masquerade as U. If W starts with $\text{ID}(U)$ and a fake value b'_U, then there is no way for her to compute the exponent a'_U corresponding to b'_U if the **Discrete Log** problem is intractible. Without a'_U, a key computation cannot be performed by W (who is pretending to be U).

FIGURE 8.9
Girault Key Agreement Protocol

---

1. U chooses r_U at random and computes

$$s_U = \alpha^{r_U} \bmod n.$$

2. U sends ID(U), p_U and s_U to V.
3. V chooses r_V at random and computes

$$s_V = \alpha^{r_V} \bmod n.$$

4. V sends ID(V), p_V and s_V to U.
5. U computes

$$K = s_V{}^{a_U} \left(p_V{}^e + \text{ID}(V)\right)^{r_U} \bmod n;$$

and V computes

$$K = s_U{}^{a_V} \left(p_U{}^e + \text{ID}(U)\right)^{r_V} \bmod n.$$

The situation is similar if W acts as an intruder-in-the-middle. W will be able to prevent U and V from computing a common key, but W is unable to duplicate the computations of either U or V. Thus the scheme provides implicit key authentication, as did the **MTI** protocol.

An attentive reader might wonder why U is required to supply the value a_U to the TA. Indeed, the TA can compute p_U directly from b_U, without knowing a_U. Actually, the important thing here is that the TA should be convinced that U knows the value of a_U before the TA computes p_U for U.

We illustrate this point by showing how the scheme can be attacked if the TA indiscriminately issues public keys p_U to users without first checking that they possess the value a_U corresponding to their b_U. Suppose W chooses a fake value a'_U and computes the corresponding value

$$b'_U = \alpha^{a'_U} \bmod n.$$

Here is how he can determine the corresponding public key

$$p'_U = (b'_U - \text{ID}(U))^d \bmod n.$$

W will compute

$$b'_W = b'_U - \text{ID}(U) + \text{ID}(W)$$

and then give b'_W and $\text{ID}(W)$ to the TA. Suppose the TA issues the public key

$$p'_W = (b'_W - \text{ID}(W))^d \bmod n$$

to W. Using the fact that

$$b'_W - \text{ID}(W) \equiv b'_U - \text{ID}(U) \pmod{n},$$

it is immediate that

$$p'_W = p'_U.$$

Now, at some later time, suppose U and V execute the protocol, and W substitutes information as follows:

$$U \quad \xrightarrow{\quad \text{ID}(U), p_U, \alpha^{r_U} \bmod n \quad} \quad W \quad \xrightarrow{\quad \text{ID}(U), p'_U, \alpha^{r'_U} \bmod n \quad} \quad V$$
$$\xleftarrow{\quad \text{ID}(V), p_V, \alpha^{r_V} \bmod n \quad} \qquad \xleftarrow{\quad \text{ID}(V), p_V, \alpha^{r_V} \bmod n \quad}$$

Now V will compute the key

$$K' = \alpha^{r'_U a_V + r_V a'_U} \bmod n,$$

whereas U will compute the key

$$K = \alpha^{r_U a_V + r_V a_U} \bmod n.$$

W can compute K' as

$$K' = s_V{}^{a'_U} (p_V{}^e + \text{ID}(V))^{r'_U} \bmod n.$$

Thus W and V share a key, but V thinks he is sharing a key with U. So W will be able to decrypt messages sent by V to U.

8.5 Notes and References

Blom presented his key predistribution scheme in [BL85]. Generalizations can be found in Blundo *et al.* [BDSHKVY93] and Beimel and Chor [BC94].

Diffie and Hellman presented their key exchange algorithm in [DH76]. The idea of key exchange was discovered independently by Merkle [ME78]. The material on authenticated key exchange is taken from Diffie, van Oorschot, and Wiener [DVW92].

Version V of **Kerberos** is described in [KN93]. For a recent descriptive article on **Kerberos**, see Schiller [SC94].

The protocols of Matsumoto, Takashima, and Imai can be found in [MTI86]. Self-certifying key distribution was introduced by Girault [GIR91]. The scheme he presented was actually a key predistribution scheme; modification to a key agreement scheme is based on [RV94].

Two recent surveys on key distribution and key agreement are Rueppel and Van Oorschot [RV94] and van Tilburg [vT93].

Exercises

8.1 Suppose the **Blom Scheme** with $k = 1$ is implemented for a set of four users, U, V, W and X. Suppose that $p = 7873$, $r_U = 2365$, $r_V = 6648$, $r_W = 1837$ and $r_X = 2186$. The secret g polynomials are as follows:

$$g_U(x) = 6018 + 6351x$$

$$g_V(x) = 3749 + 7121x$$

$$g_W(x) = 7601 + 7802x$$

$$g_X(x) = 635 + 6828x.$$

(a) Compute the key for each pair of users, verifying that each pair of users obtains a common key (that is, $K_{U,V} = K_{V,U}$, etc.).

(b) Show how W and X together can compute $K_{U,V}$.

8.2 Suppose the **Blom Scheme** with $k = 2$ is implemented for a set of five users, U, V, W, X and Y. Suppose that $p = 97$, $r_U = 14$, $r_V = 38$, $r_W = 92$, $r_X = 69$ and $r_Y = 70$. The secret g polynomials are as follows:

$$g_U(x) = 15 + 15x + 2x^2$$

$$g_V(x) = 95 + 77x + 83x^2$$

$$g_W(x) = 88 + 32x + 18x^2$$

$$g_X(x) = 62 + 91x + 59x^2$$

$$g_X(x) = 10 + 82x + 52x^2.$$

(a) Show how U and V each will compute the key $K_{U,V} = K_{V,U}$.

(b) Show how W, X and Y together can compute $K_{U,V}$.

8.3 Suppose that U and V carry out the **Diffie-Hellman Key Exchange** with $p = 27001$ and $\alpha = 101$. Suppose that U chooses $a_U = 21768$ and V chooses $a_V = 9898$. Show the computations performed by both U and V, and determine the key that they will compute.

8.4 Suppose that U and V carry out the **MTI Protocol** where $p = 30113$ and $\alpha = 52$. Suppose that U has $a_U = 8642$ and chooses $r_U = 28654$, and V has $a_V = 24673$ and chooses $r_V = 12385$. Show the computations performed by both U and V, and determine the key that they will compute.

8.5 If a passive adversary tries to compute the key K constructed by U and V by using the **MTI** protocol, then he is faced with an instance of what we might term the **MTI** problem, which we present in Figure 8.10.   Prove that any algorithm that can be used to solve the **MTI** problem can be used to solve the **Diffie-Hellman** problem, and vice versa.

8.6 Consider the **Girault Scheme** where $p = 167$, $q = 179$, and hence $n = 29893$. Suppose $\alpha = 2$ and $e = 11101$.

(a) Compute d.

(b) Given that $ID(U) = 10021$ and $a_U = 9843$, compute b_U and p_U. Given that $ID(V) = 10022$ and $a_V = 7692$, compute b_V and p_V.

(c) Show how b_U can be computed from p_U and $ID(U)$ using the public exponent e. Similarly, show how b_V can be computed from p_V and $ID(V)$.

FIGURE 8.10
The MTI problem

Problem Instance $I = (p, \alpha, \beta, \gamma, \delta, \epsilon)$, where p is prime, $\alpha \in \mathbb{Z}_p^*$ is a primitive element, and $\beta, \gamma, \delta, \epsilon \in \mathbb{Z}_p^*$.

Objective   Compute $\beta^{\log_\alpha \gamma} \delta^{\log_\alpha \epsilon} \bmod p$.

(d) Suppose that U chooses $r_U = 15556$ and V chooses $r_V = 6420$. Compute s_U and s_V, and show how U and V each compute their common key.

9

Identification Schemes

9.1 Introduction

Cryptographic methods enable many seemingly impossible problems to be solved. One such problem is the construction of secure identification schemes. There are many common, everyday situations where it is necessary to electronically "prove" one's identity. Some typical scenarios are as follows:

1. To withdraw money from an automated teller machine (or ATM), we use a card together with a four-digit personal identification number (PIN).

2. To charge purchases over the telephone to a credit card, all that is necessary is a credit card number (and the expiry date).

3. To charge long-distance telephone calls (using a calling card), one requires only a telephone number together with a four-digit PIN.

4. To do a remote login to a computer over a network, it suffices to know a valid user name and the corresponding password.

In practice, these types of schemes are not usually implemented in a secure way. In the protocols performed over the telephone, any eavesdropper can use the identifying information for their own purposes. This could include the person who is the recipient of the information; many credit card "scams" operate in this way. An ATM card is somewhat more secure, but there are still weaknesses. For example, someone monitoring the communication line can obtain all the information encoded on the card's magnetic strip, as well as the PIN. This could allow an imposter to gain access to a bank account. Finally, remote computer login is a serious problem due to the fact that user IDs and passwords are transmitted over the network in unencrypted form. Thus they are vulnerable to anyone who is monitoring the computer network.

The goal of an identification scheme is that someone "listening in" as Alice identifies herself to Bob, say, should not subsequently be able to misrepresent herself as Alice. Furthermore, we should try to guard against the possibility that

FIGURE 9.1
Challenge-and-response protocol

---

1. Bob chooses a *challenge*, x, which is a random 64-bit string. Bob sends x to Alice.

2. Alice computes
$$y = e_K(x)$$
and sends it to Bob.

3. Bob computes
$$y' = e_K(x)$$
and verifies that $y' = y$.

Bob himself might try to impersonate Alice after she has identified herself to him. In other words, Alice wants to be able to prove her identity electronically without "giving away" her identifying information.

Several such identification schemes have been discovered. One practical objective is to find a scheme that is simple enough that it can be implemented on a smart card, which is essentially a credit card equipped with a chip that can perform arithmetic computations. Hence, both the amount of computation and the memory requirements should be kept as small as possible. Such a card would be a more secure alternative to current ATM cards. However, it is important to note that the "extra" security pertains to someone monitoring the communication line. Since it is the card that is "proving" its identity, we have no extra protection against a lost card. It would still be necessary to include a PIN in order to establish that it is the real owner of the card who is initiating the identification protocol.

In later sections, we will describe some of the more popular identification schemes. But first, we give a very simple scheme that can be based on any private-key cryptosystem, e.g., **DES**. The protocol, which is described in Figure 9.1, is called a *challenge-and-response* protocol. In it, we assume that Alice is identifying herself to Bob, and Alice and Bob share a common secret key, K, which specifies an encryption function e_K.

We illustrate this protocol with a small example.

Example 9.1
Assume Alice and Bob use an encryption function which does a modular exponentiation:
$$e_K(x) = x^{101379} \bmod 167653.$$

Suppose Bob's challenge is $x = 77835$. Then Alice responds with $y = 100369$.
▯

Virtually all identification schemes are challenge-and-response protocols, but the most useful schemes do not require shared keys. This idea will be pursued in the remainder of the chapter.

9.2   The Schnorr Identification Scheme

We begin by describing the **Schnorr Identification Scheme**, which is one of the most attractive practical identification schemes. The scheme requires a trusted authority, which we denote by TA. The TA will choose parameters for the scheme as follows:

1. p is a large prime (i.e., $p \geq 2^{512}$) such that the discrete log problem in $\mathbb{Z}_p^*$ is intractible.

2. q is a large prime divisor of $p - 1$ (i.e., $q \geq 2^{140}$).

3. $\alpha \in \mathbb{Z}_p^*$ has order q (such an α can be computed as the $(p-1)/q$th power of a primitive element).

4. A *security parameter* t such that $q > 2^t$. For most practical applications, $t = 40$ will provide adequate security.

5. The TA also establishes a secure signature scheme with a secret signing algorithm sig_{TA} and a public verification algorithm ver_{TA}.

6. A secure hash function is specified. As usual, all information is to be hashed before it is signed. In order to make the protocols easier to read, we will omit the hashing steps from the descriptions of the protocols.

The parameters p, q, and α, the public verification algorithm ver_{TA} and the hash function are all made public.

A certificate will be issued to Alice by the TA. When Alice wants to obtain a certificate from the TA, the steps in Figure 9.2 are carried out. At a later time, when Alice wants to prove her identity to Bob, say, the protocol of Figure 9.3 is executed.

As mentioned above, t is a security parameter. Its purpose is to prevent an impostor posing as Alice, say Olga, from guessing Bob's challenge, r. For, if Olga guessed the correct value of r, she could choose any value for y and compute

$$\gamma = \alpha^y v^r \bmod p.$$

She would give Bob γ in step 1, and then when she receives the challenge r, she would supply the value y she has already chosen. Then γ would be verified by Bob in step 6.

FIGURE 9.2
Issuing a certificate to Alice

---

1. The TA establishes Alice's identity by means of conventional forms of identification such as a birth certificate, passport, etc. Then the TA forms a string ID(Alice) which contains her identification information.

2. Alice secretly chooses a random exponent a, where $0 \leq a \leq q - 1$. Alice computes

$$v = \alpha^{-a} \bmod p$$

and gives v to the TA.

3. The TA generates a signature

$$s = sig_{TA}(\text{ID}(\text{Alice}), v).$$

The certificate

$$\mathbf{C}(\text{Alice}) = (\text{ID}(\text{Alice}), v, s)$$

is given to Alice.

The probability that Olga will guess the value of r correctly is 2^{-t} if r is chosen at random by Bob. Thus, $t = 40$ should be a reasonable value for most applications. (But notice that Bob should choose his challenge r at random every time Alice identifies herself to him. If Bob always used the same challenge r, then Olga could impersonate Alice by the method described above.)

Basically, there are two things happening in the verification protocol. First, the signature s proves the validity of Alice's certificate. Thus Bob verifies the signature of the TA on Alice's certificate to convince himself that the certificate itself is authentic. This is essentially the same way that certificates were used in Chapter 8.

The second part of the protocol concerns the secret number a. The value a functions like a PIN in that it convinces Bob that the person carrying out the identification protocol is indeed Alice. But there is an important difference from a PIN: in the identification protocol, the value of a is not revealed. Instead Alice (or more accurately, Alice's smart card) "proves" that she/it knows the value of a in step 5 of the protocol by computing the value y in response to the challenge r issued by Bob. Since the value of a is not revealed, this technique is called a *proof of knowledge*.

FIGURE 9.3
The Schnorr identification scheme

---

1. Alice chooses a random number k, where $0 \leq k \leq q - 1$, and computes
$$\gamma = \alpha^k \bmod p.$$

2. Alice sends her certificate $\mathbf{C}(\text{Alice}) = (\text{ID}(\text{Alice}), v, s)$ and γ to Bob.

3. Bob verifies the signature of the TA by checking that $ver_{\text{TA}}(\text{ID}(\text{Alice}), v, s) = \text{true}$.

4. Bob chooses a random number r, $1 \leq r \leq 2^t$ and gives it to Alice.

5. Alice computes
$$y = k + ar \bmod q$$
and gives y to Bob.

6. Bob verifies that
$$\gamma \equiv \alpha^y v^r \pmod{p}.$$

The following congruences demonstrate that Alice will be able to prove her identity to Bob:

$$\alpha^y v^r \equiv \alpha^{k+ar} v^r \pmod{p}$$
$$\equiv \alpha^{k+ar} \alpha^{-ar} \pmod{p}$$
$$\equiv \alpha^k \pmod{p}$$
$$\equiv \gamma \pmod{p}.$$

Thus Bob will accept Alice's proof of identity (assuming he is honest), and the protocol is said to have the *completeness* property.

Here is a small (toy) example illustrating the challenge-and-response aspect of the protocol.

Example 9.2
Suppose $p = 88667$, $q = 1031$ and $t = 10$. The element $\alpha = 70322$ has order q in $\mathbb{Z}_p^{*}$. Suppose Alice's secret exponent is $a = 755$; then

$$v = \alpha^{-a} \bmod p$$
$$= 70322^{1031-755} \bmod 88667$$
$$= 13136.$$

Now suppose Alice chooses $k = 543$. Then she computes

$$\gamma = \alpha^k \bmod p$$
$$= 70322^{543} \bmod 88667$$
$$= 84109.$$

and sends γ to Bob. Suppose Bob issues the challenge $r = 1000$. Then Alice computes

$$y = k + ar \bmod q$$
$$= 543 + 755 \times 1000 \bmod 1031$$
$$= 851$$

and sends y to Bob. Bob then verifies that

$$84109 \equiv 70322^{851} 13136^{1000} \pmod{88667}.$$

So Bob believes that he is communicating with Alice. ▢

Next, let's consider how someone might try to impersonate Alice. An imposter, Olga, might try to impersonate Alice by forging a certificate

$$\mathbf{C}'(\text{Alice}) = (\text{ID}(\text{Alice}), v', s'),$$

where $v' \neq v$. But s' is supposed to be a signature of $(\text{ID}(\text{Alice}), v')$, and this is verified by Bob in step 3 of the protocol. If the signature scheme of the TA is secure, Olga will not be able to forge a signature s' which will subsequently be verified by Bob.

Another approach would be for Olga to use Alice's correct certificate, which is $\mathbf{C}(\text{Alice}) = (\text{ID}(\text{Alice}), v, s)$ (recall that certificates are not secret, and the information on a certificate is revealed each time the identification protocol is executed). But Olga will not be able to impersonate Alice unless she also knows the value of a. This is because of the "challenge" r in step 4. In step 5, Olga would have to compute y, but y is a function of a. The computation of a from v involves solving a discrete log problem, which we assume is intractible.

We can prove a more precise statement about the security of the protocol, as follows.

THEOREM 9.1
Suppose Olga knows a value γ for which she has probability $\epsilon \geq 1/2^{t-1}$ of successfully impersonating Alice in the verification protocol. Then Olga can compute a in polynomial time.

PROOF For a fraction ϵ of the 2^t possible challenges r, Olga can compute a value y which will be accepted in step 6 by Bob. Since $\epsilon \geq 1/2^{t-1}$, we have that $2^t \epsilon \geq 2$, and therefore Olga can compute values y_1, y_2, r_1 and r_2 such that

$$y_1 \not\equiv y_2 \pmod{q}$$

and

$$\gamma \equiv \alpha^{y_1} v^{r_1} \equiv \alpha^{y_2} v^{r_2} \pmod{p}.$$

It follows that

$$\alpha^{y_1 - y_2} \equiv v^{r_2 - r_1} \pmod{p}.$$

Since $v = \alpha^{-a}$, we have that

$$y_1 - y_2 \equiv a(r_1 - r_2) \pmod{q}.$$

Now, $0 < |r_2 - r_1| < 2^t$ and $q > 2^t$ is prime. Hence $\gcd(r_2 - r_1, q) = 1$, and Olga can compute

$$a = (y_1 - y_2)(r_1 - r_2)^{-1} \bmod q,$$

as desired.   ∎

The above theorem proves that anyone who has a non-negligible chance of successfully executing the identification protocol must know (or be able to compute in polynomial time) Alice's secret exponent a. This property is often referred to as *soundness*.

We illustrate with an example.

Example 9.3
Suppose we have the same parameters as in Example 9.2: $p = 88667, q = 1031$, $t = 10, \alpha = 70322, a = 755$ and $v = 13136$. Suppose Olga learns that

$$\alpha^{851} v^{1000} \equiv \alpha^{454} v^{19} \pmod{p}.$$

Then she can compute

$$a = (851 - 454)(1000 - 19)^{-1} \bmod 1031 = 755,$$

and thus discover Alice's secret exponent.   □

We have proved that the protocol is sound and complete. But soundness and completeness are not sufficient to ensure that the protocol is "secure." For example, if Alice simply revealed the value of her exponent a to prove her identity to Olga (say), the protocol would still be sound and complete. However, it would be completely insecure, since Olga could subsequently impersonate Alice.

This motivates the consideration of the secret information released to a verifier (or an observer) who takes part in the protocol (in this protocol, the secret information is the value of the exponent a). Our hope is that no information about a can be gained by Olga when Alice proves her identity, for then Olga would be able to masquerade as Alice.

In general, we could envision a situation whereby Alice proves her identity to Olga, say, on several different occasions. Perhaps Olga does not choose her challenges (i.e., the values of r) in a random way. After several executions of the protocol, Olga will try to determine the value of a so she can subsequently impersonate Alice. If Olga can determine no information about the value of a by taking part in a polynomial number of executions of the protocol and then performing a polynomial amount of computation, then we would be convinced that the protocol is secure.

It has not been proven that the **Schnorr Scheme** is secure. But in the next section, we present a modification of the **Schnorr Scheme,** due to Okamoto, that can be proved to be secure given a certain computational assumption.

The **Schnorr Scheme** was designed to be very fast and efficient, both from a computational point of view and in the amount of information that needs to be exchanged in the protocol. It is also designed to minimize the amount of computation performed by Alice. This is desirable because in many practical applications, Alice's computations will be performed by a smart card with low computing power, while Bob's computations will be performed by a more powerful computer.

For the purpose of discussion, let's assume that $\text{ID}(\text{Alice})$ is a 512-bit string. v also comprises 512 bits, and s will be 320 bits if the **DSS** is used as a signature scheme. The total size of the certificate $\mathbf{C}(\text{Alice})$ (which needs to be stored on Alice's smart card) is then 1344 bits.

Let us consider Alice's computations: step 1 requires a modular exponentiation to be performed; step 5 comprises one modular addition and one modular multiplication. It is the modular exponentiation that is computationally intensive, but this can be precomputed offline, if desired. The online computations to be performed by Alice are very modest.

It is also a simple matter to calculate the number of bits that are communicated during the protocol. We can depict the information that is communicated in the form of a diagram:

$$
\begin{array}{ccc}
 & \xrightarrow{\quad \mathbf{C}, \gamma \quad} & \\
\text{Alice} & \xleftarrow{\quad\quad r \quad\quad} & \text{Bob} \\
 & \xrightarrow{\quad\quad y \quad\quad} &
\end{array}
$$

Alice gives Bob $1344 + 512 = 1856$ bits of information in step 2; Bob gives Alice 40 bits in step 4; and Alice gives Bob 140 bits in step 6. So the communication requirements are quite modest, as well.

FIGURE 9.4
Issuing a certificate to Alice

---

1. The TA establishes Alice's identity and issues an identification string ID(Alice).

2. Alice secretly chooses two random exponents a_1, a_2, where $0 \leq a_1, a_2 \leq q - 1$. Alice computes

$$v = \alpha_1^{-a_1} \alpha_2^{-a_2} \bmod p$$

and gives v to the TA.

3. The TA generates a signature

$$s = sig_{TA}(\text{ID}(\text{Alice}), v).$$

The certificate

$$\mathbf{C}(\text{Alice}) = (\text{ID}(\text{Alice}), v, s)$$

is given to Alice.

9.3   The Okamoto Identification Scheme

In this section, we present a modification of the **Schnorr Scheme** due to Okamoto. This modification can be proved secure, assuming the intractibility of computing a particular discrete logarithm in $\mathbb{Z}_p$.

To set up the scheme, the TA chooses p and q as in the **Schnorr Scheme**. The TA also chooses two elements $\alpha_1, \alpha_2 \in \mathbb{Z}_p^*$ both having order q. The value $c = \log_{\alpha_1} \alpha_2$ is kept secret from all the participants, including Alice. We will assume that it is infeasible for anyone (even a coalition of Alice and Olga, say) to compute the value c. As before, the TA chooses a signature scheme and hash function. The certificate issued to Alice by the TA is constructed as described in Figure 9.4. The **Okamoto Identification Scheme** is presented in Figure 9.5.

Here is an example of the **Okamoto Scheme**.

Example 9.4
As in previous examples, we will take $p = 88667$, $q = 1031$, and $t = 10$. Suppose $\alpha_1 = 58902$ and $\alpha_2 = 73611$ (both α_1 and α_2 have order q in $\mathbb{Z}_p^*$). Now, suppose $a_1 = 846$ and $a_2 = 515$; then $v = 13078$.

FIGURE 9.5
The Okamoto identification scheme

---

1. Alice chooses random numbers k_1, k_2, where $0 \le k_1, k_2 \le q - 1$, and computes
$$\gamma = \alpha_1{}^{k_1} \alpha_2{}^{k_2} \bmod p.$$

2. Alice sends her certificate $C(\text{Alice}) = (\text{ID}(\text{Alice}), v, s)$ and γ to Bob.

3. Bob verifies the signature of the TA by checking that $ver_{\text{TA}}(\text{ID}(\text{Alice}), v, s) = \text{true}$.

4. Bob chooses a random number r, $1 \le r \le 2^t$ and gives it to Alice.

5. Alice computes
$$y_1 = k_1 + a_1 r \bmod q$$
and
$$y_2 = k_2 + a_2 r \bmod q$$
and gives y_1 and y_2 to Bob.

6. Bob verifies that
$$\gamma \equiv \alpha_1{}^{y_1} \alpha_2{}^{y_2} v^r \pmod{p}.$$

Suppose Alice chooses $k_1 = 899$ and $k_2 = 16$; then $\gamma = 14574$. If Bob issues the challenge $r = 489$ then Alice will respond with $y_1 = 131$ and $y_2 = 287$. Bob will verify that

$$58902^{131} 73611^{287} 13078^{489} \equiv 14574 \pmod{88667}.$$

So Bob will accept Alice's proof of identity. □

The proof that the protocol is complete (i.e., that Bob will accept Alice's proof of identity) is straightforward. The main difference between Okamoto's and Schnorr's scheme is that we can prove that the **Okamoto Scheme** is secure provided that the computation of the discrete logarithm $\log_{\alpha_1} \alpha_2$ is intractable.

The proof of security is quite subtle. Here is the general idea: As before, Alice identifies herself to Olga polynomially many times by executing the protocol. We then suppose (hoping to obtain a contradiction) that Olga is able to learn some information about the values of Alice's secret exponents a_1 and a_2. If this is so, then we will show that (with high probability) Alice and Olga together will be able to compute the discrete logarithm c in polynomial time. This contradicts the assumption made above, and proves that Olga must be unable to obtain any information about Alice's exponents by taking part in the protocol.

The first part of this procedure is similar to the soundness proof for the **Schnorr Scheme**.

THEOREM 9.2
Suppose Olga knows a value γ for which she has probability $\epsilon \geq 1/2^{t-1}$ of successfully impersonating Alice in the verification protocol. Then, in polynomial time, Olga can compute values b_1 and b_2 such that

$$v \equiv \alpha_1^{-b_1} \alpha_2^{-b_2} \pmod{p}.$$

PROOF   For a fraction ϵ of the 2^t possible challenges r, Olga can compute values y_1, y_2, z_1, z_2, r and s with $r \neq s$ and

$$\gamma \equiv \alpha_1^{y_1} \alpha_2^{y_2} v^r \equiv \alpha_1^{z_1} \alpha_2^{z_2} v^s \pmod{p}.$$

Define

$$b_1 = (y_1 - z_1)(r - s)^{-1} \bmod q$$

and

$$b_2 = (y_2 - z_2)(r - s)^{-1} \bmod q.$$

Then it is easy to check that

$$v \equiv \alpha_1^{-b_1} \alpha_2^{-b_2} \pmod{p},$$

as desired.   ∎

We now proceed to show how Alice and Olga can together compute the value of c.

THEOREM 9.3
Suppose Olga knows a value γ for which she has probability $\epsilon \geq 1/2^{t-1}$ of successfully impersonating Alice in the verification protocol. Then, with probability $1 - 1/q$, Alice and Olga can together compute $\log_{\alpha_1} \alpha_2$ in polynomial time.

PROOF   By Theorem 9.2, Olga is able to determine values b_1 and b_2 such that

$$v \equiv \alpha_1^{-b_1} \alpha_2^{-b_2} \pmod{p}.$$

Now suppose that Alice reveals the values a_1 and a_2 to Olga. Of course

$$v \equiv \alpha_1^{-a_1} \alpha_2^{-a_2} \pmod{p},$$

so it must be the case that

$$\alpha_1^{a_1 - b_1} \equiv \alpha_2^{b_2 - a_2} \pmod{p}.$$

Suppose that $(a_1, a_2) \neq (b_1, b_2)$. Then $(a_2 - b_2)^{-1} \bmod q$ exists, and the discrete log

$$c = \log_{\alpha_1} \alpha_2 = (a_1 - b_1)(b_2 - a_2)^{-1} \bmod q$$

can be computed in polynomial time.

There remains to be considered the possibility that $(a_1, a_2) = (b_1, b_2)$. If this happens, then the value of c cannot be computed as described above. However, we will argue that $(a_1, a_2) = (b_1, b_2)$ will happen only with very small probability $1/q$, so the procedure whereby Alice and Olga compute c will almost surely succeed.

Define

$$\mathcal{A} = \{(a_1', a_2') \in \mathbb{Z}_q \times \mathbb{Z}_q : \alpha_1^{-a_1'} \alpha_2^{-a_2'} \equiv \alpha_1^{-a_1} \alpha_2^{-a_2} \pmod{p}\}.$$

That is, $\mathcal{A}$ consists of all the possible ordered pairs that could be Alice's secret exponents. Observe that

$$\mathcal{A} = \{(a_1 - c\theta, a_2 + \theta) : \theta \in \mathbb{Z}_q\},$$

where $c = \log_{\alpha_1} \alpha_2$. Thus $\mathcal{A}$ consists of q ordered pairs.

The ordered pair (b_1, b_2) computed by Olga is certainly in the set $\mathcal{A}$. We will argue that the value of the pair (b_1, b_2) is independent of the value of the pair (a_1, a_2) that comprises Alice's secret exponents. Since (a_1, a_2) was originally chosen at random by Alice, it must be the case that the probability that $(a_1, a_2) = (b_1, b_2)$ is $1/q$.

So, we need to say what we mean by (b_1, b_2) being "independent" of (a_1, a_2). The idea is that Alice's pair (a_1, a_2) is one of the q possible ordered pairs in the set $\mathcal{A}$, and no information about which is the "correct" ordered pair is revealed by Alice identifying herself to Olga. (Stated informally, Olga knows that an ordered pair from $\mathcal{A}$ comprises Alice's exponents, but she has no way of telling which one.)

Let's look at the information that is exchanged during the identification protocol. Basically, in each execution of the protocol, Alice chooses a γ; Olga chooses an r; and Alice reveals y_1 and y_2 such that

$$\gamma \equiv \alpha_1^{y_1} \alpha_2^{y_2} v^r \pmod{p}.$$

Recall that Alice computes

$$y_1 = k_1 + a_1 r \bmod q$$

and

$$y_2 = k_2 + a_2 r \bmod q,$$

where

$$\gamma = \alpha_1^{k_1} \alpha_2^{k_2} \bmod p.$$

But note that k_1 and k_2 are not revealed (nor are a_1 and a_2).

The particular quadruple (γ, r, y_1, y_2) that is generated during one execution of the protocol appears to depend on Alice's ordered pair (a_1, a_2), since y_1 and y_2 are defined in terms of a_1 and a_2. But we will show that each such quadruple could equally well be generated from any other ordered pair $(a_1', a_2') \in A$. To see this, suppose $(a_1', a_2') \in A$, i.e., $a_1' = a_1 - c\theta$ and $a_2' = a_2 + \theta$, where $0 \le \theta \le q - 1$. We can express y_1 and y_2 as follows:

$$y_1 = k_1 + a_1 r$$
$$= k_1 + (a_1' + c\theta)r$$
$$= (k_1 + rc\theta) + a_1' r,$$

and

$$y_2 = k_2 + a_2 r$$
$$= k_2 + (a_2' - \theta)r$$
$$= (k_2 - r\theta) + a_2' r,$$

where all arithmetic is performed in $\mathbb{Z}_q$. That is, the quadruple (γ, r, y_1, y_2) is also consistent with the ordered pair (a_1', a_2') using the random choices $k_1' = k_1 + rc\theta$ and $k_2' = k_2 - r\theta$ to produce (the same) γ. We have already noted that the values of k_1 and k_2 are not revealed by Alice, so the quadruple (γ, r, y_1, y_2) yields no information regarding which ordered pair in A Alice is actually using for her secret exponents. This completes the proof.   ∎

This security proof is certainly quite elegant and subtle. It would perhaps be useful to recap the features of the protocol that lead to the proof of security. The basic idea involves having Alice choose two secret exponents rather than one. There are a total of q pairs in the set A that are "equivalent" to Alice's pair (a_1, a_2). The fact that leads to the ultimate contradiction is that knowledge of two different pairs in A provides an efficient method of computing the discrete logarithm c. Alice, of course, knows one pair in A; and we proved that if Olga can impersonate Alice, then Olga is able to compute a pair in A which (with high probability) is different from Alice's pair. Thus Alice and Olga together can find two pairs in A and compute c, which provides the desired contradiction.

Here is an example to illustrate the computation of $\log_{\alpha_1} \alpha_2$ by Alice and Olga.

Example 9.5
As in Example 9.4, we will take $p = 88667$, $q = 1031$ and $t = 10$, and assume that $v = 13078$.

Suppose Olga has determined that

$$\alpha_1{}^{131} \alpha_2{}^{287} v^{489} \equiv \alpha_1{}^{890} \alpha_2{}^{303} v^{199} \pmod{p}.$$

Then she can compute

$$b_1 = (131 - 890)(489 - 199)^{-1} \bmod 1031 = 456$$

and

$$b_2 = (287 - 303)(489 - 199)^{-1} \bmod 1031 = 519.$$

Now, using the values of a_1 and a_2 supplied by Alice, the value

$$c = (846 - 456)(519 - 515)^{-1} \bmod 1031 = 613$$

is computed. This value c is in fact $\log_{\alpha_1} \alpha_2$, as can be verified by calculating

$$58902^{613} \bmod 88667 = 73611.$$

$\square$

Finally, we should emphasize that, although there is no known proof that the **Schnorr Scheme** is secure (even assuming that the discrete logarithm problem is intractible), neither is there any known weakness in the scheme. Actually, the **Schnorr Scheme** might be preferred in practice to the **Okamoto Scheme** simply because it is somewhat faster.

9.4 The Guillou-Quisquater Identification Scheme

In this section, we describe another identification scheme, due to Guillou and Quisquater, that is based on **RSA**.

The set-up of the scheme is as follows: The TA chooses two primes p and q and forms the product $n = pq$. The values of p and q are secret, while n is public. As is usually the case, p and q should be chosen large enough that factoring n is intractible. Also, the TA chooses a large prime integer b which will function as a security parameter as well as being a public RSA encryption exponent; to be specific, let us suppose that b is a 40-bit prime. Finally, the TA chooses a signature scheme and hash function.

The certificate issued to Alice by the TA is constructed as described in Figure 9.6. When Alice wants to prove her identity to Bob, say, the protocol of Figure 9.7 is executed. We will prove that the **Guillou-Quisquater Scheme** is sound and complete. However, the scheme has not been proved to be secure (even assuming that the **RSA** cryptosystem is secure).

FIGURE 9.6
Issuing a certificate to Alice

---

1. The TA establishes Alice's identity and issues an identification string ID(Alice).

2. Alice secretly chooses an integer u, where $0 \leq u \leq n - 1$. Alice computes
    $$v = (u^{-1})^b \bmod n$$
    and gives v to the TA.

3. The TA generates a signature
    $$s = sig_{TA}(\text{ID}(\text{Alice}), v).$$
    The certificate
    $$\mathbf{C}(\text{Alice}) = (\text{ID}(\text{Alice}), v, s)$$
    is given to Alice.

FIGURE 9.7
The Guillou-Quisquater identification scheme

---

1. Alice chooses a random number k, where $0 \leq k \leq n - 1$ and computes
    $$\gamma = k^b \bmod n.$$

2. Alice gives Bob her certificate $\mathbf{C}(\text{Alice}) = (\text{ID}(\text{Alice}), v, s)$ and γ.

3. Bob verifies the signature of the TA by checking that $ver_{TA}(\text{ID}(\text{Alice}), v, s) = \text{true}$.

4. Bob chooses a random number r, $0 \leq r \leq b - 1$ and gives it to Alice.

5. Alice computes
    $$y = ku^r \bmod n$$
    and gives y to Bob.

6. Bob verifies that
    $$\gamma \equiv v^r y^b \pmod{n}.$$

Example 9.6

Suppose the TA chooses $p = 467$ and $q = 479$, so $n = 223693$. Suppose also that $b = 503$ and Alice's secret integer $u = 101576$. Then she will compute

$$v = (u^{-1})^b \bmod n$$
$$= (101576^{-1})^{503} \bmod 223693$$
$$= 89888.$$

Now, let's assume that Alice is proving her identity to Bob and she chooses $k = 187485$; then she gives Bob the value

$$\gamma = k^b \bmod n$$
$$= 187485^{503} \bmod 223693$$
$$= 24412.$$

Suppose Bob responds with the challenge $r = 375$. Then Alice will compute

$$y = ku^r \bmod n$$
$$= 187485 \times 101576^{375} \bmod 223693$$
$$= 93725$$

and gives it to Bob. Bob then verifies that

$$24412 \equiv 89888^{375} 93725^{503} \pmod{223693}.$$

Hence, Bob accepts Alice's proof of identity. ⬜

As is generally the case, proving completeness is quite simple:

$$v^r y^b \equiv (u^{-b})^r (ku^r)^b \pmod{n}$$
$$\equiv u^{-br} k^b u^{br} \pmod{n}$$
$$\equiv k^b \pmod{n}$$
$$\equiv \gamma \pmod{n}.$$

Now, let us consider soundness. We will prove that the scheme is sound provided that it is infeasible to compute u from v. Since v is formed from u by RSA encryption, this is a plausible assumption to make.

THEOREM 9.4
Suppose Olga knows a value γ for which she has probability $\epsilon > 1/b$ of successfully impersonating Alice in the verification protocol. Then, in polynomial time, Olga can compute u.

PROOF   For some γ, Olga can compute values y_1, y_2, r_1, r_2 with $r_1 \neq r_2$, such that

$$\gamma \equiv v^{r_1} y_1{}^b \equiv v^{r_2} y_2{}^b \pmod{n}.$$

Suppose, without loss of generality, that $r_1 > r_2$. Then we have

$$v^{r_1 - r_2} \equiv (y_2/y_1)^b \pmod{n}.$$

Since $0 < r_1 - r_2 < b$ and b is prime, $t = (r_1 - r_2)^{-1} \bmod b$ exists, and it can be computed in polynomial time by Olga using the Euclidean algorithm. Hence, we have that

$$v^{(r_1 - r_2)t} \equiv (y_2/y_1)^{bt} \pmod{n}.$$

Now,

$$(r_1 - r_2)t = \ell b + 1$$

for some positive integer ℓ, so

$$v^{\ell b + 1} \equiv (y_2/y_1)^{bt} \pmod{n},$$

or equivalently,

$$v \equiv (y_2/y_1)^{bt} (v^{-1})^{\ell b} \pmod{n}.$$

Now raise both sides of the congruence to the power $b^{-1} \bmod \phi(n)$, to get the following:

$$u^{-1} \equiv (y_2/y_1)^{t} (v^{-1})^{\ell} \pmod{n}.$$

Finally, compute the inverse modulo n of both sides of this congruence, to obtain the following formula for u:

$$u = (y_1/y_2)^{t} v^{\ell} \bmod n.$$

Olga can use this formula to compute u in polynomial time.   ∎

Example 9.7
As in the previous example, suppose that $n = 223693$, $b = 503$, $u = 101576$ and $v = 89888$. Suppose Olga has learned that

$$v^{401} 103386^b \equiv v^{375} 93725^b \pmod{n}.$$

She will first compute

$$t = (r_1 - r_2)^{-1} \bmod b$$

$$= (401 - 375)^{-1} \bmod 503$$

$$= 445.$$

FIGURE 9.8
Issuing a value u to Alice

---

1. The TA establishes Alice's identity and issues an identification string ID(Alice).
2. The TA computes

$$u = (h(\text{ID}(\text{Alice}))^{-1})^a \bmod n$$

and gives u to Alice.

Next, she calculates

$$\ell = \frac{(r_1 - r_2)t - 1}{b}$$
$$= \frac{(401 - 375)445 - 1}{503}$$
$$= 23.$$

Finally, she can obtain the secret value u as follows:

$$u = (y_1/y_2)^t v^\ell \bmod n$$
$$= (103386/93725)^{445} 89888^{23} \bmod 223693$$
$$= 101576.$$

Thus Alice's secret exponent has been compromised. ▯

9.4.1 Identity-based Identification Schemes

The **Guillou-Quisquater Identification Scheme** can be tranformed into what is known as an *identity-based* identification scheme. This basically means that certificates are not necessary. Instead, the TA computes the value of u as a function of Alice's ID string, using a public hash function h with range $\mathbb{Z}_n$. This is done as indicated in Figure 9.8. The identification protocol now works as described in Figure 9.9. The value v is computed from Alice's ID string via the public hash function h. In order to carry out the identification protocol, Alice needs to know the value of u, which can be computed only by the TA (assuming that the **RSA** cryptosystem is secure). If Olga tries to identify herself as Alice, she will not succeed because she does not know the value of u.

FIGURE 9.9
The Guillou-Quisquater identity-based identification scheme

---

1. Alice chooses a random number k, where $0 \leq k \leq n - 1$ and computes
$$\gamma = k^b \bmod n.$$

2. Alice gives $\text{ID}(\text{Alice})$ and γ to Bob.

3. Bob computes
$$v = h(\text{ID}(\text{Alice})).$$

4. Bob chooses a random number r, $0 \leq r \leq b - 1$ and gives it to Alice.

5. Alice computes
$$y = ku^r \bmod n$$
and gives y to Bob.

6. Bob verifies that
$$\gamma \equiv v^r y^b \pmod{n}.$$

9.5   Converting Identification to Signature Schemes

There is a standard method of converting an identification scheme to a signature scheme. The basic idea is to replace the verifier (Bob) by a public hash function, h. In a signature scheme obtained by this approach, the message is not hashed before it is signed; the hashing is integrated into the signing algorithm.

We illustrate this approach by converting the **Schnorr Scheme** into a signature scheme. See Figure 9.10. In practice, one would probably take the hash function h to be the **SHS**, with the result reduced modulo q. Since the **SHS** produces a bitstring of length 160 and q is a 160-bit prime, the modulo q reduction is necessary only if the message digest produced by the **SHS** exceeds q; and even in this situation it is necessary only to subtract q from the result.

In proceeding from an identification scheme to a signature scheme, we replaced a 40-bit challenge by a 160-bit message digest. 40 bits suffice for a challenge since an impostor needs to be able to guess the challenge in order to precompute a response that will be accepted. But in the context of a signature scheme, we need message digests of a much larger size, in order to prevent attacking the scheme by finding collisions in the hash function.

Other identification schemes can be converted to signature schemes in a similar fashion.

FIGURE 9.10
Schnorr Signature Scheme

Let p be a 512-bit prime such that the discrete log problem in $\mathbb{Z}_p$ is intractible, and let q be a 160-bit prime that divides $p - 1$. Let $\alpha \in \mathbb{Z}_p^*$ be a qth root of 1 modulo p. Let h be a hash function with range $\mathbb{Z}_q$. Define $\mathcal{P} = \mathbb{Z}_p^*$, $\mathcal{A} = \mathbb{Z}_p^* \times \mathbb{Z}_q$, and define

$$\mathcal{K} = \{(p, q, \alpha, a, v) : v \equiv \alpha^{-a} \pmod{p}\}.$$

The values p, q, α, and v are public, and a is secret.

For $K = (p, q, \alpha, a, v)$, and for a (secret) random number $k \in \mathbb{Z}_q^*$, define

$$sig_K(x, k) = (\gamma, y),$$

where

$$\gamma = \alpha^k \bmod p$$

and

$$y = k + ah(x, \gamma) \bmod q.$$

For $x, \gamma \in \mathbb{Z}_p^*$ and $y \in \mathbb{Z}_q$, define

$$ver(x, \gamma, y) = \text{true} \Leftrightarrow \gamma \equiv \alpha^y v^{h(x,\gamma)} \pmod{p}.$$

9.6 Notes and References

The **Schnorr Identification Scheme** is from [SC91], the **Okamoto Scheme** was presented in [OK93], and the **Guillou-Quisquater Scheme** can be found in [GQ88]. Another scheme that can be proved secure under a plausible computational assumption has been given by Brickell and McCurley [BM92].

Other popular identification schemes include the **Feige-Fiat-Shamir Scheme** [FFS88] (see also [FS87]) and Shamir's **Permuted Kernel Scheme** [SH90]. The **Feige-Fiat-Shamir Scheme** is proved secure using zero-knowledge techniques (see Chapter 13 for more information on zero-knowledge proofs).

The method of constructing signature schemes from identification schemes is due to Fiat and Shamir [FS87]. They also describe an identity-based version of their identification scheme.

Surveys on identification schemes have been published by Burmester, Desmedt, and Beth [BDB92] and de Waleffe and Quisquater [DWQ93].

Exercises

9.1 Consider the following possible identification scheme. Alice possesses a secret key $n = pq$, where p and q are prime and $p \equiv q \equiv 3 \pmod 4$. The values n and ID(Alice) are signed by the TA, as usual, and stored on Alice's certificate. When Alice wants to identify herself to Bob, say, Bob will present Alice with a random quadratic residue modulo n, say x. Then Alice will compute a square root y of x and give it to Bob. Bob then verifies that $y^2 \equiv x \pmod n$. Explain why this scheme is insecure.

9.2 Suppose Alice is using the **Schnorr Scheme** where $q = 1201, p = 122503, t = 10$ and $\alpha = 11538$.

(a) Verify that α has order q in $\mathbb{Z}_p^*$.

(b) Suppose that Alice's secret exponent is $a = 357$. Compute v.

(c) Suppose that $k = 868$. Compute γ.

(d) Suppose that Bob issues the challenge $r = 501$. Compute Alice's response y.

(e) Perform Bob's calculations to verify y.

9.3 Suppose that Alice uses the **Schnorr Scheme** with p, q, t and α as in Exercise 9.2. Now suppose that $v = 51131$, and Olga has learned that

$$\alpha^3 v^{148} \equiv \alpha^{151} v^{1077} \pmod p.$$

Show how Olga can compute Alice's secret exponent a.

9.4 Suppose that Alice is using the **Okamoto Scheme** with $q = 1201, p = 122503, t = 10, \alpha_1 = 60497$ and $\alpha_2 = 17163$.

(a) Suppose that Alice's secret exponents are $a_1 = 432$ and $a_2 = 423$. Compute v.

(b) Suppose that $k_1 = 389$ and $k_2 = 191$. Compute γ.

(c) Suppose that Bob issues the challenge $r = 21$. Compute Alice's response, y_1 and y_2.

(d) Perform Bob's calculations to verify y_1 and y_2.

9.5 Suppose that Alice uses the **Okamoto Scheme** with $p, q, t, \alpha_1,$ and α_2 as in Exercise 9.4. Suppose also that $v = 119504$.

(a) Verify that

$$\alpha_1^{70} \alpha_2^{1033} v^{877} \equiv \alpha_1^{248} \alpha_2^{883} v^{992} \pmod p.$$

(b) Use this information to compute b_1 and b_2 such that

$$\alpha_1^{-b_1} \alpha_2^{-b_2} \equiv v \pmod p.$$

(c) Now suppose that Alice reveals that $a_1 = 484$ and $a_2 = 935$. Show how Alice and Olga together will compute $\log_{\alpha_1} \alpha_2$.

9.6 Suppose that Alice is using the **Guillou-Quisquater Scheme** with $p = 503, q = 379$, and $b = 509$.

(a) Suppose that Alice's secret $u = 155863$. Compute v.

(b) Suppose that $k = 123845$. Compute γ.

(c) Suppose that Bob issues the challenge $r = 487$. Compute Alice's response, y.

(d) Perform Bob's calculations to verify y.

9.7 Suppose that Alice is using the **Guillou-Quisquater Scheme** with $n = 199543$, $b = 523$ and $v = 146152$. Suppose that Olga has discovered that

$$v^{456} 101360^b \equiv v^{257} 36056^b \pmod{n}.$$

Show how Olga can compute u.

10

Authentication Codes

10.1 Introduction

We have spent a considerable amount of time studying cryptosystems, which are used to obtain secrecy. An authentication code provides a method of ensuring the *integrity* of a message, i.e., that the message has not been tampered with and that it originated with the presumed transmitter. Our goal is to achieve this authentication capability even in the presence of an active opponent, Oscar, who can observe messages in the channel and introduce messages of his own choosing into the channel. This goal is accomplished in the "private-key" setting whereby Alice and Bob share a secret key, K, before any message is transmitted.

In this chapter, we study codes that provide authentication but no secrecy. In such a code, a key is used to compute an authentication tag which will enable Bob to check the authenticity of the message he receives. Another application of an authentication code is verify that data in a large file has not been tampered with. An authentication tag would be stored with the data; the key used to generate and verify the authenticator would be stored separately, in a "secure" area.

We should also point out that, in many respects, an authentication code is similar to a signature scheme or to a message authentication code (MAC). The main differences are as follows: The security of an authentication code is unconditional, whereas signature schemes and MACs are studied from the point of view of computational security. Also, when an authentication code (or a MAC) is used, a message can be verified only by the intended receiver. In comparison, anyone can verify a signature using a public verification algorithm.

We now give a formal definition of the terminology we use in the study of authentication codes.

DEFINITION 10.1 *An authentication code is a four-tuple (S, A, K, E), where the following conditions are satisfied:*

1. *S is a finite set of possible source states*

FIGURE 10.1
Impersonation by Oscar

$$(s, a)$$

Oscar $\xrightarrow{\hspace{5cm}}$ Bob

2. $\mathcal{A}$ *is a finite set of possible authentication tags*
3. $\mathcal{K}$, *the keyspace, is a finite set of possible keys*
4. *For each $K \in \mathcal{K}$, there is an authentication rule $e_K : \mathcal{S} \to \mathcal{A}$.*

The message set is defined to be $\mathcal{M} = \mathcal{S} \times \mathcal{A}$.

REMARK Note that a source state is analogous to a plaintext. A message consists of a plaintext with an appended authentication tag; it could be more precisely referred to as a *signed message*. Also, an authentication rule need not be an injective function. ∎

In order to transmit a (signed) message, Alice and Bob follow the following protocol. First, they jointly choose a random key $K \in \mathcal{K}$. This is done in secret, as in a private-key cryptosystem. At a later time, suppose that Alice wants to communicate a source state $s \in \mathcal{S}$ to Bob over an insecure channel. Alice computes $a = e_K(s)$ and sends the message (s, a) to Bob. When Bob receives (s, a), he computes $a' = e_K(s)$. If $a' = a$, then he accepts the message as authentic; otherwise, he rejects it.

We will study two different types of attacks that Oscar might carry out. In both of these attacks, Oscar is an "intruder-in-the-middle." These attacks described are as follows:

Impersonation

Oscar introduces a message (s, a) into the channel, hoping to have it accepted as authentic by Bob. This is depicted in Figure 10.1.

Substitution

Oscar observes a message (s, a) in the channel, and then changes it to (s', a'), where $s' \neq s$, again hoping to have it accepted as authentic by Bob. Hence, he is hoping to mislead Bob as to the source state. This is depicted in Figure 10.2.

Associated with each of these attacks is a *deception probability*, which represents the probability that Oscar will successfully deceive Bob, if he (Oscar) follows an optimal strategy. These probabilities are denoted by Pd_0 (impersonation) and Pd_1 (substitution). In order to compute Pd_0 and Pd_1, we need to

FIGURE 10.2
Substitution by Oscar

$$\text{Alice} \xrightarrow{\quad (s,a) \quad} \text{Oscar} \xrightarrow{\quad (s',a') \quad} \text{Bob}$$

specify probability distributions on S and K. These will be denoted by p_S and p_K, respectively. We assume that the authentication code and these two probability distributions are known to Oscar. The only information that Alice and Bob possess that is not known to Oscar is the value of the key, K. This is analogous to the way that we studied the unconditional security of private-key cryptosystems.

10.2   Computing Deception Probabilities

In this section, we look at the computation of deception probabilities. We begin with a small example of an authentication code.

Example 10.1
Suppose

$$S = A = \mathbb{Z}_3$$

and

$$K = \mathbb{Z}_3 \times \mathbb{Z}_3.$$

For each $(i, j) \in K$ and each $s \in S$, define

$$e_{ij}(s) = is + j \bmod 3.$$

It will be useful to study the *authentication matrix*, which tabulates all the values $e_{ij}(s)$. For each key $K \in K$ and for each $s \in S$, place the authentication tag $e_K(s)$ in row K and column s of a $|K| \times |S|$ matrix M. The array M is presented in Figure 10.3.

Suppose that the key is chosen at random, i.e., $p_K(K) = 1/9$ for each $K \in K$. We do not specify the probability distribution p_S since it turns out to be immaterial in this example.

Let's first consider an impersonation attack. Oscar will pick a source state s, and attempt to guess the "correct" authentication tag. Denote by K_0 the actual key being used (which is unknown to Oscar). Oscar will succeed in deceiving Bob if he guesses the tag $a_0 = e_{K_0}(s)$. However, for any $s \in S$ and $a \in A$, it is easy to verify that there are exactly three (out of nine) authentication rules $K \in K$

FIGURE 10.3
An authentication matrix

key	0	1	2
$(0,0)$	0	0	0
$(0,1)$	1	1	1
$(0,2)$	2	2	2
$(1,0)$	0	1	2
$(1,1)$	1	2	0
$(1,2)$	2	0	1
$(2,0)$	0	2	1
$(2,1)$	1	0	2
$(2,2)$	2	1	0

such that $e_K(s) = a$. (In other words, each symbol occurs three times in each column of the authentication matrix.) Hence, it follows that $Pd_0 = 1/3$.

Substitution is a bit more complicated to analyze. As a specific case, suppose Oscar observes the message $(0,0)$ in the channel. This does give Oscar some information about the key: he now knows that

$$K_0 \in \{(0,0), (1,0), (2,0)\}.$$

Now suppose Oscar replaces the message $(0,0)$ with the message $(1,1)$. Then, he will succeed in his deception if and only if $K_0 = (1,0)$. The probability that K_0 is the key is $1/3$, since the key is known to be in the set $\{(0,0), (1,0), (2,0)\}$.

A similar analysis can be done for any substitution that Oscar might make. In general, if Oscar observes the message (s, a), and replaces it with any message (s', a') where $s' \neq s$, then he deceives Bob with probability $1/3$. We can see this as follows. Observation of (s, a) restricts the key to one of three possibilities. Then, for each choice of (s', a'), there is one key (out of the three possible keys) under which a' is the authentication tag for s'. ⬜

Let's now discuss how to compute the deception probabilities in general. First, we consider Pd_0. As above, let K_0 denote the key chosen by Alice and Bob. For $s \in S$ and $a \in A$, define $payoff(s, a)$ to be the probability that Bob will accept the message (s, a) as being authentic. It is not difficult to see that

$$payoff(s, a) = prob(a = e_{K_0}(s))$$

$$= \sum_{\{K \in \mathcal{K} : e_K(s) = a\}} p_{\mathcal{K}}(K).$$

That is, $payoff(s, a)$ is computing by selecting the rows of the authentication matrix that have entry a in column s, and summing the probabilities of the corresponding keys.

In order to maximize his chance of success, Oscar will choose (s, a) such that $payoff(s, a)$ is a maximum. Hence,

$$Pd_0 = \max\{payoff(s, a) : s \in \mathcal{S}, a \in \mathcal{A}\}. \tag{10.1}$$

Note that Pd_0 does not depend on the probability distribution $p_{\mathcal{S}}$.

Pd_1 is more difficult to compute, and it may depend on the probability distribution $p_{\mathcal{S}}$. Let's first consider the following problem: Suppose Oscar observes the message (s, a) in the channel. Oscar will substitute some (s', a') for (s, a), where $s' \neq s$. Hence, for $s, s' \in \mathcal{S}$, $s \neq s'$, and $a, a' \in \mathcal{A}$, we define $payoff(s', a'; s, a)$ to be the probability that a substitution of (s, a) with (s', a') will succeed in deceiving Bob. Then we can compute the following:

$$
\begin{aligned}
payoff(s', a'; s, a) &= prob(a' = e_{K_0}(s') | a = e_{K_0}(s)) \\
&= \frac{prob(a' = e_{K_0}(s') \wedge a = e_{K_0}(s))}{prob(a = e_{K_0}(s))} \\
&= \frac{\displaystyle\sum_{\{K \in \mathcal{K}: e_K(s)=a, e_K(s')=a'\}} p_{\mathcal{K}}(K)}{\displaystyle\sum_{\{K \in \mathcal{K}: e_K(s)=a\}} p_{\mathcal{K}}(K)} \\
&= \frac{\displaystyle\sum_{\{K \in \mathcal{K}: e_K(s)=a, e_K(s')=a'\}} p_{\mathcal{K}}(K)}{payoff(s, a)}.
\end{aligned}
$$

The numerator of this fraction is found by selecting the rows of the authentication matrix that have the value a in column s and the value a' in column s', and summing the probabilities of the corresponding keys.

Since Oscar wants to maximize his chance of deceiving Bob, he will compute

$$p_{s,a} = \max\{payoff(s', a'; s, a) : s' \in \mathcal{S}, s \neq s', a' \in \mathcal{A}\}.$$

The quantity $p_{s,a}$ denotes the probability that Oscar can deceive Bob with a substitution, given that (s, a) is the message observed in the channel.

Now, how do we compute the deception probability Pd_1? Evidently, we have to compute a weighted average of the quantities $p_{s,a}$ with respect to the probabilities $p_{\mathcal{M}}(s, a)$ of observing messages (s, a) in the channel. That is, we calculate Pd_1 to be

$$Pd_1 = \sum_{(s,a) \in \mathcal{M}} p_{\mathcal{M}}(s, a) p_{s,a}. \tag{10.2}$$

FIGURE 10.4
An authentication matrix

key	1	2	3	4
1	1	1	1	2
2	2	2	1	2
3	1	2	2	1

The probability distribution $p_{\mathcal{M}}$ is as follows:

$$p_{\mathcal{M}}(s,a) = p_{\mathcal{S}}(s) \times p_{\mathcal{K}}(a|s)$$

$$= p_{\mathcal{S}}(s) \times \sum_{\{K \in \mathcal{K}: e_K(s)=a\}} p_{\mathcal{K}}(K)$$

$$= p_{\mathcal{S}}(s) \times payoff(s,a).$$

In Example 10.1,
$$payoff(s,a) = 1/3$$

for all s, a, so $Pd_0 = 1/3$. Also, it can be checked that

$$payoff(s',a';s,a) = 1/3$$

for all $s, s', a, a', s \neq s'$. Hence, $Pd_1 = 1/3$ for any probability distribution $p_{\mathcal{S}}$. (In general, though, Pd_1 will depend on $p_{\mathcal{S}}$.)
Let's look at the computation of Pd_0 and Pd_1 for a less "regular" example.

Example 10.2
Consider the authentication matrix of Figure 10.4. Suppose the probability distributions on $\mathcal{S}$ and $\mathcal{K}$ are
$$p_{\mathcal{S}}(i) = 1/4,$$

$1 \leq i \leq 4$; and
$$p_{\mathcal{K}}(1) = 1/2, p_{\mathcal{K}}(2) = p_{\mathcal{K}}(3) = 1/4.$$

The values $payoff(s,a)$ are as follows:

$$payoff(1,1) = 3/4$$
$$payoff(1,2) = 1/4$$
$$payoff(2,1) = 1/2$$

$$payoff\,(2,2) = 1/2$$
$$payoff\,(3,1) = 3/4$$
$$payoff\,(3,2) = 1/4$$
$$payoff\,(4,1) = 1/4$$
$$payoff\,(4,2) = 3/4.$$

Hence, $Pd_0 = 3/4$. Oscar's optimal impersonation strategy is to place any of the messages $(1,1)$, $(3,1)$ or $(4,2)$ into the channel.

Now we turn to the computation of Pd_1. First, we present the various values $payoff\,(s',a';s,a)$ in the form of a matrix. The entry in row (s,a) and column (s',a') is the value $payoff\,(s',a';s,a)$.

	$(1,1)$	$(1,2)$	$(2,1)$	$(2,2)$	$(3,1)$	$(3,2)$	$(4,1)$	$(4,2)$
$(1,1)$			$2/3$	$1/3$	$2/3$	$1/3$	$1/3$	$2/3$
$(1,2)$			0	1	1	0	1	0
$(2,1)$	1	0			0	1	0	1
$(2,2)$	$1/2$	$1/2$			$1/2$	$1/2$	$1/2$	$1/2$
$(3,1)$	$2/3$	$1/3$	$2/3$	$1/3$			0	1
$(3,2)$	1	0	0	1			1	0
$(4,1)$	1	0	0	1	0	1		
$(4,2)$	$2/3$	$1/3$	$2/3$	$1/3$	1	0		

Thus we have $p_{1,1} = 2/3$, $p_{2,2} = 1/2$, and $p_{s,a} = 1$ for all other s,a. It is then a simple matter to evaluate $Pd_1 = 7/8$. An optimal substitution strategy for Oscar is as follows:

$$(1,1) \rightarrow (2,1)$$
$$(1,2) \rightarrow (2,2)$$
$$(2,1) \rightarrow (1,1)$$
$$(2,2) \rightarrow (1,1)$$
$$(3,1) \rightarrow (4,2)$$
$$(3,2) \rightarrow (1,1)$$
$$(4,1) \rightarrow (1,1)$$
$$(4,2) \rightarrow (3,1).$$

This strategy indeed yields $Pd_1 = 7/8$.   ▯

The computation of Pd_1 in Example 10.2 is straightforward but lengthy. We can in fact simplify the computation of Pd_1 by observing that we divide by the quantity $payoff(s, a)$ in the computation of $p_{s,a}$, and then later multiply by $payoff(s, a)$ in the computation of Pd_1. Of course, these two operations cancel each other out. Suppose we define

$$q_{s,a} = \max \left\{ \sum_{\{K \in \mathcal{K}: e_K(s)=a, e_K(s')=a'\}} p_K(K) : s' \in \mathcal{S}, s' \neq s, a' \in \mathcal{A} \right\}$$

for all s, a. Then we have the following more concise formula for Pd_1:

$$Pd_1 = \sum_{(s,a) \in \mathcal{M}} p_S(s) q_{s,a}. \tag{10.3}$$

10.3 Combinatorial Bounds

We have seen that the security of an authentication code is measured by the deception probabilities. Hence, we want to construct codes so that these probabilities are as small as possible. But other considerations are also important. Let's consider the various objectives that we might strive for in an authentication code:

1. The deception probabilities Pd_0 and Pd_1 must be small enough to obtain the desired level of security.

2. The number of source states must be large enough so that we can communicate the desired information by appending an authentication tag to one source state.

3. The size of the key space should be minimized, since the value of the key must be communicated over a secure channel. (Note that the key must be changed every time a message is communicated, as is done with the **One-time Pad**.)

In this section, we determine lower bounds on the deception probabilities, which will be computed in terms of other parameters of the code. Recall that we have defined an authentication code to consist of a four-tuple $(\mathcal{S}, \mathcal{A}, \mathcal{K}, \mathcal{E})$. Throughout this section, we will denote $|\mathcal{A}| = \ell$.

Suppose we fix a source state $s \in \mathcal{S}$. Then we can compute:

$$\sum_{a \in \mathcal{A}} payoff(s, a) = \sum_{a \in \mathcal{A}} \sum_{\{K \in \mathcal{K}: e_K(s)=a\}} p_K(K)$$

$$= \sum_{K \in \mathcal{K}} p_K(K)$$

$$= 1.$$

Hence, for every $s \in \mathcal{S}$, there exists an authentication tag $a(s)$ such that

$$payoff(s, a(s)) \geq \frac{1}{\ell}.$$

The following theorem follows easily.

THEOREM 10.1
Suppose $(\mathcal{S}, \mathcal{A}, \mathcal{K}, \mathcal{E})$ is an authentication code. Then $Pd_0 \geq 1/\ell$, where $\ell = |\mathcal{A}|$. Further, $Pd_0 = 1/\ell$ if and only if

$$\sum_{\{K \in \mathcal{K}: e_K(s)=a\}} p_{\mathcal{K}}(K) = \frac{1}{\ell} \tag{10.4}$$

for every $s \in \mathcal{S}$, $a \in \mathcal{A}$.

Now, we turn our attention to substitution. Suppose we fix s, a and s', where $s' \neq s$. Then we have the following:

$$\sum_{a' \in \mathcal{A}} payoff(s', a'; s, a) = \sum_{a' \in \mathcal{A}} \frac{\sum_{\{K \in \mathcal{K}: e_K(s)=a, e_K(s')=a'\}} p_{\mathcal{K}}(K)}{\sum_{\{K \in \mathcal{K}: e_K(s)=a\}} p_{\mathcal{K}}(K)}$$

$$= \frac{\sum_{\{K \in \mathcal{K}: e_K(s)=a\}} p_{\mathcal{K}}(K)}{\sum_{\{K \in \mathcal{K}: e_K(s)=a\}} p_{\mathcal{K}}(K)}$$

$$= 1.$$

So, there exists an authentication tag $a'(s', s, a)$ such that

$$payoff(s', a'(s', s, a); s, a) \geq \frac{1}{\ell}.$$

The next theorem follows as a consequence.

THEOREM 10.2
Suppose $(\mathcal{S}, \mathcal{A}, \mathcal{K}, \mathcal{E})$ is an authentication code. Then $Pd_1 \geq 1/\ell$, where $\ell = |\mathcal{A}|$. Further, $Pd_1 = 1/\ell$ if and only if

$$\frac{\sum_{\{K \in \mathcal{K}: e_K(s)=a, e_K(s')=a'\}} p_{\mathcal{K}}(K)}{\sum_{\{K \in \mathcal{K}: e_K(s)=a\}} p_{\mathcal{K}}(K)} = \frac{1}{\ell} \tag{10.5}$$

for every $s, s' \in \mathcal{S}, s' \neq s, a, a' \in \mathcal{A}$.

PROOF We have

$$Pd_1 = \sum_{(s,a)\in\mathcal{M}} p_{\mathcal{M}}(s,a)p_{s,a}$$

$$\geq \sum_{(s,a)\in\mathcal{M}} \frac{p_{\mathcal{M}}(s,a)}{\ell}$$

$$= \frac{1}{\ell}.$$

Further, equality occurs if and only if $p_{s,a} = 1/\ell$ for every (s,a). But this is in turn equivalent to the condition that $payoff(s',a';s,a) = 1/\ell$ for every (s,a). ∎

Combining Theorems 10.1 and 10.2, we get the following:

THEOREM 10.3
Suppose $(\mathcal{S},\mathcal{A},\mathcal{K},\mathcal{E})$ is an authentication code, where $\ell = |\mathcal{A}|$. Then $Pd_0 = Pd_1 = 1/\ell$ if and only if

$$\sum_{\{K\in\mathcal{K}:e_K(s)=a,e_K(s')=a'\}} p_{\mathcal{K}}(K) = \frac{1}{\ell^2} \tag{10.6}$$

for every $s,s' \in \mathcal{S}, s' \neq s, a,a' \in \mathcal{A}$.

PROOF Equations (10.4) and (10.5) imply Equation (10.6). Conversely, Equation (10.6) implies Equations (10.4) and (10.5). ∎

If the keys are equiprobable, then we obtain the following corollary:

COROLLARY 10.4
Suppose $(\mathcal{S},\mathcal{A},\mathcal{K},\mathcal{E})$ is an authentication code where $\ell = |\mathcal{A}|$, and keys are chosen equiprobably. Then $Pd_0 = Pd_1 = 1/\ell$ if and only if

$$|\{K \in \mathcal{K} : e_K(s) = a, e_K(s') = a'\}| = \frac{|\mathcal{K}|}{\ell^2}, \tag{10.7}$$

for every $s,s' \in \mathcal{S}, s' \neq s, a,a' \in \mathcal{A}$.

10.3.1 Orthogonal Arrays

In this section, we look at the connections between authentication codes and certain combinatorial structures called orthogonal arrays. First, we give a definition.

FIGURE 10.5
An $OA(3, 3, 1)$

$$\begin{pmatrix} 0 & 0 & 0 \\ 1 & 1 & 1 \\ 2 & 2 & 2 \\ 0 & 1 & 2 \\ 1 & 2 & 0 \\ 2 & 0 & 1 \\ 0 & 2 & 1 \\ 1 & 0 & 2 \\ 2 & 1 & 0 \end{pmatrix}$$

DEFINITION 10.2 *An orthogonal array $OA(n, k, \lambda)$ is a $\lambda n^2 \times k$ array of n symbols, such that in any two columns of the array every one of the possible n^2 pairs of symbols occurs in exactly λ rows.*

Orthogonal arrays are well-studied structures in combinatorial design theory, and are equivalent to other structures such as transversal designs, mutually orthogonal Latin squares and nets.

In Figure 10.5, we present an orthogonal array $OA(3, 3, 1)$ which is obtained from the authentication matrix of Figure 10.3. Any orthogonal array $OA(n, k, \lambda)$ can be used to construct an authentication code with $Pd_0 = Pd_1 = 1/n$, as stated in the following theorem.

THEOREM 10.5

Suppose there is an orthogonal array $OA(n, k, \lambda)$. Then there is an authentication code $(\mathcal{S}, \mathcal{A}, \mathcal{K}, \mathcal{E})$, where $|\mathcal{S}| = k$, $|\mathcal{A}| = n$, $|\mathcal{K}| = \lambda n^2$ and $Pd_0 = Pd_1 = 1/n$.

PROOF Use each row of the orthogonal array as an authentication rule with equal probability $1/(\lambda n^2)$. The correspondences are as follows:

orthogonal array	authentication code
row	authentication rule
column	source state
symbol	authentication tag

Since Equation (10.7) is satisfied, we can apply Corollary 10.4, obtaining a code with the stated properties.   ∎

10.3.2 Constructions and Bounds for OAs

Suppose that we construct an authentication code from an $OA(n, k, \lambda)$. The parameter n determines the number of authenticators (i.e., the security of the code), while the parameter k determines the number of source states the code can accommodate. The parameter λ relates only to the number of keys, which is λn^2. Of course, the case $\lambda = 1$ is most desirable, but we will see that it is sometimes necessary to use orthogonal arrays with higher values of λ.

Suppose we want to construct an authentication code with a specified source set $\mathcal{S}$, and a specified security level ϵ (i.e., so that $Pd_0 \leq \epsilon$ and $Pd_1 \leq \epsilon$). An appropriate orthogonal array will satisfy the following conditions:

1. $n \geq 1/\epsilon$
2. $k \geq |\mathcal{S}|$ (observe that we can always delete one or more columns from an orthogonal array and the resulting array is still an orthogonal array, so we do not require $k = |\mathcal{S}|$)
3. λ is minimized, subject to the two previous conditions being satisfied.

Let's first consider orthogonal arrays with $\lambda = 1$. For a given value of n, we are interested in maximizing the number of columns. Here is a necessary condition for existence:

THEOREM 10.6
Suppose there exists an $OA(n, k, 1)$. *Then* $k \leq n + 1$.

PROOF   Let A be an $OA(n, k, 1)$ on symbol set $X = \{0, 1, \ldots, n-1\}$. Suppose π is a permutation of X, and we permute the symbols in any column of A according to the permutation π. The result is again an $OA(n, k, 1)$. Hence, by applying a succession of permutations of this type, we can assume without loss of generality that the first row of A is $(00 \ldots 0)$.

We next show that each symbol must occur exactly n times in each column of A. Choose two columns, say c and c', and let x be any symbol. Then for each symbol x', there is a unique row of A in which x occurs in column c and x' occurs in column c'. Letting x' vary over X, we see that x occurs exactly n times in column c.

Now, since the first row is $(00 \ldots 0)$, we have exhausted all occurrences of ordered pairs $(0, 0)$. Hence, no other row contains more than one occurrence of 0. Now, let us count the number of rows containing at least one 0: the total is $1 + k(n-1)$. But this total cannot exceed the total number of rows in A, which is n^2. Hence, $1 + k(n-1) \leq n^2$, so $k \leq n+1$, as desired.   ∎

We now present a construction for orthogonal arrays with $\lambda = 1$ in which $k = n$. This is, in fact, the construction that was used to obtain the orthogonal array presented in Figure 10.5.

THEOREM 10.7

Suppose p is prime. Then there exists an orthogonal array $\mathrm{OA}(p, p, 1)$.

PROOF The array will be a $p^2 \times p$ array, where the rows are indexed by $\mathbb{Z}_p \times \mathbb{Z}_p$ and the columns are indexed by $\mathbb{Z}_p$. The entry in row (i, j) and column x is defined to be $ix + j \bmod p$.

Suppose we choose two columns, $x, y, x \neq y$, and two symbols a, b. We want to find a (unique) row (i, j) such that a occurs in column x and b occurs in column y of row (i, j). Hence, we want to solve the two equations

$$a = ix + j$$
$$b = iy + j$$

for the unknowns i and j (where all arithmetic is done in the field $\mathbb{Z}_p$). But this system has the unique solution

$$i = (a - b)(x - y)^{-1} \bmod p$$
$$j = a - ix \bmod p.$$

Hence, we have an orthogonal array.   ∎

We remark that any $\mathrm{OA}(n, n, 1)$ can be extended by one column to form an $\mathrm{OA}(n, n + 1, 1)$ (see the Exercises). Hence, using Theorem 10.7, we can obtain an infinite class of OA's that meet the bound of Theorem 10.6 with equality.

Theorem 10.6 tells us that $\lambda > 1$ if $k > n + 1$. We will prove a more general result that places a lower bound on λ as a function of n and k. First, however, we derive an important inequality that we will use in the proof.

LEMMA 10.8

Suppose $b_1, \ldots, b_m$ are real numbers. Then

$$m \sum_{i=1}^{m} b_i^{\,2} \geq \left(\sum_{i=1}^{m} b_i \right)^2.$$

PROOF Apply Jensen's Inequality (Theorem 2.5) with $f(x) = -x^2$ and $a_i = 1/m, 1 \leq i \leq m$. The function f is continuous and concave, so we obtain

$$-\sum_{i=1}^{m} \frac{b_i^{\,2}}{m} \leq -\left(\sum_{i=1}^{m} \frac{b_i}{m} \right)^2,$$

which simplifies to give the desired result.   ∎

THEOREM 10.9
Suppose there exists an $\mathrm{OA}(n, k, \lambda)$. *Then*

$$\lambda \geq \frac{k(n-1)+1}{n^2}.$$

PROOF Let A be an $\mathrm{OA}(n, k, \lambda)$ on symbol set $X = \{0, 1, \ldots, n-1\}$, where, without loss of generality, the first row of A is $(00\ldots0)$ (as in Theorem 10.6).

Let us denote the set of rows of A by $\mathcal{R}$, let r_1 denote the first row, and let $\mathcal{R}_1 = \mathcal{R}\backslash\{r_1\}$. For any row r of A, denote by x_r the number of occurrences of 0 in row r. It is easy to count the total number of occurrences of 0 in $\mathcal{R}_1$. Since each symbol must occur exactly λn times in each column of A, we have that

$$\sum_{r \in \mathcal{R}_1} x_r = k(\lambda n - 1).$$

Now, the number of times the ordered pair $(0, 0)$ occurs in rows in $\mathcal{R}_1$ is

$$\sum_{r \in \mathcal{R}_1} x_r(x_r - 1) = \sum_{r \in \mathcal{R}_1} x_r^2 - \sum_{r \in \mathcal{R}_1} x_r$$

$$= \sum_{r \in \mathcal{R}_1} x_r^2 - k(\lambda n - 1).$$

Applying Lemma 10.8, we obtain

$$\sum_{r \in \mathcal{R}_1} x_r^2 \geq \frac{(k(\lambda n - 1))^2}{\lambda n^2 - 1},$$

and hence

$$\sum_{r \in \mathcal{R}_1} x_r(x_r - 1) \geq \frac{(k(\lambda n - 1))^2}{\lambda n^2 - 1} - k(\lambda n - 1).$$

On the other hand, in any given pair of columns, the ordered pair $(0, 0)$ occurs in exactly λ rows. Since there are $k(k-1)$ ordered pairs of columns, it follows that the exact number of occurrences of the ordered pair $(0, 0)$ in rows in $\mathcal{R}_1$ is $(\lambda - 1)k(k-1)$. We therefore have

$$(\lambda - 1)k(k-1) \geq \frac{(k(\lambda n - 1))^2}{\lambda n^2 - 1} - k(\lambda n - 1),$$

and hence

$$((\lambda - 1)k(k-1) + k(\lambda n - 1))(\lambda n^2 - 1) \geq (k(\lambda n - 1))^2.$$

If we divide out a factor of k, we get

$$(\lambda k - k - \lambda + \lambda n)(\lambda n^2 - 1) \geq k(\lambda n - 1)^2.$$

Expanding, we have

$$\lambda^2 kn^2 - \lambda kn^2 - \lambda^2 n^2 + \lambda^2 n^3 - \lambda k + k + \lambda - \lambda n \geq \lambda^2 kn^2 - 2\lambda kn + k.$$

This simplifies to give

$$-\lambda^2 n^2 + \lambda^2 n^3 \geq \lambda kn^2 + \lambda k - \lambda + \lambda n - 2\lambda kn,$$

or

$$\lambda^2(n^3 - n^2) \geq \lambda(k(n-1)^2 + n - 1).$$

Finally, taking out a factor of $\lambda(n-1)$, we obtain

$$\lambda n^2 \geq k(n-1) + 1,$$

which is the desired bound.   ∎

Our next result establishes the existence of an infinite class of orthogonal arrays that meet the above bound with equality.

THEOREM 10.10
Suppose p is prime and $d \geq 2$ is an integer. Then there is an orthogonal array $OA(p, (p^d - 1)/(p - 1), p^{d-2})$.

PROOF Denote by $(\mathbb{Z}_p)^d$ the vector space of all d-tuples over $\mathbb{Z}_p$. We will construct A, an $OA(p, (p^d - 1)/(p - 1), p^{d-2})$ in which the rows and columns are indexed by certain vectors in $(\mathbb{Z}_p)^d$. The entries of A will be elements of $\mathbb{Z}_p$. The set of rows is defined to be $\mathcal{R} = (\mathbb{Z}_p)^d$; the set of columns is

$$\mathcal{C} = \{(c_1, \ldots, c_d) \in (\mathbb{Z}_p)^d : \exists j, 0 \leq j \leq d - 1, c_1 = \ldots = c_j = 0, c_{j+1} = 1\}.$$

$\mathcal{R}$ consists of all vectors in $(\mathbb{Z}_p)^d$, so $|\mathcal{R}| = p^d$. $\mathcal{C}$ consists of all non-zero vectors that have the first non-zero coordinate equal to 1. Observe that

$$|\mathcal{C}| = \frac{p^d - 1}{p - 1},$$

and that no two vectors in $\mathcal{C}$ are scalar multiples of each other.

Now, for each $\bar{r} \in \mathcal{R}$ and each $\bar{c} \in \mathcal{C}$, define

$$A(\bar{r}, \bar{c}) = \bar{r} \cdot \bar{c},$$

where $\cdot$ denotes the inner product of two vectors (reduced modulo p).

We prove that A is the desired orthogonal array. Let $\bar{b}, \bar{c} \in \mathcal{C}$ be two distinct columns, and let $x, y \in \mathbb{Z}_p$. We will count the number of rows $\bar{r}$ such that $A(\bar{r}, \bar{b}) = x$ and $A(\bar{r}, \bar{c}) = y$. Denote $\bar{r} = (r_1, r_2, \ldots, r_d)$, $\bar{b} = (b_1, b_2, \ldots, b_d)$

FIGURE 10.6
An $OA(2, 7, 2)$

$$\begin{pmatrix}
0 & 0 & 0 & 0 & 0 & 0 & 0 \\
1 & 0 & 1 & 0 & 1 & 0 & 1 \\
0 & 1 & 1 & 0 & 0 & 1 & 1 \\
1 & 1 & 0 & 0 & 1 & 1 & 0 \\
0 & 0 & 0 & 1 & 1 & 1 & 1 \\
1 & 0 & 1 & 1 & 0 & 1 & 0 \\
0 & 1 & 1 & 1 & 1 & 0 & 0 \\
1 & 1 & 0 & 1 & 0 & 0 & 1
\end{pmatrix}$$

and $\bar{c} = (c_1, c_2, \ldots, c_d)$. The two equations $\bar{r} \cdot \bar{b} = x, \bar{r} \cdot \bar{c} = y$ can be written as two linear equations in $\mathbb{Z}_p$:

$$b_1 r_1 + \ldots + b_d r_d = x$$
$$c_1 r_1 + \ldots + c_d r_d = y.$$

This is a system of two linear equations in the d unknowns $r_1, \ldots r_d$. Since $\bar{b}$ and $\bar{c}$ are not scalar multiples, the two equations are linearly independent. Hence, this system has a solution space of dimension $d - 2$. That is, the number of solutions (i.e., the number of rows in which x occurs in column $\bar{b}$ and y occurs in column $\bar{c}$) is p^{d-2}, as desired. $\blacksquare$

Let's carry out a small example of this construction.

Example 10.3
Suppose we take $p = 2, d = 3$. Then we will construct an $OA(2, 7, 2)$. We have

$$\mathcal{R} = \{000, 001, 010, 011, 100, 101, 110, 111\}$$

and

$$\mathcal{C} = \{001, 010, 011, 100, 101, 110, 111\}.$$

The orthogonal array in Figure 10.6 results. $\square$

10.3.3 Characterizations of Authentication Codes

To this point, we have studied authentication codes obtained from orthogonal arrays. Then we looked at necessary existence conditions and constructions for orthogonal arrays. One might wonder whether there are better alternatives to the orthogonal array approach. However, two characterization theorems tell us that this is not the case if we restrict our attention to authentication codes in which the deception probabilities are as small as possible.

We first prove the following partial converse to Theorem 10.5:

THEOREM 10.11
Suppose $(\mathcal{S}, \mathcal{A}, \mathcal{K}, \mathcal{E})$ is an authentication code where $|\mathcal{A}| = n$ and $Pd_0 = Pd_1 = 1/n$. Then $|\mathcal{K}| \geq n^2$. Further, $|\mathcal{K}| = n^2$ if and only if there is an orthogonal array $\mathrm{OA}(n, k, 1)$ where $|\mathcal{S}| = k$, and $p_{\mathcal{K}}(K) = 1/n^2$ for every key $K \in \mathcal{K}$.

PROOF Fix two (arbitrary) source states s and s', $s \neq s'$, and consider Equation (10.6). For each ordered pair (a, a') of authentication tags, define

$$\mathcal{K}_{a,a'} = \{K \in \mathcal{K} : e_K(s) = a, e_K(s') = a'\}.$$

Then $|\mathcal{K}_{a,a'}| > 0$ for every pair (a, a'). Also, the n^2 sets $\mathcal{K}_{a,a'}$ are disjoint. Hence, $|\mathcal{K}| \geq n^2$.

Now, suppose that $|\mathcal{K}| = n^2$. Then $|\mathcal{K}_{a,a'}| = 1$ for every pair (a, a'), and Equation (10.6) tells us that $p_{\mathcal{K}}(K) = 1/n^2$ for every key $K \in \mathcal{K}$.

It remains to show that the authentication matrix forms an orthogonal array $\mathrm{OA}(n, k, 1)$. Consider the columns indexed by the source states s and s'. Since $|\mathcal{K}_{a,a'}| = 1$ for every (a, a'), we have every ordered pair occurring exactly once in these two columns. Since, s and s' are arbitrary, we see that every ordered pair occurs exactly once in any two columns. ∎

The following characterization is more difficult; we state it without proof.

THEOREM 10.12
Suppose $(\mathcal{S}, \mathcal{A}, \mathcal{K}, \mathcal{E})$ is an authentication code where $|\mathcal{S}| = k$, $|\mathcal{A}| = n$ and $Pd_0 = Pd_1 = 1/n$. Then $|\mathcal{K}| \geq k(n-1) + 1$. Further, $|\mathcal{K}| = k(n-1) + 1$ if and only if there is an orthogonal array $\mathrm{OA}(n, k, \lambda)$, where $\lambda = (k(n-1) + 1)/n^2$, and $p_{\mathcal{K}}(K) = 1/(k(n-1) + 1)$ for every key $K \in \mathcal{K}$.

REMARK Notice that Theorem 10.10 provides an infinite class of orthogonal arrays that meet the bound of Theorem 10.12 with equality. ∎

10.4 Entropy Bounds

In this section, we use entropy techniques to obtain bounds on the deception probabilities. The first of these is a bound on Pd_0.

THEOREM 10.13
Suppose that $(\mathcal{S}, \mathcal{A}, \mathcal{K}, \mathcal{E})$ is an authentication code. Then

$$\log Pd_0 \geq H(\mathbf{K}|\mathbf{M}) - H(\mathbf{K}).$$

PROOF From Equation (10.1), we have

$$Pd_0 = \max\{payoff(s, a) : s \in \mathcal{S}, a \in \mathcal{A}\}.$$

Since the maximum of the values $payoff(s, a)$ is greater than their weighted average, we obtain

$$Pd_0 \geq \sum_{s \in \mathcal{S}, a \in \mathcal{A}} p_{\mathcal{M}}(s, a) payoff(s, a).$$

Hence, by Jensen's inequality (Theorem 2.5), we have

$$\log Pd_0 \geq \log \sum_{s \in \mathcal{S}, a \in \mathcal{A}} p_{\mathcal{M}}(s, a) payoff(s, a)$$

$$\geq \sum_{s \in \mathcal{S}, a \in \mathcal{A}} p_{\mathcal{M}}(s, a) \log payoff(s, a).$$

Recalling from Section 10.2 that

$$p_{\mathcal{M}}(s, a) = p_{\mathcal{S}}(s) \times payoff(s, a),$$

we see that

$$\log Pd_0 \geq \sum_{s \in \mathcal{S}, a \in \mathcal{A}} p_{\mathcal{S}}(s) payoff(s, a) \log payoff(s, a).$$

Now, we observe that $payoff(s, a) = p_{\mathcal{A}}(a|s)$ (i.e., the probability that a is the authenticator, given that s is the source state). Hence,

$$\log Pd_0 \geq \sum_{s \in \mathcal{S}, a \in \mathcal{A}} p_{\mathcal{S}}(s) p_{\mathcal{A}}(a|s) \log p_{\mathcal{A}}(a|s)$$

$$= -H(\mathbf{A}|\mathbf{S}),$$

by the definition of conditional entropy. We complete the proof by showing that $-H(\mathbf{A}|\mathbf{S}) = H(\mathbf{K}|\mathbf{M}) - H(\mathbf{K})$. This follows from basic entropy identities. On one hand, we have

$$H(\mathbf{K}, \mathbf{A}, \mathbf{S}) = H(\mathbf{K}|\mathbf{A}, \mathbf{S}) + H(\mathbf{A}|\mathbf{S}) + H(\mathbf{S}).$$

On the other hand, we compute

$$H(\mathbf{K}, \mathbf{A}, \mathbf{S}) = H(\mathbf{A}|\mathbf{K}, \mathbf{S}) + H(\mathbf{K}, \mathbf{S})$$
$$= H(\mathbf{K}) + H(\mathbf{S}),$$

where we use the facts that $H(\mathbf{A}|\mathbf{K}, \mathbf{S}) = 0$ since the key and source state uniquely determine the authenticator, and $H(\mathbf{K}, \mathbf{S}) = H(\mathbf{K}) + H(\mathbf{S})$ since the source and key are independent events.

Equating the two expressions for $H(\mathbf{K}, \mathbf{A}, \mathbf{S})$, we obtain

$$-H(\mathbf{A}|\mathbf{S}) = H(\mathbf{K}|\mathbf{A}, \mathbf{S}) - H(\mathbf{K}).$$

But a message $m = (s, a)$ is defined to consist of a source state and an authenticator (i.e., $\mathcal{M} = \mathcal{S} \times \mathcal{A}$). Hence, $H(\mathbf{K}|\mathbf{A}, \mathbf{S}) = H(\mathbf{K}|\mathbf{M})$ and the proof is complete. ∎

There is a similar bound for Pd_1 which we will not prove here. It is as follows:

THEOREM 10.14
Suppose that $(\mathcal{S}, \mathcal{A}, \mathcal{K}, \mathcal{E})$ *is an authentication code. Then*

$$\log Pd_1 \geq H(\mathbf{K}|\mathbf{M}^2) - H(\mathbf{K}|\mathbf{M}).$$

We need to define what we mean by the random variable $\mathbf{M}^2$. Suppose we authenticate two distinct source states using the same key K. In this way, we obtain an ordered pair of messages $(m_1, m_2) \in \mathcal{M} \times \mathcal{M}$. In order to define a probability distribution on $\mathcal{M} \times \mathcal{M}$, it is necessary to define a probability distribution on $\mathcal{S} \times \mathcal{S}$, with the stipulation that $p_{\mathcal{S} \times \mathcal{S}}(s, s) = 0$ for every $s \in \mathcal{S}$ (that is, we do not allow source states to be repeated). The probability distributions on $\mathcal{K}$ and $\mathcal{S} \times \mathcal{S}$ will induce a probability distribution on $\mathcal{M} \times \mathcal{M}$, in the same way that the probability distributions on $\mathcal{K}$ and $\mathcal{S}$ induce a probability distribution on $\mathcal{M}$.

As an illustration of the two bounds, we consider our basic orthogonal array construction and show that the bounds of Theorems 10.13 and 10.14 are both met with equality. First, it is clear that

$$H(\mathbf{K}) = \log \lambda n^2,$$

since each of the λn^2 authentication rules are chosen with equal probability. Let's next turn to the computation of $H(\mathbf{K}|\mathbf{M})$. If any message $m = (s, a)$ is observed, this restricts the possible keys to a subset of size λn. Each of these λn keys is equally likely. Hence, $H(\mathbf{K}|m) = \log \lambda n$, for any message m. Then, we get the following:

$$H(\mathbf{K}|\mathbf{M}) = \sum_{m \in \mathcal{M}} p_{\mathcal{M}}(m) H(\mathbf{K}|m)$$

$$= \sum_{m \in \mathcal{M}} p_{\mathcal{M}}(m) \log \lambda n$$

$$= \log \lambda n.$$

Thus we have

$$H(\mathbf{K}|\mathbf{M}) - H(\mathbf{K}) = \log \lambda n - \log \lambda n^2 = -\log n = \log Pd_0,$$

so the bound is met with equality.

If we observe two messages which have been produced using the same key (and different source states), then the number of possible keys is reduced to λ. Using similar reasoning as above, we have that $H(\mathbf{K}|\mathbf{M}^2) = \log \lambda$. Then

$$H(\mathbf{K}|\mathbf{M}^2) - H(\mathbf{K}|\mathbf{M}) = \log \lambda - \log \lambda n = -\log n = \log Pd_1,$$

so this bound is also met with equality.

10.5 Notes and References

Authentication codes were invented in 1974 by Gilbert, MacWilliams, and Sloane [GMS74]. Much of the theory of authentication codes was developed by Simmons, who proved many fundamental results in the area. Two useful survey articles by Simmons are [SI92] and [SI88]. Another good survey is Massey [MA86].

The connections between orthogonal arrays and authentication codes has been addressed by several researchers. The treatment here is based on three papers by Stinson [ST88], [ST90] and [ST92]. Orthogonal arrays have been studied for over 45 years by researchers in statistics and in combinatorial design theory. For example, the bound in Theorem 10.9 was first proved by Plackett and Berman in 1945 in [PB45]. Many interesting results on orthogonal arrays can be found in various textbooks on combinatorial design theory such as Beth, Jungnickel, and Lenz [BJL85].

Finally, the use of entropy techniques in the study of authentication codes was introduced by Simmons. The bound of Theorem 10.13 was first proved in Simmons [SI85]; a proof of Theorem 10.14 can be found in Walker [WA90].

Exercises

10.1 Compute Pd_0 and Pd_1 for the following authentication code, represented in matrix form:

key	1	2	3	4
1	1	1	2	3
2	1	2	3	1
3	2	1	3	1
4	2	3	1	2
5	3	2	1	3
6	3	3	2	1

The probability distributions on S and $\mathcal{K}$ are as follows:

$$p_S(1) = p_S(4) = 1/6, p_S(2) = p_S(3) = 1/3$$
$$p_\mathcal{K}(1) = p_\mathcal{K}(6) = 1/4, p_\mathcal{K}(2) = p_\mathcal{K}(3) = p_\mathcal{K}(4) = p_\mathcal{K}(5) = 1/8.$$

What are the optimal impersonation and substitution strategies?

10.2 We have seen a construction for an orthogonal array $OA(p, p, 1)$ when p is prime. Prove that this $OA(p, p, 1)$ can always be extended by one extra column to form an $OA(p, p + 1, 1)$. Illustrate your construction in the case $p = 5$.

10.3 Suppose A is an $OA(n_1, k, \lambda_1)$ on symbol set $\{1, \ldots, n_1\}$ and suppose B is an $OA(n_2, k, \lambda_2)$ on symbol set $\{1, \ldots, n_2\}$. We construct C, an $OA(n_1 n_2, k, \lambda_1 \lambda_2)$ on symbol set $\{1, \ldots, n_1\} \times \{1, \ldots, n_2\}$, as follows: for each row $r_1 = (x_1, \ldots, x_k)$ of A and for each row $s_1 = (y_1, \ldots, y_k)$ of B, define a row

$$t_1 = ((x_1, y_1), \ldots, (x_k, y_k))$$

of C. Prove that C is indeed an $OA(n_1 n_2, k, \lambda_1 \lambda_2)$.

10.4 Construct an orthogonal array $OA(3, 13, 3)$.

10.5 Write a computer program to compute $H(\mathbf{K})$, $H(\mathbf{K}|\mathbf{M})$ and $H(\mathbf{K}|\mathbf{M}^2)$ for the authentication code from Exercise 10.1. The probability distribution on sequences of two sources is as follows:

$$p_{S^2}(1, 2) = p_{S^2}(1, 3) = p_{S^2}(1, 4) = 1/18$$
$$p_{S^2}(2, 1) = p_{S^2}(2, 3) = p_{S^2}(2, 4) = 1/9$$
$$p_{S^2}(3, 1) = p_{S^2}(3, 2) = p_{S^2}(3, 4) = 1/9$$
$$p_{S^2}(4, 1) = p_{S^2}(4, 2) = p_{S^2}(4, 3) = 1/18$$

Compare the entropy bounds for Pd_0 and Pd_1 with the actual values you computed in Exercise 10.1.

HINT To compute $p_\mathcal{K}(k|m)$, use Bayes' formula

$$p_\mathcal{K}(k|m) = \frac{p_\mathcal{M}(m|k) p_\mathcal{K}(k)}{p_\mathcal{M}(m)}.$$

We already know how to calculate $p_\mathcal{M}(m)$. To compute $p_\mathcal{M}(m|k)$, write $m = (s, a)$, and then observe that $p_\mathcal{M}(m|k) = p_S(s)$ if $e_k(s) = a$, and $p_\mathcal{M}(m|k) = 0$ otherwise.

To compute $p_\mathcal{K}(k|m_1, m_2)$, use Bayes' formula

$$p_\mathcal{K}(k|m_1, m_2) = \frac{p_{\mathcal{M}^2}(m_1, m_2|k) p_\mathcal{K}(k)}{p_{\mathcal{M}^2}(m_1, m_2)}.$$

$p_{\mathcal{M}^2}(m_1, m_2)$ can be calculated as follows: write $m_1 = (s_1, a_1)$ and $m_2 = (s_2, a_2)$. Then

$$p_{\mathcal{M}^2}(m_1, m_2) = p_{S^2}(s_1, s_2) \times \sum_{\{K \in \mathcal{K} : e_k(s_1) = a_1, e_k(s_2) = a_2\}} p_\mathcal{K}(K).$$

(Note the similarity with the computation of $p(m)$.) To compute $p_{\mathcal{M}^2}(m_1, m_2|k)$, observe that $p_{\mathcal{M}^2}(m_1, m_2|k) = p_{S^2}(s_1, s_2)$ if $e_k(s_1) = a_1$ and $e_k(s_2) = a_2$, and $p_{\mathcal{M}^2}(m_1, m_2|k) = 0$, otherwise.

11

Secret Sharing Schemes

11.1 Introduction: The Shamir Threshold Scheme

In a bank, there is a vault which must be opened every day. The bank employs three senior tellers, but they do not trust the combination to any individual teller. Hence, we would like to design a system whereby any two of the three senior tellers can gain access to the vault, but no individual teller can do so. This problem can be solved by means of a *secret sharing scheme*, the topic of this chapter.

Here is an interesting "real-world" example of this situation: According to *Time Magazine*[1], control of nuclear weapons in Russia involves a similar "two-out-of-three" access mechanism. The three parties involved are the President, the Defense Minister and the Defense Ministry.

We first study a special type of secret sharing scheme called a threshold scheme. Here is an informal definition.

DEFINITION 11.1 *Let t, w be positive integers, $t \leq w$. A (t, w)-threshold scheme is a method of sharing a key K among a set of w participants (denoted by $\mathcal{P}$), in such a way that any t participants can compute the value of K, but no group of $t - 1$ participants can do so.*

Note that the examples described above are $(2, 3)$-threshold schemes.

The value of K is chosen by a special participant called the *dealer*. The dealer is denoted by D and we assume $D \notin \mathcal{P}$. When D wants to share the key K among the participants in $\mathcal{P}$, he gives each participant some partial information called a *share*. The shares should be distributed secretly, so no participant knows the share given to another participant.

At a later time, a subset of participants $B \subseteq \mathcal{P}$ will pool their shares in an attempt to compute the key K. (Alternatively, they could give their shares to a trusted authority which will perform the computation for them.) If $|B| \geq t$, then

---

[1] Time Magazine, May 4, 1992, p. 13

FIGURE 11.1
The Shamir (t, w)-threshold scheme in $\mathbb{Z}_p$

Initialization Phase

1. D chooses w distinct, non-zero elements of $\mathbb{Z}_p$, denoted x_i, $1 \leq i \leq w$ (this is where we require $p \geq w + 1$). For $1 \leq i \leq w$, D gives the value x_i to P_i. The values x_i are public.

Share Distribution

2. Suppose D wants to share a key $K \in \mathbb{Z}_p$. D secretly chooses (independently at random) $t - 1$ elements of $\mathbb{Z}_p$, $a_1, \ldots, a_{t-1}$.

3. For $1 \leq i \leq w$, D computes $y_i = a(x_i)$, where

$$a(x) = K + \sum_{j=1}^{t-1} a_j x^j \bmod p.$$

4. For $1 \leq i \leq w$, D gives the share y_i to P_i.

---

they should be able to compute the value of K as a function of the shares they collectively hold; if $|B| < t$, then they should not be able to compute K.

We will use the following notation. Let

$$\mathcal{P} = \{P_i : 1 \leq i \leq w\}$$

be the set of w participants. $\mathcal{K}$ is the *key set* (i.e., the set of all possible keys); and $\mathcal{S}$ is the *share set* (i.e., the set of all possible shares).

In this section, we present a method of constructing a (t, w)-threshold scheme, called the **Shamir Threshold Scheme**, which was invented in 1979. Let $\mathcal{K} = \mathbb{Z}_p$, where $p \geq w + 1$ is prime. Also, let $\mathcal{S} = \mathbb{Z}_p$. Hence, the key will be an element of $\mathbb{Z}_p$, as will be each share given to a participant. The Shamir threshold scheme is presented in Figure 11.1. In this scheme, the dealer constructs a random polynomial $a(x)$ of degree at most $t - 1$ in which the constant term is the key, K. Every participant P_i obtains a point (x_i, y_i) on this polynomial.

Let's look at how a subset B of t participants can reconstruct the key. This is basically accomplished by means of polynomial interpolation. We will describe a couple of methods of doing this.

Suppose that participants $P_{i_1}, \ldots, P_{i_t}$ want to determine K. They know that

$$y_{i_j} = a(x_{i_j}),$$

$1 \leq j \leq t$, where $a(x) \in \mathbb{Z}_p[x]$ is the (secret) polynomial chosen by D. Since

$a(x)$ has degree at most $t - 1$, $a(x)$ can be written as

$$a(x) = a_0 + a_1x + \ldots + a_{t-1}x^{t-1},$$

where the coefficients $a_0, \ldots, a_{t-1}$ are unknown elements of $\mathbb{Z}_p$, and $a_0 = K$ is the key. Since $y_{i_j} = a(x_{i_j})$, $1 \leq j \leq t$, B can obtain t linear equations in the t unknowns $a_0, \ldots, a_{t-1}$, where all arithmetic is done in $\mathbb{Z}_p$. If the equations are linearly independent, there will be a unique solution, and a_0 will be revealed as the key.

Here is a small example to illustrate.

Example 11.1
Suppose that $p = 17$, $t = 3$, and $w = 5$; and the public x-co-ordinates are $x_i = i$, $1 \leq i \leq 5$. Suppose that $B = \{P_1, P_3, P_5\}$ pool their shares, which are respectively 8, 10, and 11. Writing the polynomial $a(x)$ as

$$a(x) = a_0 + a_1x + a_2x^2,$$

and computing $a(1)$, $a(3)$ and $a(5)$, the following three linear equations in $\mathbb{Z}_{17}$ are obtained:

$$a_0 + a_1 + a_2 = 8$$
$$a_0 + 3a_1 + 9a_2 = 10$$
$$a_0 + 5a_1 + 8a_2 = 11.$$

This system does have a unique solution in $\mathbb{Z}_{17}$: $a_0 = 13$, $a_1 = 10$, and $a_2 = 2$. The key is therefore $K = a_0 = 13$.   ▯

Clearly, it is important that the system of t linear equations has a unique solution, as in Example 11.1. We show now that this is always the case. In general, we have

$$y_{i_j} = a(x_{i_j}),$$

$1 \leq j \leq t$, where

$$a(x) = a_0 + a_ix + \ldots + a_{t-1}x^{t-1}$$

and

$$a_0 = K.$$

The system of linear equations (in $\mathbb{Z}_p$) is the following:

$$
\begin{aligned}
a_0 + a_1x_{i_1} + a_2x_{i_1}{}^2 + \ldots + a_{t-1}x_{i_1}{}^{t-1} &= y_{i_1} \\
a_0 + a_1x_{i_2} + a_2x_{i_2}{}^2 + \ldots + a_{t-1}x_{i_2}{}^{t-1} &= y_{i_2} \\
&\vdots \\
a_0 + a_1x_{i_t} + a_2x_{i_t}{}^2 + \ldots + a_{t-1}x_{i_t}{}^{t-1} &= y_{i_t}.
\end{aligned}
$$

This can be written in matrix form as follows:

$$
\begin{pmatrix}
1 & x_{i_1} & x_{i_1}{}^2 & \cdots & x_{i_1}{}^{t-1} \\
1 & x_{i_2} & x_{i_2}{}^2 & \cdots & x_{i_2}{}^{t-1} \\
\vdots & \vdots & \vdots & & \vdots \\
1 & x_{i_t} & x_{i_t}{}^2 & \cdots & x_{i_t}{}^{t-1}
\end{pmatrix}
\begin{pmatrix}
a_0 \\ a_1 \\ \vdots \\ a_{t-1}
\end{pmatrix}
=
\begin{pmatrix}
y_{i_1} \\ y_{i_2} \\ \vdots \\ y_{i_t}
\end{pmatrix}.
$$

Now, the coefficient matrix A is a so-called Vandermonde matrix. There is a well-known formula for the determinant of a Vandermonde matrix, namely

$$
\det A = \prod_{1 \le j < k \le t} (x_{i_k} - x_{i_j}) \bmod p.
$$

Recall that the x_i's are all distinct, so no term $x_{i_j} - x_{i_k}$ in this product is equal to zero. The product is computed in $\mathbb{Z}_p$, where p is prime, which is a field. Since the product of non-zero terms in a field is always non-zero, we have that $\det A \ne 0$. Since the determinant of the coefficient matrix is non-zero, the system has a unique solution over the field $\mathbb{Z}_p$. This establishes that any group of t participants will be able to recover the key in this threshold scheme.

What happens if a group of $t - 1$ participants attempt to compute K? Proceeding as above, they will obtain a system of $t - 1$ equations in t unknowns. Suppose they hypothesize a value y_0 for the key. Since the key is $a_0 = a(0)$, this will yield a tth equation, and the coefficient matrix of the resulting system of t equations in t unknowns will again be a Vandermonde matrix. As before, there will be a unique solution. Hence, for every hypothesized value y_0 of the key, there is a unique polynomial $a_{y_0}(x)$ such that

$$
y_{i_j} = a_{y_0}(x_{i_j}),
$$

$1 \le j \le t - 1$, and such that

$$
y_0 = a_{y_0}(0).
$$

Hence, no value of the key can be ruled out, and thus a group of $t - 1$ participants can obtain no information about the key.

We have analyzed the Shamir scheme from the point of view of solving systems of linear equations over $\mathbb{Z}_p$. There is an alternative method, based on the Lagrange interpolation formula for polynomials. The Lagrange interpolation formula is an explicit formula for the (unique) polynomial $a(x)$ of degree at most t that we computed above. The formula is as follows:

$$
a(x) = \sum_{j=1}^{t} y_{i_j} \prod_{1 \le k \le t,\, k \ne j} \frac{x - x_{i_k}}{x_{i_j} - x_{i_k}}.
$$

It is easy to verify the correctness of this formula by substituting $x = x_{i_j}$: all terms in the summation vanish except for the jth term, which is y_{i_j}. Thus, we have a polynomial of degree at most $t - 1$ which contains the t ordered pairs

(x_{i_j}, y_{i_j}), $1 \leq j \leq t$. We already proved above that this polynomial is unique, so the interpolation formula does yield the correct polynomial.

A group B of t participants can compute $a(x)$ by using the interpolation formula. But a simplification is possible, since the participants in B do not need to know the whole polynomial $a(x)$. It is sufficient for them to compute the constant term $K = a(0)$. Hence, they can compute the following expression, which is obtained by substituting $x = 0$ into the Lagrange interpolation formula:

$$K = \sum_{j=1}^{t} y_{i_j} \prod_{1 \leq k \leq t, k \neq j} \frac{x_{i_k}}{x_{i_k} - x_{i_j}}.$$

Suppose we define

$$b_j = \prod_{1 \leq k \leq t, k \neq j} \frac{x_{i_k}}{x_{i_k} - x_{i_j}},$$

$1 \leq j \leq t$. (Note that these values b_j can be precomputed, if desired, and their values are not secret.) Then we have

$$K = \sum_{j=1}^{t} b_j y_{i_j}.$$

Hence, the key is a linear combination of the t shares.

To illustrate this approach, let's recompute the key from Example 11.1.

Example 11.1   *(Cont.)*
The participants $\{P_1, P_3, P_5\}$ can compute b_1, b_2, and b_3 according to the formula given above. For example, they would obtain

$$b_1 = \frac{x_3 x_5}{(x_3 - x_1)(x_5 - x_1)} \bmod 17$$

$$= 3 \times 5 \times (-2)^{-1} \times (-4)^{-1} \bmod 17$$

$$= 4.$$

Similarly, $b_2 = 3$ and $b_3 = 11$. Then, given shares $8, 10$, and 11 (respectively), they would obtain

$$K = 4 \times 8 + 3 \times 10 + 11 \times 11 \bmod 17 = 13,$$

as before.   □

The last topic of this section is a simplified construction for threshold schemes in the special case $w = t$. This construction will work for any key set $\mathcal{K} = \mathbb{Z}_m$ with $\mathcal{S} = \mathbb{Z}_m$. (For this scheme, it is not required that m be prime, and it is not necessary that $m \geq w + 1$.) If D wants to share the key $K \in \mathbb{Z}_m$, he carries out the protocol of Figure 11.2.

FIGURE 11.2
A (t, t)-threshold scheme in $\mathbb{Z}_m$

---

1. D secretly chooses (independently at random) $t - 1$ elements of $\mathbb{Z}_m$, $y_1, \ldots, y_{t-1}$.
2. D computes
$$y_t = K - \sum_{i=1}^{t-1} y_i \bmod m.$$
3. For $1 \leq i \leq t$, D gives the share y_i to P_i.

Observe that the t participants can compute K by the formula

$$K = \sum_{i=1}^{t} y_i \bmod m.$$

Can $t-1$ participants compute K? Clearly, the first $t-1$ participants cannot do so, since they receive $t - 1$ independent random numbers as their shares. Consider the $t - 1$ participants in the set $\mathcal{P} \setminus \{P_i\}$, where $1 \leq i \leq t - 1$. These $t - 1$ participants possess the shares

$$y_1, \ldots, y_{i-1}, y_{i+1}, \ldots, y_{t-1}$$

and

$$K - \sum_{i=1}^{t-1} y_i.$$

By summing their shares, they can compute $K - y_i$. However, they do not know the random value y_i, and hence they have no information as to the value of K. Consequently, we have a (t, t)-threshold scheme.

11.2 Access Structures and General Secret Sharing

In the previous section, we desired that any t of the w participants should be able to determine the key. A more general situation is to specify exactly which subsets of participants should be able to determine the key and which should not. Let Γ be a set of subsets of $\mathcal{P}$; the subsets in Γ are those subsets of participants that should be able to compute the key. Γ is called an *access structure* and the subsets in Γ are called *authorized subsets*.

Let $\mathcal{K}$ be the key set and let $\mathcal{S}$ be the share set. As before, when a dealer D wants to share a key $K \in \mathcal{K}$, he will give each participant a share from $\mathcal{S}$. At a later time a subset of participants will attempt to determine K from the shares they collectively hold.

DEFINITION 11.2   *A perfect secret sharing scheme realizing the access structure Γ is a method of sharing a key K among a set of w participants (denoted by $\mathcal{P}$), in such a way that the following two properties are satisfied:*

1. *If an authorized subset of participants $B \subseteq \mathcal{P}$ pool their shares, then they can determine the value of K.*

2. *If an unauthorized subset of participants $B \subseteq \mathcal{P}$ pool their shares, then they can determine nothing about the value of K.*

Observe that a (t, w)-threshold scheme realizes the access structure

$$\{B \subseteq \mathcal{P} : |B| \geq t\}.$$

Such an access structure is called a *threshold* access structure. We showed in the previous section that the Shamir scheme is a perfect scheme realizing the threshold access structure.

We study the unconditional security of secret sharing schemes. That is, we do not place any limit on the amount of computation that can be performed by an unauthorized subset of participants.

Suppose that $B \in \Gamma$ and $B \subseteq C \subseteq \mathcal{P}$. Suppose the subset C wants to determine K. Since B is an authorized subset, it can already determine K. Hence, the subset C can determine K by ignoring the shares of the participants in $C \backslash B$. Stated another way, a superset of an authorized set is again an authorized set. What this says is that the access structure should satisfy the *monotone* property:

$$\text{if } B \in \Gamma \text{ and } B \subseteq C \subseteq \mathcal{P}, \text{ then } C \in \Gamma.$$

In the remainder of this chapter, we will assume that all access structures are monotone.

If Γ is an access structure, then $B \in \Gamma$ is a *minimal* authorized subset if $A \notin \Gamma$ whenever $A \subseteq B$, $A \neq B$. The set of minimal authorized subsets of Γ is denoted Γ_0 and is called the *basis* of Γ. Since Γ consists of all subsets of $\mathcal{P}$ that are supersets of a subset in the basis Γ_0, Γ is determined uniquely as a function of Γ_0. Expressed mathematically, we have

$$\Gamma = \{C \subseteq \mathcal{P} : B \subseteq C, B \in \Gamma_0\}.$$

We say that Γ is the *closure* of Γ_0 and write

$$\Gamma = cl(\Gamma_0).$$

Example 11.2
Suppose $\mathcal{P} = \{P_1, P_2, P_3, P_4\}$ and

$$\Gamma_0 = \{\{P_1, P_2, P_4\}, \{P_1, P_3, P_4\}, \{P_2, P_3\}\}.$$

Then
$$\Gamma = \Gamma_0 \bigcup \{\{P_1, P_2, P_3\}, \{P_2, P_3, P_4\}, \{P_1, P_2, P_3, P_4\}\}.$$

Conversely, given this access structure Γ, it is easy to see that Γ_0 consists of the minimal subsets in Γ. ▯

In the case of a (t, w)-threshold access structure, the basis consists of all subsets of (exactly) t participants.

11.3 The Monotone Circuit Construction

In this section, we will give a conceptually simple and elegant construction due to Benaloh and Leichter that shows that any (monotone) access structure can be realized by a perfect secret sharing scheme. The idea is to first build a monotone circuit that "recognizes" the access structure, and then to build the secret sharing scheme from the description of the circuit. We call this the *monotone circuit construction*.

Suppose we have a boolean circuit **C**, with w boolean inputs, $x_1, \ldots, x_w$ (corresponding to the w participants $P_1, \ldots, P_w$), and one boolean output, y. The circuit consists of "or" gates and "and" gates; we do not allow any "not" gates. Such a circuit is called a *monotone* circuit. The reason for this nomenclature is that changing any input x_i from "0" (false) to "1" (true) can never result in the output y changing from "1" to "0." The circuit is permitted to have arbitrary fan-in, but we require fan-out equal to 1 (that is, a gate can have arbitrarily many input wires, but only one output wire).

If we specify boolean values for the w inputs of such a monotone circuit, we can define

$$B(x_1, \ldots, x_w) = \{P_i : x_i = 1\},$$

i.e., the subset of $\mathcal{P}$ corresponding to the true inputs. Suppose **C** is a monotone circuit, and define

$$\Gamma(\mathbf{C}) = \{B(x_1, \ldots, x_w) : \mathbf{C}(x_1, \ldots, x_w) = 1\},$$

where $\mathbf{C}(x_1, \ldots, x_w)$ denotes the output of **C**, given inputs $x_1, \ldots, x_w$. Since the circuit **C** is monotone, it follows that $\Gamma(\mathbf{C})$ is a monotone set of subsets of $\mathcal{P}$.

It is easy to see that there is a one-to-one correspondence between monotone circuits of this type and boolean formulae which contain the operators $\wedge$ ("and") and $\vee$ ("or"), but do not contain any negations.

If Γ is a monotone set of subsets of $\mathcal{P}$, then it is easy to construct a monotone circuit $\mathbf{C}$ such that $\Gamma(\mathbf{C}) = \Gamma$. One way to do this is as follows. Let Γ_0 be the basis of Γ. Then construct the *disjunctive normal form boolean formula*

$$\bigvee_{B \in \Gamma_0} \left(\bigwedge_{P_i \in B} P_i \right).$$

In Example 11.2, where

$$\Gamma_0 = \{\{P_1, P_2, P_4\}, \{P_1, P_3, P_4\}, \{P_2, P_3\}\},$$

we would obtain the boolean formula

$$(P_1 \wedge P_2 \wedge P_4) \vee (P_1 \wedge P_3 \wedge P_4) \vee (P_2 \wedge P_3). \tag{11.1}$$

Each clause in the boolean formula corresponds to an "and" gate of the associated monotone circuit; the final disjunction corresponds to an "or" gate. The number of gates in the circuit is $|\Gamma_0| + 1$.

Suppose $\mathbf{C}$ is any monotone circuit that recognizes Γ (note that $\mathbf{C}$ need not be the circuit described above.) We describe an algorithm which enables D, the dealer, to construct a perfect secret sharing scheme that realizes Γ. This scheme will use as a building block the (t, t)-schemes constructed in Figure 11.2. Hence, we take the key set to be $\mathcal{K} = \mathbb{Z}_m$ for some integer m.

The algorithm proceeds by assigning a value $f(W) \in \mathcal{K}$ to every wire W in the circuit $\mathbf{C}$. Initially, the output wire W_{out} of the circuit is assigned the value K, the key. The algorithm iterates a number of times, until every wire has a value assigned to it. Finally, each participant P_i is given the list of values $f(W)$ such that W is an input wire of the circuit which receives input x_i.

A description of the construction is given in Figure 11.3. Note that, whenever a gate G is an "and" gate having (say) t input wires, we share the "key" $f(W_G)$ among the input wires using a (t, t)-threshold scheme.

Let's carry out this procedure for the access structure of Example 11.2, using the circuit corresponding to the boolean formula (11.1).

Example 11.3

We illustrate the construction in Figure 11.4. Suppose K is the key. The value K is given to each of the three input wires of the final "or" gate. Next, we consider the "and" gate corresponding to the clause $P_1 \wedge P_2 \wedge P_4$. The three input wires are assigned values $a_1, a_2, K - a_1 - a_2$, respectively, where all arithmetic is done in $\mathbb{Z}_m$. In a similar way, the three input wires corresponding to $P_1 \wedge P_3 \wedge P_4$ are assigned values $b_1, b_2, K - b_1 - b_2$. Finally, the two input wires corresponding to $P_2 \wedge P_3$ are assigned values $c_1, K - c_1$. Note that a_1, a_2, b_1, b_2 and c_1 are all independent random values in $\mathbb{Z}_m$. If we look at the shares that the four participants receive, we have the following:

FIGURE 11.3
The monotone circuit construction

1. $f(W_{out}) = K$
2. **while** there exists a wire W such that $f(W)$ is not defined **do**
3.   find a gate G of $\mathbf{C}$ such that $f(W_G)$ is defined, where W_G is the output wire of G, but $f(W)$ is not defined for any of the input wires of G
4.   **if** G is an "or" gate **then**
5. $f(W) = f(W_G)$ for every input wire W of G
6.   **else** (G is an "and" gate)
7.   let the input wires of G be $W_1, \ldots, W_t$
8.   choose (independently at random) $t - 1$ elements of $\mathbb{Z}_m$, denoted by $y_{G,1}, \ldots, y_{G,t-1}$
9.   compute

$$y_{G,t} = f(W_G) - \sum_{i=1}^{t-1} y_{G,i} \bmod m$$

10.   **for** $1 \le i \le t$ **do**
11. $f(W_i) = y_{G,i}$

1. P_1 receives a_1, b_1.
2. P_2 receives a_2, c_1.
3. P_3 receives $b_2, K - c_1$.
4. P_4 receives $K - a_1 - a_2, K - b_1 - b_2$.

Thus, every participant receives two elements of $\mathbb{Z}_m$ as his or her share.

Let's prove that the scheme is perfect. First, we verify that each basis subset can compute K. The authorized subset $\{P_1, P_2, P_4\}$ can compute

$$K = a_1 + a_2 + (K - a_1 - a_2) \bmod m.$$

The subset $\{P_1, P_3, P_4\}$ can compute

$$K = b_1 + b_2 + (K - b_1 - b_2) \bmod m.$$

Finally, the subset $\{P_2, P_3\}$ can compute

$$K = c_1 + (K - c_1) \bmod m.$$

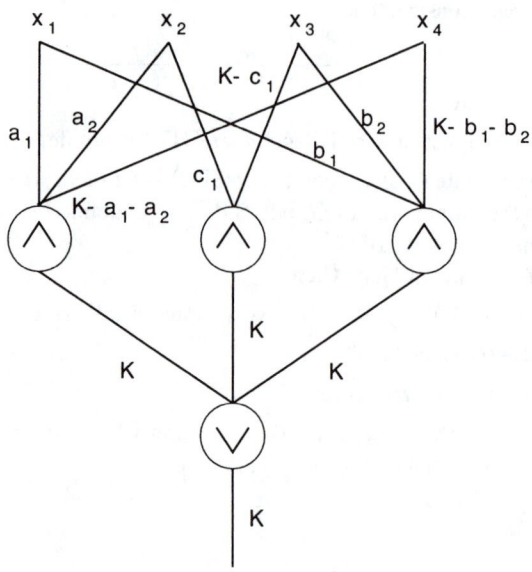

FIGURE 11.4
A monotone circuit

Thus any authorized subset can compute K, so we turn our attention to the unauthorized subsets. Note that we do not need to look at all the unauthorized subsets. For, if B_1 and B_2 are both unauthorized subsets, $B_1 \subseteq B_2$, and B_2 cannot compute K, then neither can B_1 compute K. Define a subset $B \subseteq \mathcal{P}$ to be a *maximal* unauthorized subset if $B_1 \in \Gamma$ for all $B_1 \supseteq B$, $B_1 \neq B$. It follows that it suffices to verify that none of the maximal unauthorized subsets can determine any information about K. Here, the maximal unauthorized subsets are

$$\{P_1, P_2\}, \{P_1, P_3\}, \{P_1, P_4\}, \{P_2, P_4\}, \{P_3, P_4\}.$$

In each case, it is easy to see that K cannot be computed, either because some necessary piece of "random" information is missing, or because all the shares possessed by the subset are random. For example, the subset $\{P_1, P_2\}$ possesses only the random values a_1, b_1, a_2, c_1. As another example, the subset $\{P_3, P_4\}$ possesses the shares $b_2, K - c_1, K - a_1 - a_2, K - b_1 - b_2$. Since the values of c_1, a_1, a_2, and b_1 are unknown random values, K cannot be computed. In each possible case, an unauthorized subset has no information about the value of K.
$\square$

We can obtain a different scheme realizing the same access structure by using a different circuit. We illustrate by returning again to the access structure of Example 11.2.

Example 11.4
Suppose we convert the formula (11.1) to the so-called conjunctive normal form:

$$(P_1 \vee P_2) \wedge (P_1 \vee P_3) \wedge (P_2 \vee P_3) \wedge (P_2 \vee P_4) \wedge (P_3 \vee P_4). \qquad (11.2)$$

(The reader can verify that this formula is equivalent to the formula (11.1).) If we implement the scheme using the circuit corresponding to formula (11.2), then we obtain the following:

1. P_1 receives a_1, a_2.
2. P_2 receives a_1, a_3, a_4.
3. P_3 receives $a_2, a_3, K - a_1 - a_2 - a_3 - a_4$.
4. P_4 receives $a_4, K - a_1 - a_2 - a_3 - a_4$.

We leave the details for the reader to check. ⬜

We now prove that the monotone circuit construction always produces a perfect secret sharing scheme.

THEOREM 11.1
Let $\mathbf{C}$ be any monotone boolean circuit. Then the monotone circuit construction yields a perfect secret sharing scheme realizing the access structure $\Gamma(\mathbf{C})$.

PROOF We proceed by induction on the number of gates in the circuit $\mathbf{C}$. If $\mathbf{C}$ contains only one gate, then the result is fairly trivial: If $\mathbf{C}$ consists of one "or" gate, then every participant will be given the key. This scheme realizes the access structure consisting of all non-empty subsets of participants. If $\mathbf{C}$ consists of a single "and" gate with t inputs, then the scheme is the (t, t)-threshold scheme presented in Figure 11.2.

Now, as an induction assumption, suppose that there is an integer $j > 1$ such that, for all circuits $\mathbf{C}$ with fewer than j gates, the construction produces a scheme that realizes $\Gamma(\mathbf{C})$. Let $\mathbf{C}$ be a circuit on j gates. Consider the "last" gate, G, in the circuit; again, G could be either an "or" gate or an "and" gate. Let's first consider the case where G is an "or" gate. Denote the input wires to G by W_i, $1 \leq i \leq t$. These t input wires are the outputs of t sub-circuits of $\mathbf{C}$, which we denote $\mathbf{C}_i$, $1 \leq i \leq t$. Corresponding to each $\mathbf{C}_i$, we have a (sub-)scheme that realizes the access structure $\Gamma_{\mathbf{C}_i}$, by induction. Now, it is easy to see that

$$\Gamma(\mathbf{C}) = \bigcup_{i=1}^{t} \Gamma_{\mathbf{C}_i}.$$

Since every W_i is assigned the key K, it follows that the scheme realizes $\Gamma(\mathbf{C})$, as desired.

The analysis is similar if G is an "and" gate. In this situation, we have

$$\Gamma(\mathbf{C}) = \bigcap_{i=1}^{t} \Gamma_{\mathbf{C}_i}.$$

Since the key K is shared among the t wires W_i using a (t, t)-threshold scheme, it follows again that the scheme realizes $\Gamma(\mathbf{C})$. This completes the proof. ∎

Of course, when an authorized subset, B, wants to compute the key, the participants in B need to know the circuit used by D to distribute shares, and which shares correspond to which wires of the circuit. All this information will be public knowledge. Only the actual values of the shares are secret. The algorithm for reconstructing the key involves combining shares according to the circuit, with the stipulation that an "and" gate corresponds to summing the values on the input wires modulo m (provided these values are all known), and an "or" gate involves choosing the value on any input wire (with the understanding that all these values will be identical).

11.4 Formal Definitions

In this section, we will give formal mathematical definitions of a (perfect) secret sharing scheme. We represent a secret sharing scheme by a set of distribution rules. A *distribution rule* is a function

$$f : \mathcal{P} \to \mathcal{S}.$$

A distribution rule represents a possible distribution of shares to the participants, where $f(P_i)$ is the share given to P_i, $1 \leq i \leq w$.

Now, for each $K \in \mathcal{K}$, let $\mathcal{F}_K$ be a set of distribution rules. $\mathcal{F}_K$ will be distribution rules corresponding to the key having the value K. The sets of distribution rules $\mathcal{F}_K$ are public knowledge.

Next, define

$$\mathcal{F} = \bigcup_{K \in \mathcal{K}} \mathcal{F}_K.$$

$\mathcal{F}$ is the complete set of distribution rules of the scheme. If $K \in \mathcal{K}$ is the value of the key that D wishes to share, then D will choose a distribution rule $f \in \mathcal{F}_K$, and use it to distribute shares.

This is a completely general model in which we can study secret sharing schemes. Any of our existing schemes can be described in this setting by determining the possible distribution rules which the scheme will use. The fact that this model is mathematically precise makes it easier to give definitions and to present proofs.

It is useful to develop conditions which ensure that a set of distribution rules for a scheme realizes a specified access structure. This will involve looking at certain probability distributions, as we did previously when studying the concept of perfect secrecy. To begin with, we suppose that there is a probability distribution $p_{\mathcal{K}}$ on $\mathcal{K}$. Further, for every $K \in \mathcal{K}$, D will choose a distribution rule in $\mathcal{F}_K$ according to a probability distribution $p_{\mathcal{F}_K}$.

Given these probability distributions, it is straightforward to compute the probability distribution on the list of shares given to any subset of participants, B (authorized or unauthorized). This is done as follows. Suppose $B \subseteq \mathcal{P}$. Define

$$\mathcal{S}(B) = \{f|_B : f \in \mathcal{F}\},$$

where the function $f|_B$ denotes the restriction of the distribution rule f to B. That is, $f|_B : B \to S$ is defined by

$$f|_B(P_i) = f(P_i)$$

for all $P_i \in B$. Thus, $\mathcal{S}(B)$ is the set of possible distributions of shares to the participants in B.

The probability distribution on $\mathcal{S}(B)$, denoted $p_{\mathcal{S}(B)}$, is computed as follows: Let $f_B \in \mathcal{S}(B)$. Then

$$p_{\mathcal{S}(B)}(f_B) = \sum_{K \in \mathcal{K}} p_{\mathcal{K}}(K) \sum_{\{f \in \mathcal{F}_K : f|_B = f_B\}} p_{\mathcal{F}_K}(f).$$

Also,

$$p_{\mathcal{S}(B)}(f_B|K) = \sum_{\{f \in \mathcal{F}_K : f|_B = f_B\}} p_{\mathcal{F}_K}(f),$$

for all $f_B \in \mathcal{S}(B)$ and $K \in \mathcal{K}$.

Here now is a formal definition of a perfect secret sharing scheme.

DEFINITION 11.3 *Suppose Γ is an access structure and $\mathcal{F} = \cup_{K \in \mathcal{K}}\mathcal{F}_K$ is a set of distribution rules. Then $\mathcal{F}$ is a perfect secret sharing scheme realizing the access structure Γ provided that the following two properties are satisfied:*

1. *For any authorized subset of participants $B \subseteq \mathcal{P}$, there do not exist two distribution rules $f \in \mathcal{F}_K$ and $f' \in \mathcal{F}_{K'}$ with $K \neq K'$, such that $f|_B = f'|_B$. (That is, any distribution of shares to the participants in an authorized subset B determines the value of the key.)*

2. *For any unauthorized subset of participants $B \subseteq \mathcal{P}$ and for any distribution of shares $f_B \in S_B$, $p_{\mathcal{K}}(K|f_B) = p_{\mathcal{K}}(K)$ for every $K \in \mathcal{K}$. (That is, the conditional probability distribution on $\mathcal{K}$, given a distribution of shares f_B to an unauthorized subset B, is the same as the a priori probability distribution on $\mathcal{K}$. In other words, the distribution of shares to B provides no information as to the value of the key.)*

$\mathcal{F}_0$

P_1	P_2	P_3	P_4
0	0	0	0
0	1	1	3
0	2	3	1
0	3	2	2
1	0	4	0
1	1	5	3
1	2	7	1
1	3	6	2
2	4	0	0
2	5	1	3
2	6	3	1
2	7	2	2
3	4	4	0
3	5	5	3
3	6	7	1
3	7	6	2

$\mathcal{F}_1$

P_1	P_2	P_3	P_4
0	0	1	1
0	1	0	2
0	2	2	0
0	3	3	3
1	0	5	1
1	1	4	2
1	2	6	0
1	3	7	3
2	4	1	1
2	5	0	0
2	6	2	2
2	7	3	3
3	4	5	1
3	5	4	2
3	6	6	0
3	7	7	3

FIGURE 11.5
Distribution rules for a secret sharing scheme

Observe that the second property in Definition 11.3 is very similar to the concept of perfect secrecy; this similarity is why the resulting secret sharing scheme is termed "perfect."

Note that the probability $p_K(K|f_B)$ can be computed from probability distributions exhibited above using Bayes' theorem:

$$p_K(K|f_B) = \frac{p_{S(B)}(f_B|K)p_K(K)}{p_{S(B)}(f_B)}.$$

Let us now illustrate these definitions by looking at a small example.

Example 11.5
We will present the distribution rules for the scheme constructed in Example 11.4 when it is implemented in $\mathbb{Z}_2$. Each of $\mathcal{F}_0$ and $\mathcal{F}_1$ contains 16 equiprobable distribution rules. For conciseness, we replace a binary k-tuple by an integer between 0 and $2^k - 1$. If this is done, then $\mathcal{F}_0$ and $\mathcal{F}_1$ are as depicted in Figure 11.5, where each row represents a distribution rule.

This yields a perfect scheme for any probability distribution p_K on the keys. We will not perform all the verifications here, but we will look at a couple of typcial cases to illustrate the use of the two properties in Definition 11.3.

The subset $\{P_2, P_3\}$ is an authorized subset. Thus the shares that P_2 and P_3 receive should (together) determine a unique key. It can easily be checked that any distribution of shares to these two participants occurs in a distribution rule in at most one of the sets $\mathcal{F}_0$ and $\mathcal{F}_1$. For example, if P_2 has the share 3 and P_3 has the share 6, then the distribution rule must be the eighth rule in $\mathcal{F}_0$ and thus the key is 0.

On the other hand, $B = \{P_1, P_2\}$ is an unauthorized subset. It is not too hard to see that any distribution of shares to these two participants occurs in exactly one distribution rule in $\mathcal{F}_0$ and in exactly one distribution rule in $\mathcal{F}_1$. That is,

$$p_{\mathcal{S}(B)}(f_B|K) = \frac{1}{16}$$

for any $f_B \in \mathcal{S}(B)$ and for $K = 0, 1$. Next, we compute

$$p_{\mathcal{S}(B)}(f_B) = \sum_{K \in \mathcal{K}} p_{\mathcal{K}}(K) \sum_{\{f \in \mathcal{F}_K : f|_B = f_B\}} p_{\mathcal{F}_K}(f)$$

$$= \sum_{K=0}^{1} p_{\mathcal{K}}(K) \times \frac{1}{16}$$

$$= \frac{1}{16}.$$

Now, we use Bayes' theorem to compute $p_{\mathcal{K}}(K|f_B)$:

$$p_{\mathcal{K}}(K|f_B) = \frac{p_{\mathcal{S}(B)}(f_B|K)p_{\mathcal{K}}(K)}{p_{\mathcal{S}(B)}(f_B)}$$

$$= \frac{\frac{1}{16} \times p_{\mathcal{K}}(K)}{\frac{1}{16}}$$

$$= p_{\mathcal{K}}(K),$$

so the second property is satisfied for this subset B.

Similar computations can be performed for other authorized and unauthorized sets, and in each case the appropriate property is satisfied. Hence we have a perfect secret sharing scheme. $\Box$

11.5 Information Rate

The results of Section 11.3 prove that any monotone access structure can be realized by a perfect secret sharing scheme. We now want to consider the efficiency of the resulting schemes. In the case of a (t, w)-threshold scheme, we can construct a circuit corresponding to the disjunctive normal form boolean formula which will have $1 + \binom{w}{t}$ gates. Each participant will receive $\binom{w-1}{t-1}$ elements of $\mathbb{Z}_m$ as his or her share. This seems very inefficient, since a Shamir (t, w)-threshold scheme enables a key to be shared by giving each participant only one "piece" of information.

In general, we measure the efficiency of a secret sharing scheme by the information rate, which we define now.

DEFINITION 11.4   *Suppose we have a perfect secret sharing scheme realizing an access structure Γ. The information rate for P_i is the ratio*

$$\rho_i = \frac{\log_2 |\mathcal{K}|}{\log_2 |\mathcal{S}(P_i)|}.$$

(Note that $\mathcal{S}(P_i)$ denotes the set of possible shares that P_i might receive; of course $\mathcal{S}(P_i) \subseteq \mathcal{S}$.) The information rate of the scheme is denoted by ρ and is defined as

$$\rho = \min\{\rho_i : 1 \leq i \leq w\}.$$

The motivation for this definition is as follows. Since the key K comes from a finite set $\mathcal{K}$, we can think of K as being represented by a bit-string of length $\log_2 |\mathcal{K}|$, by using a binary encoding, for example. In a similar way, a share given to P_i can be represented by a bit-string of length $\log_2 |\mathcal{S}(P_i)|$. Intuitively, P_i receives $\log_2 |\mathcal{S}(P_i)|$ bits of information (in his or her share), but the information content of the key is $\log_2 |\mathcal{K}|$ bits. Thus ρ_i is the ratio of the number of bits in a share to the number of bits in the key.

Example 11.6
Let's look at the two schemes from Section 11.2. The scheme produced in Example 11.3 has

$$\rho = \frac{\log_2 m}{\log_2 m^2} = \frac{1}{2}.$$

However, in Example 11.4, we get a scheme with

$$\rho = \frac{\log_2 m}{\log_2 m^3} = \frac{1}{3}.$$

Hence, the first implementation is preferable.   ▯

In general, if we construct a scheme from a circuit $\mathbf{C}$ using the monotone circuit construction, then the information rate can be computed as indicated in the following theorem.

THEOREM 11.2

Let $\mathbf{C}$ be any monotone boolean circuit. Then there is a perfect secret sharing scheme realizing the access structure $\Gamma(\mathbf{C})$ having information rate

$$\rho = \max\{1/r_i : 1 \leq i \leq w\},$$

where r_i denotes the number of input wires to $\mathbf{C}$ carrying the input x_i.

With respect to threshold access structures, we observe that the Shamir scheme will have information rate 1, which we show below is the optimal value. In contrast, an implementation of a (t, w)-threshold scheme using a disjunctive normal form boolean circuit will have information rate $1/\binom{w-1}{t-1}$, which is much lower (and therefore inferior) if $1 < t < w$.

Obviously, a high information rate is desirable. The first general result we prove is that $\rho \leq 1$ in any scheme.

THEOREM 11.3

In any perfect secret sharing scheme realizing an access structure Γ, $\rho \leq 1$.

PROOF Suppose we have a a perfect secret sharing scheme that realizes the access structure Γ. Let $B \in \Gamma_0$ and choose any participant $P_j \in B$. Define $B' = B \backslash \{P_j\}$. Let $g \in \mathcal{S}(B)$. Now, $B' \notin \Gamma$, so the distribution of shares $g|_{B'}$ provides no information about the key. Hence, for each $K \in \mathcal{K}$, there is a distribution rule $g^K \in \mathcal{F}_K$ such that $g^K|_{B'} = g|_{B'}$. Since $B \in \Gamma$, it must be the case that $g^K(P_j) \neq g^{K'}(P_j)$ if $K \neq K'$. Hence, $|\mathcal{S}(P_j)| \geq |\mathcal{K}|$, and thus $\rho \leq 1$. ∎

Since $\rho = 1$ is the optimal situation, we refer to such a scheme an *ideal* scheme. The Shamir schemes are ideal schemes. In the next section, we present a construction for ideal schemes that generalizes the Shamir schemes.

11.6 The Brickell Vector Space Construction

In this section, we present a construction for certain ideal schemes known as the *Brickell vector space construction*.

Suppose Γ is an access structure, and let $(\mathbb{Z}_p)^d$ denote the vector space of all d-tuples over $\mathbb{Z}_p$, where p is prime and $d \geq 2$. Suppose there exists a function

$$\phi : \mathcal{P} \to (\mathbb{Z}_p)^d$$

which satisfies the property

$$(1, 0, \ldots, 0) \in \langle \phi(P_i) : P_i \in B \rangle \Leftrightarrow B \in \Gamma. \tag{11.3}$$

FIGURE 11.6
The Brickell scheme

Initialization Phase

1. For $1 \leq i \leq w$, D gives the vector $\phi(P_i) \in (\mathbb{Z}_p)^d$ to P_i. These vectors are public.

Share Distribution

2. Suppose D wants to share a key $K \in \mathbb{Z}_p$. D secretly chooses (independently at random) $d - 1$ elements of $\mathbb{Z}_p$, $a_2, \ldots, a_d$.

3. For $1 \leq i \leq w$, D computes $y_i = \bar{a} \cdot \phi(P_i)$, where

$$\bar{a} = (K, a_2, \ldots, a_d).$$

4. For $1 \leq i \leq w$, D gives the share y_i to P_i.

In other words, the vector $(1, 0, \ldots, 0)$ can be expressed as a linear combination of the vectors in the set $\{\phi(P_i) : P_i \in B\}$ if and only if B is an authorized subset.

Now, suppose there is a function ϕ that satisfies Property (11.3). (In general, finding such a function is often a matter of trial and error, though we will see some explicit constructions of suitable functions ϕ for certain access structures a bit later.) We are going to construct an ideal secret sharing scheme with $\mathcal{K} = \mathcal{S}(P_i) = \mathbb{Z}_p$, $1 \leq i \leq w$. The distribution rules of the scheme are as follows: for every vector $\bar{a} = (a_1, \ldots, a_d) \in \mathbb{Z}_p{}^d$, define a distribution rule $f_{\bar{a}} \in \mathcal{F}_{a_1}$, where

$$f_{\bar{a}}(x) = \bar{a} \cdot \phi(x)$$

for every $x \in \mathcal{P}$, and the operation "·" is the inner product modulo p.

Note that each $\mathcal{F}_K$ contains p^{d-1} distribution rules. We will suppose that each probability distribution $p_{\mathcal{F}_K}$ is equiprobable: $p_{\mathcal{F}_K}(f) = 1/p^{d-1}$ for every $f \in \mathcal{F}_K$. The Brickell scheme is presented in Figure 11.6.

We have the following result.

THEOREM 11.4
Suppose ϕ satisfies Property (11.3). Then the sets of distribution rules $\mathcal{F}_K$, $K \in \mathcal{K}$, comprise an ideal scheme that realizes Γ.

PROOF   First, we will show that if B is an authorized subset, then the participants in B can compute K. Since

$$(1, 0, \ldots, 0) \in \langle \phi(P_i) : P_i \in B \rangle,$$

we can write

$$(1, 0, \ldots, 0) = \sum_{\{i : P_i \in B\}} c_i \phi(P_i),$$

where each $c_i \in \mathbb{Z}_p$. Denote by s_i the share given to P_i. Then

$$s_i = \bar{a} \cdot \phi(P_i),$$

where $\bar{a}$ is an unknown vector chosen by D and

$$K = a_1 = \bar{a} \cdot (1, 0, \ldots, 0).$$

By the linearity of the inner product operation,

$$K = \sum_{\{i : P_i \in B\}} c_i \bar{a} \cdot \phi(P_i).$$

Thus, it is a simple matter for the participants in B to compute

$$K = \sum_{\{i : P_i \in B\}} c_i s_i.$$

What happens if B is not an authorized subset? Denote by e the dimension of the subspace $\langle \phi(P_i) : P_i \in B \rangle$ (note that $e \leq |B|$). Choose any $K \in \mathcal{K}$, and consider the system of equations:

$$\phi(P_i) \cdot \bar{a} = s_i, \forall P_i \in B$$

$$(1, 0, \ldots, 0) \cdot \bar{a} = K.$$

This is a system of linear equations in the d unknowns $a_1, \ldots, a_d$. The coefficient matrix has rank $e + 1$, since

$$(1, 0, \ldots, 0) \notin \langle \phi(P_i) : P_i \in B \rangle.$$

Provided the system of equations is consistent, the solution space has dimension $d - e - 1$ (independent of the value of K). It will then follow that there are precisely p^{d-e-1} distribution rules in each $\mathcal{F}_K$ that are consistent with any possible distribution of shares to B. By a similar computation as was performed in Example 11.5, we see that $p_{\mathcal{K}}(K|f_B) = p_{\mathcal{K}}(K)$ for every $K \in \mathcal{K}$, where $f_B(P_i) = s_i$ for all $P_i \in B$.

Why is the system consistent? The first $|B|$ equations are consistent, since the vector $\bar{a}$ chosen by D is a solution. Since

$$(1, 0, \ldots, 0) \notin \langle \phi(P_i) : P_i \in B \rangle$$

(as mentioned above), the last equation is consistent with the first $|B|$ equations. This completes the proof. ∎

It is interesting to observe that the Shamir (t, w)-threshold scheme is a special case of the vector space construction. To see this, define $d = t$ and let

$$\phi(P_i) = (1, x_i, x_i^2, \ldots, x_i^{t-1})$$

for $1 \leq i \leq w$, where x_i is the x-coordinate given to P_i. The resulting scheme is equivalent to the Shamir scheme; we leave the details to the reader to check.

Here is another general result that is easy to prove. It concerns access structures that have as a basis a collection of pairs of participants that forms a complete multipartite graph. A graph $G = (V, E)$ with vertex set V and edge set E is defined to be a *complete multipartite graph* if the vertex set V can be partitioned into subsets $V_1, \ldots, V_\ell$ such that $\{x, y\} \in E$ if and only if $x \in V_i, y \in V_j$, where $i \neq j$. The sets V_i are called *parts*. The complete multipartite graph is denoted by $K_{n_1, \ldots, n_\ell}$ if $|V_i| = n_i, 1 \leq i \leq \ell$. A complete multipartite graph $K_{1, \ldots, 1}$ (with ℓ parts) is in fact a *complete graph* and is denoted K_ℓ.

THEOREM 11.5
Suppose $G = (V, E)$ is a complete multipartite graph. Then there is an ideal scheme realizing the access structure $cl(E)$ on participant set V.

PROOF   Let $V_1, \ldots, V_\ell$ be the parts of G. Let $x_1, \ldots, x_\ell$ be distinct elements of $\mathbb{Z}_p$, where $p \geq \ell$. Let $d = 2$. For every participant $v \in V_i$, define $\phi(v) = (x_i, 1)$. It is straightforward to verify Property (11.3). By Theorem 11.4, we have an ideal scheme.   ∎

To illustrate the application of these constructions, we will consider the possible access structures for up to four participants. Note that it suffices to consider only the access structures in which the basis cannot be partitioned into two non-empty subsets on disjoint participant sets. (For example, $\Gamma_0 = \{\{P_1, P_2\}, \{P_3, P_4\}\}$ can be partitioned as $\{\{P_1, P_2\}\} \cup \{\{P_3, P_4\}\}$ so we do not consider it.) We list the non-isomorphic access structures of this type on two, three, and four participants in Table 11.1 (the quantities ρ^* are defined in Section 11.7).

Of these 18 access structures, we can already obtain ideal schemes for ten of them using the constructions we have at our disposal now. These ten access structures are either threshold access structures or have a basis which is a complete multipartite graph, so Theorem 11.5 can be applied. One such access structure is # 9, whose basis is the complete multipartite graph $K_{1,1,2}$. We illustrate in the following example.

Example 11.7
For access structure # 9, take $d = 2$, $p \geq 3$, and define ϕ as follows:

$$\phi(P_1) = (0, 1)$$

TABLE 11.1
Access structures for at most four participants

	w	subsets in Γ_0	ρ^*	comments
1.	2	$P_1 P_2$	1	$(2,2)$-threshold
2.	3	$P_1 P_2, P_2 P_3$	1	$\Gamma_0 \cong K_{1,2}$
3.	3	$P_1 P_2, P_2 P_3, P_1 P_3$	1	$(2,3)$-threshold
4.	3	$P_1 P_2 P_3$	1	$(3,3)$-threshold
5.	4	$P_1 P_2, P_2 P_3, P_3 P_4$	$2/3$	
6.	4	$P_1 P_2, P_1 P_3, P_1 P_4$	1	$\Gamma_0 \cong K_{1,3}$
7.	4	$P_1 P_2, P_1 P_4, P_2 P_3, P_3 P_4$	1	$\Gamma_0 \cong K_{2,2}$
8.	4	$P_1 P_2, P_2 P_3, P_2 P_4, P_3 P_4$	$2/3$	
9.	4	$P_1 P_2, P_1 P_3, P_1 P_4, P_2 P_3, P_2 P_4$	1	$\Gamma_0 \cong K_{1,1,2}$
10.	4	$P_1 P_2, P_1 P_3, P_1 P_4, P_2 P_3, P_2 P_4, P_3 P_4$	1	$(2,4)$-threshold
11.	4	$P_1 P_2 P_3, P_1 P_4$	1	
12.	4	$P_1 P_3 P_4, P_1 P_2, P_2 P_3$	$2/3$	
13.	4	$P_1 P_3 P_4, P_1 P_2, P_2 P_3, P_2 P_4$	$2/3$	
14.	4	$P_1 P_2 P_3, P_1 P_2 P_4$	1	
15.	4	$P_1 P_2 P_3, P_1 P_2 P_4, P_3 P_4$	1	
16.	4	$P_1 P_2 P_3, P_1 P_2 P_4, P_1 P_3 P_4$	1	
17.	4	$P_1 P_2 P_3, P_1 P_2 P_4, P_1 P_3 P_4, P_2 P_3 P_4$	1	$(3,4)$-threshold
18.	4	$P_1 P_2 P_3 P_4$	1	$(4,4)$-threshold

$$\phi(P_2) = (1,1)$$
$$\phi(P_3) = (2,1)$$
$$\phi(P_4) = (2,1).$$

Applying Theorem 11.5, an ideal scheme results. ⬚

Eight access structures remain to be considered. It is possible to use *ad hoc* applications of the vector space construction to construct ideal schemes for four of these: # 11, # 14, # 15 and # 16. We present the constructions for # 11 and # 14 here.

Example 11.8
For access structure # 11, take $d = 3$, $p \geq 3$, and define ϕ as follows:

$$\phi(P_1) = (0,1,0)$$
$$\phi(P_2) = (1,0,1)$$
$$\phi(P_3) = (0,1,-1)$$
$$\phi(P_4) = (1,1,0).$$

First, we have

$$\phi(P_4) - \phi(P_1) = (1,1,0) - (0,1,0)$$
$$= (1,0,0).$$

Also,

$$\phi(P_2) + \phi(P_3) - \phi(P_1) = (1,0,1) + (0,1,-1) - (0,1,0)$$
$$= (1,0,0).$$

Hence,

$$(1,0,0) \in \langle \phi(P_1), \phi(P_2), \phi(P_3) \rangle$$

and

$$(1,0,0) \in \langle \phi(P_1), \phi(P_4) \rangle.$$

Now, it suffices to show that

$$(1,0,0) \notin \langle \phi(P_i) : P_i \in B \rangle$$

if B is a maximal unauthorized subset. There are three such subsets B to be considered: $\{P_1, P_2\}$, $\{P_1, P_3\}$, and $\{P_2, P_3, P_4\}$. In each case, we need to establish that a system of linear equations has no solution. For example, suppose that

$$(1,0,0) = a_2\phi(P_2) + a_3\phi(P_3) + a_4\phi(P_4),$$

where $a_2, a_3, a_4 \in \mathbb{Z}_p$. This is equivalent to the system

$$a_2 + a_4 = 1$$
$$a_3 + a_4 = 0$$
$$a_2 - a_3 = 0.$$

The system is easily seen to have no solution. We leave the other two subsets B for the reader to consider. ▯

Example 11.9

For access structure # 14, take $d = 3$, $p \geq 2$ and define ϕ as follows:

$$\phi(P_1) = (0,1,0)$$
$$\phi(P_2) = (1,0,1)$$
$$\phi(P_3) = (0,1,1)$$
$$\phi(P_4) = (0,1,1).$$

Again, Property (11.3) is satisfied and hence an ideal scheme results. ▯

Constructions of ideal schemes for the access structures # 15 and # 16 are left as exercises. In the next section, we will show that the remaining four access structures cannot be realized by ideal schemes.

11.7 An Upper Bound on the Information Rate

Four access structures remain to be considered: # 5, # 8, # 12, and # 13. We will see in this section that in each case, there does not exist a scheme having information rate $\rho > 2/3$.

Denote by $\rho^* = \rho^*(\Gamma)$ the maximum information rate for any perfect secret sharing scheme realizing a specified access structure Γ. The first result we present is an entropy bound that will lead to an upper bound on ρ^* for certain access structures. We have defined a probability distribution $p_{\mathcal{K}}$ on $\mathcal{K}$; the entropy of this probability distribution is denoted $H(\mathbf{K})$. We have also denoted by $p_{S(B)}$ the probability distribution on the shares given to a subset $B \subseteq \mathcal{P}$. We will denote the entropy of this probability distribution by $H(\mathbf{B})$.

We begin by giving yet another definition of perfect secret sharing schemes, this time using the language of entropy. This definition is equivalent to Definition 11.3.

DEFINITION 11.5 *Suppose Γ is an access structure and $\mathcal{F}$ is a set of distribution rules. Then $\mathcal{F}$ is a perfect secret sharing scheme realizing the access structure Γ provided that the following two properties are satisfied:*

1. *For any authorized subset of participants $B \subseteq \mathcal{P}$, $H(\mathbf{K}|\mathbf{B}) = 0$.*

2. *For any unauthorized subset of participants $B \subseteq \mathcal{P}$, $H(\mathbf{K}|\mathbf{B}) = H(\mathbf{K})$.*

We will require several entropy identities and inequalities. Some of these results were given in Section 2.3 and the rest are proved similarly, so we state them without proof in the following Lemma.

LEMMA 11.6
Let $\mathbf{X}$, $\mathbf{Y}$ and $\mathbf{Z}$ be random variables. Then the following hold:

$$H(\mathbf{XY}) = H(\mathbf{X}|\mathbf{Y}) + H(\mathbf{Y}) \tag{11.4}$$

$$H(\mathbf{XY}|\mathbf{Z}) = H(\mathbf{X}|\mathbf{YZ}) + H(\mathbf{Y}|\mathbf{Z}) \tag{11.5}$$

$$H(\mathbf{XY}|\mathbf{Z}) = H(\mathbf{Y}|\mathbf{XZ}) + H(\mathbf{X}|\mathbf{Z}) \tag{11.6}$$

$$H(\mathbf{X}|\mathbf{Y}) \geq 0 \tag{11.7}$$

$$H(\mathbf{X}|\mathbf{Z}) \geq H(\mathbf{X}|\mathbf{YZ}) \tag{11.8}$$

$$H(\mathbf{XY}|\mathbf{Z}) \geq H(\mathbf{Y}|\mathbf{Z}) \tag{11.9}$$

We next prove two preliminary entropy lemmas for secret sharing schemes.

LEMMA 11.7
Suppose Γ is an access structure and $\mathcal{F}$ is a set of distribution rules realizing Γ.
Suppose $B \notin \Gamma$ and $A \cup B \in \Gamma$, where $A, B \subseteq \mathcal{P}$. Then

$$H(\mathbf{A}|\mathbf{B}) = H(\mathbf{K}) + H(\mathbf{A}|\mathbf{BK}).$$

PROOF   From Equations 11.5 and 11.6, we have that

$$H(\mathbf{AK}|\mathbf{B}) = H(\mathbf{A}|\mathbf{BK}) + H(\mathbf{K}|\mathbf{B})$$

and

$$H(\mathbf{AK}|\mathbf{B}) = H(\mathbf{K}|\mathbf{AB}) + H(\mathbf{A}|\mathbf{B}),$$

so

$$H(\mathbf{A}|\mathbf{BK}) + H(\mathbf{K}|\mathbf{B}) = H(\mathbf{K}|\mathbf{AB}) + H(\mathbf{A}|\mathbf{B}).$$

Since, by Property 2 of Definition 11.5, we have

$$H(\mathbf{K}|\mathbf{B}) = H(\mathbf{K}),$$

and, by Property 1 of Definition 11.5, we have

$$H(\mathbf{K}|\mathbf{AB}) = 0,$$

the result follows.   ∎

LEMMA 11.8
Suppose Γ is an access structure and $\mathcal{F}$ is a set of distribution rules realizing Γ.
Suppose $A \cup B \notin \Gamma$, where $A, B \subseteq \mathcal{P}$. Then $H(\mathbf{A}|\mathbf{B}) = H(\mathbf{A}|\mathbf{BK})$.

PROOF   As in Lemma 11.7, we have that

$$H(\mathbf{A}|\mathbf{BK}) + H(\mathbf{K}|\mathbf{B}) = H(\mathbf{K}|\mathbf{AB}) + H(\mathbf{A}|\mathbf{B}).$$

Since

$$H(\mathbf{K}|\mathbf{B}) = H(\mathbf{K})$$

and

$$H(\mathbf{K}|\mathbf{AB}) = H(\mathbf{K}),$$

the result follows.   ∎

We now prove the following important theorem.

THEOREM 11.9
Suppose Γ is an access structure such that

$$\{W, X\}, \{X, Y\}, \{W, Y, Z\} \in \Gamma$$

and

$$\{W, Y\}, \{X\}, \{W, Z\} \notin \Gamma.$$

Let $\mathcal{F}$ be any perfect secret sharing scheme realizing Γ. Then $H(\mathbf{XY}) \geq 3H(\mathbf{K})$.

PROOF We establish a sequence of inequalities:

$$
\begin{aligned}
H(\mathbf{K}) &= H(\mathbf{Y}|\mathbf{WZ}) - H(\mathbf{Y}|\mathbf{WZK}) && \text{by Lemma 11.7} \\
&\leq H(\mathbf{Y}|\mathbf{WZ}) && \text{by (11.7)} \\
&\leq H(\mathbf{Y}|\mathbf{W}) && \text{by (11.8)} \\
&= H(\mathbf{Y}|\mathbf{WK}) && \text{by Lemma 11.8} \\
&\leq H(\mathbf{XY}|\mathbf{WK}) && \text{by (11.9)} \\
&= H(\mathbf{X}|\mathbf{WK}) + H(\mathbf{Y}|\mathbf{WXK}) && \text{by (11.5)} \\
&\leq H(\mathbf{X}|\mathbf{WK}) + H(\mathbf{Y}|\mathbf{XK}) && \text{by (11.8)} \\
&= H(\mathbf{X}|\mathbf{W}) - H(\mathbf{K}) + H(\mathbf{Y}|\mathbf{X}) - H(\mathbf{K}) && \text{by Lemma 11.7} \\
&\leq H(\mathbf{X}) - H(\mathbf{K}) + H(\mathbf{Y}|\mathbf{X}) - H(\mathbf{K}) && \text{by (11.7)} \\
&= H(\mathbf{XY}) - 2H(\mathbf{K}) && \text{by (11.4).}
\end{aligned}
$$

Hence, the result follows. ∎

COROLLARY 11.10
Suppose that Γ is an access structure that satisfies the hypotheses of Theorem 11.9. Suppose the $|\mathcal{K}|$ keys are equally probable. Then $\rho \leq 2/3$.

PROOF Since the keys are equiprobable, we have

$$H(\mathbf{K}) = \log_2 |\mathcal{K}|.$$

Also, we have that

$$H(\mathbf{XY}) \leq H(\mathbf{X}) + H(\mathbf{Y})$$
$$\leq \log_2 |\mathcal{S}(X)| + \log_2 |\mathcal{S}(Y)|.$$

By Theorem 11.9, we have that

$$H(\mathbf{XY}) \geq 3H(\mathbf{K}).$$

Hence it follows that

$$\log_2 |\mathcal{S}(X)| + \log_2 |\mathcal{S}(Y)| \geq 3 \log_2 |\mathcal{K}|.$$

Now, by the definition of information rate, we have

$$\rho \leq \frac{\log_2 |\mathcal{K}|}{\log_2 |\mathcal{S}(X)|}$$

and

$$\rho \leq \frac{\log_2 |\mathcal{K}|}{\log_2 |\mathcal{S}(Y)|}.$$

It follows that

$$3 \log_2 |\mathcal{K}| \leq \log_2 |\mathcal{S}(X)| + \log_2 |\mathcal{S}(Y)|$$
$$\leq \frac{\log_2 |\mathcal{K}|}{\rho} + \frac{\log_2 |\mathcal{K}|}{\rho}$$
$$= 2 \frac{\log_2 |\mathcal{K}|}{\rho}.$$

Hence, $\rho \leq 2/3$.   ∎

For the access structures # 5, # 8, # 12, and # 13, the hypotheses of Theorem 11.9 are satisfied. Hence, $\rho^* \leq 2/3$ for these four access structures.

We also have the following result concerning ρ^* in the case where the access structure has a basis Γ_0 which is a graph. The proof involves showing that any connected graph which is not a multipartite graph contains an induced subgraph on four vertices that is isomorphic to the basis of access structure # 5 or # 8. If $G = (V, E)$ is a graph with vertex set V and edge set E, and $V_1 \subseteq V$, then the *induced subgraph* $G[V_1]$ is defined to be the graph (V_1, E_1), where

$$E_1 = \{uv \in E, u, v \in V_1\}.$$

THEOREM 11.11
Suppose G is a connected graph that is not a complete multipartite graph. Let $\Gamma(G)$ be the access structure that is the closure of E, where E is the edge set of G. Then $\rho^*(\Gamma(G)) \leq 2/3$.

PROOF   We will first prove that any connected graph that is not a complete multipartite graph must contain four vertices w, x, y, z such that the induced subgraph $G[w, x, y, z]$ is isomorphic to either the basis of access structure # 5 or # 8.

Let G^C denote the complement of G. Since G is not a complete multipartite graph, there must exist three vertices x, y, z such that $xy, yz \in E(G^C)$ and $xz \in E(G)$. Define

$$d = \min\{d_G(y, x), d_G(y, z)\},$$

where d_G denotes the length of a shortest path (in G) between two vertices. Then $d \geq 2$. Without loss of generality, we can assume that $d = d_G(y, x)$ by symmetry. Let

$$y_0, y_1, \ldots, y_{d-1}, x$$

be a path in G, where $y_0 = y$. We have that

$$y_{d-2}z, y_{d-2}x \in E(G^C)$$

and

$$y_{d-2}y_{d-1}, y_{d-1}x, xz \in E(G).$$

It follows that $G[y_{d-2}, y_{d-1}, x, z]$ is isomorphic to the basis of access structure # 5 or # 8, as desired.

So, we can assume that we have found four vertices w, x, y, z such that the induced subgraph $G[w, x, y, z]$ is isomorphic to either the basis of access structure # 5 or # 8. Now, let $\mathcal{F}$ be any scheme realizing the access structure $\Gamma(G)$. If we restrict the domain of the distribution rules to $\{w, x, y, z\}$, then we obtain a scheme $\mathcal{F}'$ realizing access structure # 5 or # 8. It is also obvious that $\rho(\mathcal{F}') \geq \rho(\mathcal{F})$. Since $\rho(\mathcal{F}') \leq 2/3$, it follows that $\rho(\mathcal{F}) \leq 2/3$. This completes the proof. ∎

Since $\rho^* = 1$ for complete multipartite graphs, Theorem 11.11 tells us that it is never the case that $2/3 < \rho^* < 1$ for any access structure that is the closure of the edge set of a connected graph.

11.8 The Decomposition Construction

We still have four access structures in Table 11.1 to consider. Of course, we can use the monotone circuit construction to produce schemes for these access structures. However, by this method, the best we can do is to obtain information rate $\rho = 1/2$ in each case. We can get $\rho = 1/2$ in cases # 5 and # 12 by using a disjunctive normal form boolean circuit. For cases # 8 and # 13, a disjunctive normal form boolean circuit will yield $\rho = 1/3$, but other monotone circuits exist which allow us to attain $\rho = 1/2$. But in fact, it is possible to construct schemes with $\rho = 2/3$ for each of these four access structures, by employing constructions that use ideal schemes as building blocks in the construction of larger schemes.

We present a construction of this type called the "decomposition construction." First, we need to define an important concept.

DEFINITION 11.6 *Suppose Γ is an access structure having basis Γ_0. Let $\mathcal{K}$ be a specified key set. An ideal $\mathcal{K}$-decomposition of Γ_0 consists of a set $\{\Gamma_1, \ldots \Gamma_n\}$ such that the following properties are satisfied:*

1. $\Gamma_k \subseteq \Gamma_0$ *for* $1 \leq k \leq n$
2. $\bigcup_{k=1}^{n} \Gamma_k = \Gamma_0$
3. *for* $1 \leq k \leq n$, *there exists an ideal scheme with key set* $\mathcal{K}$, *on the subset of participants*

$$\mathcal{P}_k = \bigcup_{B \in \Gamma_k} B,$$

for the access structure having basis Γ_k.

Given an ideal $\mathcal{K}$-decomposition of an access structure Γ, we can easily construct a perfect secret sharing scheme, as described in the following theorem.

THEOREM 11.12

Suppose Γ *is an access structure having basis* Γ_0. *Let* $\mathcal{K}$ *be a specified key set, and suppose* $\{\Gamma_1, \ldots \Gamma_n\}$ *is an ideal* $\mathcal{K}$-decomposition of Γ. *For every participant* P_i, *define*

$$R_i = |\{k : P_i \in \mathcal{P}_k\}|.$$

Then there exists a perfect secret sharing scheme realizing Γ, *having information rate* $\rho = 1/R$, *where*

$$R = \max\{R_i : 1 \leq i \leq w\}.$$

PROOF For $1 \leq k \leq n$, we have an ideal scheme realizing the access structure with basis Γ_k, with key set $\mathcal{K}$, having $\mathcal{F}^k$ as its set of distribution rules. We will construct a scheme realizing Γ, with key set $\mathcal{K}$. The set of distribution rules $\mathcal{F}$ is constructed according to the following recipe. Suppose D wants to share a key K. Then, for $1 \leq k \leq n$, he chooses a random distribution rule $f^k \in \mathcal{F}_K^k$ and distributes the resulting shares to the participants in $\mathcal{P}_k$.

We omit the proof that the scheme is perfect. However, it is easy to compute the information rate of the resulting scheme. Since each of the component schemes is ideal, it follows that

$$|\mathcal{S}(P_i)| = |\mathcal{K}|^{R_i},$$

for $1 \leq i \leq w$. So

$$\rho_i = \frac{1}{R_i},$$

and

$$\rho = \frac{1}{\max\{R_i : 1 \leq i \leq w\}},$$

which is what we were required to prove.   ∎

Although Theorem 11.12 is useful, it is often much more useful to employ a generalization in which we have ℓ ideal $\mathcal{K}$-decompositions of Γ_0 instead of just one. Each of the ℓ decompositions is used to share a key chosen from $\mathcal{K}$. Thus, we build a scheme with key set $\mathcal{K}^\ell$. The construction of the scheme and its information rate are as stated in the following theorem.

THEOREM 11.13 *(Decomposition Construction)*
Suppose Γ is an access structure having basis Γ_0, and $\ell \geq 1$ is an integer. Let $\mathcal{K}$ be a specified key set, and for $1 \leq j \leq \ell$, suppose that $\mathcal{D}_j = \{\Gamma_{j,1}, \ldots \Gamma_{j,n_j}\}$ is an ideal decomposition of Γ_0. Let $\mathcal{P}_{j,k}$ denote the participant set for the access structure $\Gamma_{j,k}$. For every participant P_i, define

$$R_i = \sum_{j=1}^{\ell} |\{k : P_i \in \mathcal{P}_{j,k}\}|.$$

Then there exists a perfect secret sharing scheme realizing Γ, having information rate $\rho = \ell/R$, where
$$R = \max\{R_i : 1 \leq i \leq w\}.$$

PROOF For $1 \leq j \leq \ell$ and $1 \leq k \leq n$, we have an ideal scheme realizing the access structure with basis $\Gamma_{j,k}$, with key set $\mathcal{K}$, having $\mathcal{F}^{j,k}$ as its set of distribution rules. We construct a scheme realizing Γ, with key set $\mathcal{K}^{\ell}$. The set of distribution rules $\mathcal{F}$ is constructed according to the following recipe. Suppose D wants to share a key $K = (K_1, \ldots, K_{\ell})$. Then for $1 \leq j \leq \ell$ and $1 \leq k \leq n$, he chooses a random distribution rule $f^{j,k} \in \mathcal{F}^{j,k}_{K_j}$ and distributes the resulting shares to the participants in $\mathcal{P}_{j,k}$.

The information rate can be computed in a manner similar to that of Theorem 11.12. ∎

Let's look at a couple of examples.

Example 11.10
Consider access structure # 5. The basis is a graph that is not a complete multi-partite graph. Therefore we know from Theorem 11.11 that $\rho^* \leq 2/3$.

Let p be prime, and consider the following two ideal $\mathbb{Z}_p$-decompositions:

$$\mathcal{D}_1 = \{\Gamma_{1,1}, \Gamma_{1,2}\},$$

where

$$\Gamma_{1,1} = \{\{P_1, P_2\}\}$$
$$\Gamma_{1,2} = \{\{P_2, P_3\}, \{P_3, P_4\}\},$$

and

$$\mathcal{D}_2 = \{\Gamma_{2,1}, \Gamma_{2,2}\},$$

where

$$\Gamma_{2,1} = \{\{P_1, P_2\}, \{P_2, P_3\}\}$$
$$\Gamma_{2,2} = \{\{P_3, P_4\}\}.$$

Each decomposition consists of a K_2 and a $K_{1,2}$, so they are indeed ideal $\mathbb{Z}_p$-decompositions. Either of them yields a scheme with $\rho = 1/2$. However, if we "combine" them by applying Theorem 11.13 with $\ell = 2$, then we get a scheme with $\rho = 2/3$, which is optimal.

One implementation of the scheme, using Theorem 11.5, is as follows. D will choose four random elements (independently) from $\mathbb{Z}_p$, say b_{11}, b_{12}, b_{21}, and b_{22}. Given a key $(K_1, K_2) \in (\mathbb{Z}_p)^2$, D distributes shares as follows:

1. P_1 receives b_{11}, b_{21}.

2. P_2 receives $b_{11} + K_1, b_{12}, b_{21} + K_2$.

3. P_3 receives $b_{12} + K_1, b_{21}, b_{22}$.

4. P_4 receives $b_{12}, b_{22} + K_2$.

(All arithmetic is performed in $\mathbb{Z}_p$.)   ⬚

Example 11.11
Consider access structure # 8. Again, $\rho^* \leq 2/3$ by Theorem 11.11, and two suitable ideal compositions will yield an (optimal) scheme with $\rho = 2/3$.

Take $\mathcal{K} = \mathbb{Z}_p$ for any prime $p \geq 3$, and define two ideal $\mathcal{K}$-decompositions to be:

$$\mathcal{D}_1 = \{\Gamma_{1,1}, \Gamma_{1,2}\},$$

where

$$\Gamma_{1,1} = \{\{P_1, P_2\}\}$$
$$\Gamma_{1,2} = \{\{P_2, P_3\}, \{P_2, P_4\}, \{P_3, P_4\}\},$$

and

$$\mathcal{D}_2 = \{\Gamma_{2,1}, \Gamma_{2,2}\},$$

where

$$\Gamma_{2,1} = \{\{P_1, P_2\}, \{P_2, P_3\}, \{P_2, P_4\}\}$$
$$\Gamma_{2,2} = \{\{P_3, P_4\}\}.$$

$\mathcal{D}_1$ consists of a K_2 and a K_3, and $\mathcal{D}_2$ consists of a K_2 and a $K_{1,3}$, so both are ideal $\mathcal{K}$-decompositions. Applying Theorem 11.13 with $\ell = 2$, we get a scheme with $\rho = 2/3$.

One implementation, using Theorem 11.5, is as follows. D will choose four random elements (independently) from $\mathbb{Z}_p$, say b_{11}, b_{12}, b_{21}, and b_{22}. Given a key $(K_1, K_2) \in (\mathbb{Z}_p)^2$, D distributes shares as follows:

1. P_1 receives $b_{11} + K_1, b_{21} + K_2$.

2. P_2 receives b_{11}, b_{12}, b_{21}.

3. P_3 receives $b_{12} + K_1, b_{21} + K_2, b_{22}$.

4. P_4 receives $b_{12} + 2K_1, b_{21} + K_2, b_{22} + K_2$.

(All arithmetic is performed in $\mathbb{Z}_p$.) □

To this point, we have explained all the information in Table 11.1 except for the values of ρ^* for access structures # 12 and # 13. These values arise from a more general version of the decomposition construction which we do not describe here; see the notes below.

11.9 Notes and References

Threshold schemes were invented independently by Blakley [BL79] and Shamir [SH79]. Secret sharing for general access structures was first studied in Ito, Saito, and Nishizeki [ISN87]; we based Section 11.2 on the approach of Benaloh and Leichter [BL90]. The vector space construction is due to Brickell [BR89A]. The entropy bound of Section 11.7 is proved in Capocelli *et al.* [CDGV93], and some of the other material from this section is found in Blundo *et al.* [BDSV93].

In this chapter, we have emphasized a linear-algebraic and combinatorial approach to secret sharing. Some interesting connections with matroid theory can be found in Brickell and Davenport [BD91]. Secret sharing schemes can also be constructed using geometric techniques. Simmons has done considerable research in this direction; we refer to [S192A] for an overview of geometric techniques in secret sharing. Further discussion of these topics, as well as constructions for schemes having information rate 2/3 for access structures # 12 and # 13, can be found in the expository paper by Stinson [ST92A].

Exercises

11.1 Write a computer program to compute the key for a Shamir (t, w)-threshold scheme implemented in $\mathbb{Z}_p$. That is, given t public x-coordinates, $x_1, x_2, \ldots, x_t$, and t y-coordinates $y_1, \ldots, y_t$, compute the resulting key. Use the Lagrange interpolation method, as it is easier to program.

(a) Test your program if $p = 31847$, $t = 5$ and $w = 10$, with the following shares:

x_1	=	413	y_1	=	25439
x_2	=	432	y_2	=	14847
x_3	=	451	y_3	=	24780
x_4	=	470	y_4	=	5910
x_5	=	489	y_5	=	12734
x_6	=	508	y_1	=	12492
x_7	=	527	y_2	=	12555
x_8	=	546	y_3	=	28578
x_9	=	565	y_4	=	20806
x_{10}	=	584	y_5	=	21462

Verify that the same key is computed by using several different subsets of five shares.

(b) Having determined the key, compute the share that would be given to a participant with x-coordinate 10000. (Note that this can be done without computing the whole secret polynomial $a(x)$.)

11.2 A dishonest dealer might distribute "bad" shares for a Shamir threshold scheme, i.e., shares for which different t-subsets determine different keys. Given all w shares, we could test the consistency of the shares by computing the key for every one of the $\binom{w}{t}$ t-subsets of participants, and verifying that the same key is computed in each case. Can you describe a more efficient method of testing the consistency of the shares?

11.3 For access structures having the following bases, use the monotone circuit construction to construct a secret sharing scheme with information rate $\rho = 1/3$.

(a) $\Gamma_0 = \{\{P_1, P_2\}, \{P_2, P_3\}, \{P_2, P_4\}, \{P_3, P_4\}\}$.

(b) $\Gamma_0 = \{\{P_1, P_3, P_4\}, \{P_1, P_2\}, \{P_2, P_3\}, \{P_2, P_4\}\}$.

(c) $\Gamma_0 = \{\{P_1, P_2\}, \{P_1, P_3\}, \{P_2, P_3, P_4\}, \{P_2, P_4, P_5\}, \{P_3, P_4, P_5\}\}$.

11.4 Use the vector space construction to obtain ideal schemes for access structures having the following bases:

(a) $\Gamma_0 = \{\{P_1, P_2, P_3\}, \{P_1, P_2, P_4\}, \{P_3, P_4\}\}$.

(b) $\Gamma_0 = \{\{P_1, P_2, P_3\}, \{P_1, P_2, P_4\}, \{P_1, P_3, P_4\}\}$.

(c) $\Gamma_0 = \{\{P_1, P_2\}, \{P_1, P_3\}, \{P_2, P_3\}, \{P_1, P_4, P_5\}, \{P_2, P_4, P_5\}\}$.

11.5 Use the decomposition construction to obtain schemes with specified information rates for access structures having the following bases:

(a) $\Gamma_0 = \{\{P_1, P_3, P_4\}, \{P_1, P_2\}, \{P_2, P_3\}\}, \rho = 3/5$.

(b) $\Gamma_0 = \{\{P_1, P_3, P_4\}, \{P_1, P_2\}, \{P_2, P_3\}, \{P_2, P_4\}\}, \rho = 4/7$.

12

Pseudo-random Number Generation

12.1 Introduction and Examples

There are many situations in cryptography where it is important to be able to generate random numbers, bit-strings, etc. For example, cryptographic keys are to be generated at random from a specified keyspace, and many protocols require random numbers to be generated during their execution. Generating random numbers by means of coin tosses or other physical processes is time-consuming and expensive, so in practice it is common to use a *pseudo-random bit generator* (or PRBG). A PRBG starts with a short random bit-string (a "seed") and expands it into a much longer "random-looking" bit-string. Thus a PRBG reduces the amount of random bits that are required in an application.

More formally, we have the following definition.

DEFINITION 12.1 *Let k, ℓ be positive integers such that $\ell \geq k + 1$ (where ℓ is a specified polynomial function of k). A (k, ℓ)-pseudo-random bit generator (more briefly, a (k, ℓ)-PRBG) is a function $f : (\mathbb{Z}_2)^k \to (\mathbb{Z}_2)^\ell$ that can be computed in polynomial time (as a function of k). The input $s_0 \in (\mathbb{Z}_2)^k$ is called the seed, and the output $f(s_0) \in (\mathbb{Z}_2)^\ell$ is called a pseudo-random bit-string.*

The function f is deterministic, so the bit-string $f(s_0)$ is dependent only on the seed. Our goal is that the pseudo-random bit-string $f(s_0)$ should "look like" truly random bits, given that the seed is chosen at random. Giving a precise definition is quite difficult, but we will try to give an intuitive description of the concept later in this chapter.

One motivating example for studying this type PRBG is as follows. Recall the concept of perfect secrecy that we studied in Chapter 2. One realization of perfect secrecy is the **One-time Pad**, where the plaintext and the key are both bit-strings of a specified length, and the ciphertext is constructed by taking the bitwise exclusive-or of the plaintext and the key. The practical difficulty of the **One-time Pad** is that the key, which must be randomly generated and communicated over a

FIGURE 12.1
Linear Congruential Generator

Let $M \geq 2$ be an integer, and let $1 \leq a, b \leq M - 1$. Define $k = \lceil \log_2 M \rceil$ and let $k + 1 \leq \ell \leq M - 1$.

For a seed s_0, where $0 \leq s_0 \leq M - 1$, define

$$s_i = (a s_{i-1} + b) \bmod M$$

for $1 \leq i \leq \ell$, and then define

$$f(s_0) = (z_1, z_2, \ldots, z_\ell),$$

where

$$z_i = s_i \bmod 2,$$

$1 \leq i \leq \ell$. Then f is a (k, ℓ)-**Linear Congruential Generator**.

secure channel, must be as long as the plaintext in order to ensure perfect secrecy. PRBGs provide a possible way of alleviating this problem. Suppose Alice and Bob agree on a PRBG and communicate a seed over the secure channel. Alice and Bob can then both compute the same string of pseudo-random bits, which will be used as a **One-time Pad**. Thus the seed functions as a key, and the PBRG can be thought of as a keystream generator for a stream cipher.

We now present some well-known PRBGs to motivate and illustrate some of the concepts we will be studying. First, we observe that a linear feedback shift register, as described in Section 1.1.7, can be thought of as a PRBG. Given a k-bit seed, an LFSR of degree k can be used to produce as many as $2^k - k - 1$ further bits before repeating. The PRBG obtained from an LFSR is very insecure: we already observed in Section 1.2.5 that knowledge of any $2k$ consecutive bits suffice to allow the seed to be determined, and hence the entire sequence can be reconstructed by an opponent. (Although we have not yet defined security of a PRBG, it should be clear that the existence of an attack of this type means that the generator is insecure!)

Another well-known (but insecure) PRBG, called the **Linear Congruential Generator**, is presented in Figure 12.1. Here is a very small example to illustrate.

Example 12.1
We can obtain a $(5, 10)$-PRBG by taking $M = 31$, $a = 3$ and $b = 5$ in the **Linear Congruential Generator**. If we consider the mapping $s \mapsto 3s + 5 \bmod 31$, then

TABLE 12.1
Bit-strings produced by the linear congruential generator

seed	sequence
0	1010001101
1	0100110101
2	1101010001
3	0001101001
4	1100101101
5	0100011010
6	1000110010
7	0101000110
8	1001101010
9	1010011010
10	0110010110
11	1010100011
12	0011001011
13	1111111111
14	0011010011
15	1010100011
16	0110100110
17	1001011010
18	0101101010
19	0101000110
20	1000110100
21	0100011001
22	1101001101
23	0001100101
24	1101010001
25	0010110101
26	1010001100
27	0110101000
28	1011010100
29	0011010100
30	0110101000

$13 \mapsto 13$, and the other 30 residues are permuted in a cycle of length 30, namely 0, 5, 20, 3, 14, 16, 22, 9, 1, 8, 29, 30, 2, 11, 7, 26, 21, 6, 23, 12, 10, 4, 17, 25, 18, 28, 27, 24, 15, 19. If the seed is anything other than 13, then the seed specifies a starting point in this cycle, and the next 10 elements, reduced modulo 2, form the pseudo-random sequence.

The 31 possible pseudo-random bit-strings produced by this generator are illustrated in Table 12.1. $\square$

We can use some concepts developed in earlier chapters to consrtruct PRBGs.

FIGURE 12.2
RSA Generator

Let p, q to be two $(k/2)$-bit primes, and define $n = pq$. Let b be chosen such that $\gcd(b, \phi(n)) = 1$. As always, n and b are public while p and q are secret.

A seed s_0 is any element of $\mathbb{Z}_n{}^*$, so s_0 has k bits. For $i \geq 1$, define

$$s_{i+1} = s_i{}^b \bmod n,$$

and then define

$$f(s_0) = (z_1, z_2, \ldots, z_\ell),$$

where

$$z_i = s_i \bmod 2,$$

$1 \leq i \leq \ell$. Then f is a (k, ℓ)-**RSA Generator**.

For example, the output feedback mode of DES, as described in Section 3.4.1, can be thought of as a PRBG; moreover, it appears to be computationally secure.

Another approach in constructing very fast PRBGs is to combine LFSRs in some way that the output looks less linear. One such method, due to Coppersmith, Krawczyk and Mansour, is called the **Shrinking Generator**. Suppose we have two LFSRs, one of degree k_1 and one of k_2. We will require a total of $k_1 + k_2$ bits as our seed, in order to initialize both LFSRs. The first LFSR will produce a sequence of bits, say $a_1, a_2, \ldots$, and the second produces a sequence of bits $b_1, b_2, \ldots$. Then we define a sequence of pseudo-random bits $z_1, z_2, \ldots$ by the rule

$$z_i = a_{i_k},$$

where i_k is the position of the kth 1 in the sequence $b_1, b_2, \ldots$. These pseudo-random bits comprise a subsequence of the bits produced by the first LFSR. This method of pseudo-random bit generation is very fast and is resistent to various known attacks, but there does not seem to be any way to prove that it is secure.

In the rest of this chapter, we will investigate PRBGs that can be proved to be secure given some plausible computational assumption. There are PRBGs based on the fundamental problems of factoring (as it relates to the **RSA** public-key cryptosystem) and the **Discrete Logarithm** problem. A PRBG based on the **RSA** encryption function is shown in Figure 12.2, and a PRBG based on the **Discrete Logarithm** problem is discussed in the exercises.

We now give an example of the **RSA Generator**.

TABLE 12.2
Bits produced by RSA generator

i	s_i	z_i
0	75634	
1	31483	1
2	31238	0
3	51968	0
4	39796	0
5	28716	0
6	14089	1
7	5923	1
8	44891	1
9	62284	0
10	11889	1
11	43467	1
12	71215	1
13	10401	1
14	77444	0
15	56794	0
16	78147	1
17	72137	1
18	89592	0
19	29022	0
20	13356	0

Example 12.2
Suppose $n = 91261 = 263 \times 347$, $b = 1547$, and $s_0 = 75364$. The first 20 bits produced by the **RSA Generator** are computed as shown in Table 12.2. Hence the bit-string resulting from this seed is

$$10000111011110011000.$$

□

12.2   Indistinguishable Probability Distributions

There are two main objectives of a pseudo-random number generator: it should be fast (i.e., computable in polynomial time as a function of k) and it should be secure. Of course, these two requirements are often conflicting. The PRBGs based on linear congruences or linear feedback shift registers are indeed very fast. These PRBGs are quite useful in simulations, but they are very insecure for cryptographic applications.

Let us now try to make precise the idea of a PRBG being "secure." Intuitively, a string of k^m bits produced by a PRBG should look "random." That is, it should be impossible in an amount of time that is polynomial in k (equivalently, polynomial in ℓ) to distinguish a string of ℓ pseudo-random bits produced by a PRBG from a string of ℓ truly random bits.

This motivates the idea of distinguishability of probability distributions. Here is a definition of this concept.

DEFINITION 12.2 *Suppose p_0 and p_1 are two probability distributions on the set $(\mathbb{Z}_2)^\ell$ of bit-strings of length ℓ. Let $\mathbf{A} : (\mathbb{Z}_2)^\ell \to \{0,1\}$ be a probabilistic algorithm that runs in polynomial time (as a function of ℓ). Let $\epsilon > 0$. For $j = 0, 1$, define*

$$E_\mathbf{A}(p_j) = \sum_{(z_1,\ldots,z_\ell) \in (\mathbb{Z}_2)^\ell} p_j(z_1,\ldots,z_\ell) \times p(\mathbf{A}(z_1,\ldots,z_\ell) = 1|(z_1,\ldots,z_\ell)).$$

We say that $\mathbf{A}$ is an ϵ-distinguisher of p_0 and p_1 provided that

$$|E_\mathbf{A}(p_0) - E_\mathbf{A}(p_1)| \geq \epsilon,$$

and we say that p_0 and p_1 are ϵ-distinguishable if there exists an ϵ-distinguisher of p_0 and p_1.

REMARK   If $\mathbf{A}$ is a deterministic algorithm, then the conditional probabilities

$$p(\mathbf{A}(z_1,\ldots,z_\ell) = 1|(z_1,\ldots,z_\ell))$$

always have the value 0 or 1.   ∎

The intuition behind this definition is as follows. The algorithm $\mathbf{A}$ tries to decide if a bit-string $(z_1,\ldots,z_\ell)$ of length ℓ is more likely to have arisen from probability distribution p_1 or from probability distribution p_0. This algorithm may use random numbers if desired, i.e., it can be probabilistic. The output $\mathbf{A}(z_1,\ldots,z_\ell)$ represents the algorithm's guess as to which of these two probability distributions is more likely to have produced $(z_1,\ldots,z_\ell)$. The quantity $E_\mathbf{A}(p_j)$ represents the average (i.e., expected) value of the output of $\mathbf{A}$ over the probability distribution p_j, for $j = 0, 1$. This is computed by summing over all possible sequences $(z_1,\ldots,z_\ell)$ the product of the probability of the ℓ-tuple $(z_1,\ldots,z_\ell)$ and the probability that $\mathbf{A}$ answers "1" when given $(z_1,\ldots,z_\ell)$ as input. $\mathbf{A}$ is an ϵ-distinguisher provided that the values of these two expectations are at least ϵ apart.

The relevance to PRBGs is as follows. Consider the sequence of ℓ bits produced by the PRBG. There are 2^ℓ possible sequences, and if the bits were chosen independently at random, each of these 2^ℓ sequences would occur with equal probability $1/2^\ell$. Thus a truly random sequence corresponds to an equiprobable distribution on the set of all bit-strings of length ℓ. Suppose we denote this probability distribution by p_0.

Now, consider sequences produced by the PRBG. Suppose a k-bit seed is chosen at random, and then the PRBG is used to obtain a bit-string of length ℓ. Then we obtain a probability distribution on the set of all bit-strings of length ℓ, which we denote by p_1. (For the purposes of illustration, suppose we make the simplifying assumption that no two seeds give rise to the same sequence of bits. Then, of the 2^ℓ possible sequences, 2^k sequences each occur with probability $1/2^k$, and the remaining $2^\ell - 2^k$ sequences never occur. So, in this case, the probability distribution p_1 is very non-uniform.)

Even though the two probability distributions p_0 and p_1 may be quite different, it is still conceivable that they might be ϵ-distinguishable only for small values of ϵ. This is our objective in constructing PRBGs.

Example 12.3
Suppose that a PRBG only produces sequences in which exactly $\ell/2$ bits have the value 0 and $\ell/2$ bits have the value 1. Define the function $\mathbf{A}$ by

$$\mathbf{A}(z_1,\ldots,z_\ell) = \begin{cases} 1 & \text{if } (z_1,\ldots,z_\ell) \text{ has } \ell/2 \text{ bits equal to } 0 \\ 0 & \text{otherwise.} \end{cases}$$

In this case, the algorithm $\mathbf{A}$ is deterministic. It is not hard to see that

$$E_{\mathbf{A}}(p_0) = \frac{\binom{\ell}{\ell/2}}{2^\ell}$$

and

$$E_{\mathbf{A}}(p_1) = 1.$$

It can be shown that

$$\lim_{\ell \to \infty} \frac{\binom{\ell}{\ell/2}}{2^\ell} = 0.$$

Hence, for any fixed value of $\epsilon < 1$, p_0 and p_1 are ϵ-distinguishable if ℓ is sufficiently large. $\qquad\square$

12.2.1 Next Bit Predictors

Another useful concept in studying PRBGs is that of a next bit predictor, which works as follows. Let f be a (k, ℓ)-PRBG. Suppose we have a probabilistic algorithm $\mathbf{B}_i$, which takes as input the first $i-1$ bits produced by f (given an unknown seed), say $z_1,\ldots,z_{i-1}$, and attempts to predict the next bit z_i. The value i can be any value such that $0 \le i \le \ell - 1$. We say that $\mathbf{B}_i$ is an ϵ-*next bit predictor* if

$\mathbf{B}_i$ can predict the ith bit of a pseudo-random sequence with probability at least $1/2 + \epsilon$, where $\epsilon > 0$.

We can give a more precise formulation of this concept in terms of probability distributions, as follows. We have already defined the probability distribution p_1 on $(\mathbb{Z}_2)^\ell$ induced by the PRBG f. We can also look at the probability distributions induced by f on any of the ℓ pseudo-random output bits (or indeed on any subset of these ℓ output bits). So, for $1 \le i \le \ell$, we will can think of the ith pseudo-random output bit as a random variable that we will denote by $\mathbf{z}_i$.

In view of these definitions, we have the following characterization of a next bit predictor.

THEOREM 12.1
Let f be a (k, ℓ)-PRBG. Then the probabilistic algorithm $\mathbf{B}_i$ is an ϵ-next bit predictor for f if and only if

$$\sum_{(z_1,\dots,z_{i-1})\in(\mathbb{Z}_2)^{i-1}} p_1(z_1,\dots,z_{i-1}) \times p(\mathbf{z}_i = \mathbf{B}_i|(z_1,\dots,z_{i-1})) \ge \frac{1}{2} + \epsilon.$$

PROOF   The probability of correctly predicting the ith bit is computed by summing over all possible $(i-1)$-tuples $(z_1,\dots,z_{i-1})$ the product of the probability that the $(i-1)$-tuple $(z_1,\dots,z_{i-1})$ is produced by the PRBG and the probability that the ith bit is predicted correctly given the $(i-1)$-tuple $(z_1,\dots,z_{i-1})$.   ∎

The reason for the expression $1/2 + \epsilon$ in this definition is that any predicting algorithm can predict the next bit of a random sequence with probability $1/2$. If a sequence is not random, then it may be possible to predict the next bit with higher probability. (Note that it is unnecessary to consider algorithms that predict the next bit with probability less than $1/2$, because in this case an algorithm that replaces every prediction z by $1 - z$ will predict the next bit with probability greater than $1/2$.)

We illustrate these ideas by producing a next-bit predictor for the **Linear Congruential Generator** of Example 12.1.

Example 12.1   (Cont.)
For any i such that $1 \le i \le 9$, Define $\mathbf{B}_i(z) = 1 - z$. That is, $\mathbf{B}_i$ predicts that a 0 is most likely to be followed by a 1, and vice versa. It is not hard to compute from Table 12.1 that each of these predictors $\mathbf{B}_i$ is a $\frac{9}{62}$-next bit predictor (i.e., they predict the next bit correctly with probability 20/31).   ☐

We can use a next bit predictor to construct a distinguishing algorithm $\mathbf{A}$, as shown in Figure 12.3. The input to algorithm $\mathbf{A}$ is a sequence of bits, $z_1,\dots,z_\ell$, and $\mathbf{A}$ calls the algorithm $\mathbf{B}_i$ as a subroutine.

FIGURE 12.3
Constructing a distinguisher from a next bit predictor

> Input: an ℓ-tuple $(z_1, \ldots, z_\ell)$
> 1. compute $z := \mathbf{B}_i(z_1, \ldots, z_{i-1})$
> 2. **if** $z = z_i$ **then**
> $\qquad \mathbf{A}(z_1, \ldots, z_\ell) = 1$
> **else**
> $\qquad \mathbf{A}(z_1, \ldots, z_\ell) = 0.$

THEOREM 12.2
Suppose $\mathbf{B}_i$ is an ϵ-next bit predictor for the (k, ℓ)-PRBG f. Let p_1 be the probability distribution induced on $(\mathbb{Z}_2)^\ell$ by f, and let p_0 be the uniform probability distribution on $(\mathbb{Z}_2)^\ell$. Then $\mathbf{A}$, as described in Figure 12.3, is an ϵ-distinguisher of p_1 and p_0.

PROOF   First, observe that

$$\mathbf{A}(z_1, \ldots, z_\ell) = 1 \Leftrightarrow \mathbf{B}_i(z_1, \ldots, z_{i-1}) = z_i.$$

Also, the output of $\mathbf{A}$ is independent of the values of $z_{i+1}, \ldots, z_\ell$. Thus we can compute as follows:

$$
\begin{aligned}
E_\mathbf{A}(p_1) &= \sum_{(z_1,\ldots,z_\ell)\in(\mathbb{Z}_2)^\ell} p_1(z_1, \ldots, z_\ell) \times p(\mathbf{A} = 1|(z_1, \ldots, z_\ell)) \\
&= \sum_{(z_1,\ldots,z_i)\in(\mathbb{Z}_2)^i} p_1(z_1, \ldots, z_i) \times p(\mathbf{A} = 1|(z_1, \ldots, z_i)) \\
&= \sum_{(z_1,\ldots,z_i)\in(\mathbb{Z}_2)^i} p_1(z_1, \ldots, z_i) \times p(\mathbf{B}_i = z_i|(z_1, \ldots, z_i)) \\
&= \sum_{(z_1,\ldots,z_{i-1})\in(\mathbb{Z}_2)^{i-1}} p_1(z_1, \ldots, z_{i-1}) \times p(z_i = \mathbf{B}_i|(z_1, \ldots, z_{i-1})) \\
&\geq \frac{1}{2} + \epsilon.
\end{aligned}
$$

On the other hand, any predictor $\mathbf{B}_i$ will predict the ith bit of a truly random sequence with probability $1/2$. Then, it is not difficult to see that $E_\mathbf{A}(p_0) = 1/2$. Hence $|E_\mathbf{A}(p_0) - E_\mathbf{A}(p_1)| \geq \epsilon$, as desired.   ∎

One of the main results in the theory of pseudo-random bit generators, due to Yao, is that a next bit predictor is a *universal* test. That is, a PRBG is "secure" if and only if there does not exist an ϵ-next bit predictor except for very small values of ϵ. Theorem 12.2 proves the implication in one direction. To prove the converse, we need to show how the existence of a distinguisher implies the existence of a next bit predictor. This is done in Theorem 12.3.

THEOREM 12.3

Suppose $\mathbf{A}$, *is an ϵ-distinguisher of p_1 and p_0, where p_1 is the probability distribution induced on $(\mathbb{Z}_2)^\ell$ by the (k, ℓ)-PRBG f, and p_0 is the uniform probability distribution on $(\mathbb{Z}_2)^\ell$. Then for some i, $1 \leq i \leq \ell - 1$, there exists an ϵ/ℓ-next bit predictor $\mathbf{B}_i$ for f.*

PROOF For $0 \leq i \leq \ell$, define q_i to be a probability distribution on $(\mathbb{Z}_2)^\ell$ where the first i bits are generated using f, and the remaining $\ell - i$ bits are generated at random. Thus $q_0 = p_0$ and $q_\ell = p_1$. We are given that

$$|E_\mathbf{A}(q_0) - E_\mathbf{A}(q_\ell)| \geq \epsilon.$$

By the triangle inequality, we have that

$$|E_\mathbf{A}(q_0) - E_\mathbf{A}(q_\ell)| \leq \sum_{i=1}^{\ell} |E_\mathbf{A}(q_{i-1}) - E_\mathbf{A}(q_i)|.$$

Hence, it follows that there is at least one value i, $1 \leq i \leq \ell$, such that

$$|E_\mathbf{A}(q_{i-1}) - E_\mathbf{A}(q_i)| \geq \frac{\epsilon}{\ell}.$$

Without loss of generality, we will assume that

$$E_\mathbf{A}(q_{i-1}) - E_\mathbf{A}(q_i) \geq \frac{\epsilon}{\ell}.$$

We are going to construct an ith bit predictor (for this specified value of i). The predicting algorithm is probabilistic in nature and is presented in Figure 12.4. Here is the idea behind this construction. The predicting algorithm in fact produces an ℓ-tuple according to the probability distribution q_{i-1}, given that $z_1, \ldots, z_{i-1}$ are generated by the PRBG. If $\mathbf{A}$ answers "0," then it thinks that the ℓ-tuple was most likely generated according to the probability distribution q_i. Now q_{i-1} and q_i differ only in that the ith bit is generated at random in q_{i-1}, whereas it is generated according to the PRBG in q_i. Hence, when $\mathbf{A}$ answers "0," it thinks that the ith bit, z_i, is what would be produced by the PRBG. Hence, in this case we take z_i as our prediction of the ith bit. If $\mathbf{A}$ answers "1," it thinks that z_i is random, so we take $1 - z_i$ as our prediction of the ith bit.

We need to compute the probability that the ith bit is predicted correctly. Observe that if $\mathbf{A}$ answers "0," then the prediction is correct with probability

$$p_1(z_i|(z_1, \ldots, z_{i-1})),$$

FIGURE 12.4
Constructing a next bit predictor from a distinguisher

Input: an $(i-1)$-tuple $(z_1, \ldots, z_{i-1})$
1. choose $(z_i, \ldots, z_\ell) \in (\mathbb{Z}_2)^{\ell-i+1}$ at random
2. compute $z := \mathbf{A}(z_1, \ldots, z_\ell)$
3. define $\mathbf{B}_i(z_1, \ldots, z_{i-1}) = (z + z_i) \bmod 2$.

where p_1 is the probability distribution induced by the PRBG. If $\mathbf{A}$ answers "1," then the prediction is correct with probability

$$1 - p_1(z_i | (z_1, \ldots, z_{i-1})).$$

For brevity, we denote $\mathbf{z} = (z_1, \ldots, z_\ell)$. In our computation, we will make use of the fact that

$$q_{i-1}(\mathbf{z}) \times p_1(z_i | (z_1, \ldots, z_{i-1})) = \frac{q_i(\mathbf{z})}{2}.$$

This can be proved easily as follows:

$$q_{i-1}(z_1, \ldots, z_\ell) \times p_1(z_i | (z_1, \ldots, z_{i-1}))$$
$$= q_{i-1}(z_1, \ldots, z_{i-1}) \times \frac{1}{2^{\ell-i+1}} \times p_1(z_i | (z_1, \ldots, z_{i-1}))$$
$$= q_i(z_1, \ldots, z_i) \times \frac{1}{2^{\ell-i+1}}$$
$$= \frac{q_i(z_1, \ldots, z_\ell)}{2}.$$

Now we can perform our main computation:

$$p(\mathbf{z}_i = \mathbf{B}_i(z_1, \ldots, z_{i-1}))$$
$$= \sum_{\mathbf{z} \in (\mathbb{Z}_2)^\ell} q_{i-1}(\mathbf{z})[p(\mathbf{A} = 0 | \mathbf{z}) \times p_1(z_i | (z_1, \ldots, z_{i-1}))$$
$$\qquad + p(\mathbf{A} = 1 | \mathbf{z}) \times (1 - p_1(z_i | (z_1, \ldots, z_{i-1})))]$$
$$= \sum_{\mathbf{z} \in (\mathbb{Z}_2)^\ell} \frac{q_i(\mathbf{z})}{2} \times p(\mathbf{A} = 0 | \mathbf{z}) + \sum_{\mathbf{z} \in (\mathbb{Z}_2)^\ell} q_{i-1}(\mathbf{z}) \times p(\mathbf{A} = 1 | \mathbf{z})$$
$$\qquad - \sum_{\mathbf{z} \in (\mathbb{Z}_2)^\ell} \frac{q_i(\mathbf{z})}{2} \times p(\mathbf{A} = 1 | \mathbf{z})$$

$$= \frac{1 - E_{\mathbf{A}}(q_i)}{2} + E_{\mathbf{A}}(q_{i-1}) - \frac{E_{\mathbf{A}}(q_i)}{2}$$

$$= \frac{1}{2} + E_{\mathbf{A}}(q_{i-1}) - E_{\mathbf{A}}(q_i)$$

$$\geq \frac{1}{2} + \frac{\epsilon}{\ell},$$

which was what we wanted to prove. ∎

12.3 The Blum-Blum-Shub Generator

In this section we describe one of the most popular PRBGs, due to Blum, Blum, and Shub. First, we review some results on Jacobi symbols from Section 4.5 and other number-theoretic facts from other parts of Chapter 4.

Suppose p and q are two distinct primes, and let $n = pq$. Recall that the Jacobi symbol

$$\left(\frac{x}{n}\right) = \begin{cases} 0 & \text{if } \gcd(x, n) > 1 \\ 1 & \text{if } \left(\frac{x}{p}\right) = \left(\frac{x}{q}\right) = 1 \text{ or if } \left(\frac{x}{p}\right) = \left(\frac{x}{q}\right) = -1 \\ -1 & \text{if one of } \left(\frac{x}{p}\right) \text{ and } \left(\frac{x}{q}\right) \text{ is 1 and the other is } -1. \end{cases}$$

Denote the quadratic residues modulo n by $\mathrm{QR}(n)$. That is,

$$\mathrm{QR}(n) = \{x^2 \bmod n : x \in \mathbb{Z}_n^*\}.$$

Recall that x is a quadratic residue modulo n if and only if

$$\left(\frac{x}{p}\right) = \left(\frac{x}{q}\right) = 1.$$

Define

$$\tilde{\mathrm{QR}}(n) = \left\{x \in \mathbb{Z}_n^* \backslash \mathrm{QR}(n) : \left(\frac{x}{n}\right) = 1\right\}.$$

Thus

$$\tilde{\mathrm{QR}}(n) = \left\{x \in \mathbb{Z}_n^* : \left(\frac{x}{p}\right) = \left(\frac{x}{q}\right) = -1\right\}.$$

An element $x \in \tilde{\mathrm{QR}}(n)$ is called a *pseudo-square* modulo n.

The **Blum-Blum-Shub Generator**, as well as some other cryptographic systems, is based on the **Quadratic Residues** problem defined in Figure 12.5. (In Chapter 4, we defined the **Quadratic Residues** problem modulo a prime and showed that it is easy to solve; here we have a composite modulus.) Observe that

FIGURE 12.5
Quadratic Residues

Problem Instance   A positive integer n that is the product of two un-known primes p and q, and an integer $x \in \mathbb{Z}_n^*$ such that $\left(\frac{x}{n}\right) = 1$.

Question   Is x a quadratic residue modulo n?

FIGURE 12.6
Blum-Blum-Shub Generator

Let p, q to be two $(k/2)$-bit primes such that $p \equiv q \equiv 3 \bmod 4$, and define $n = pq$. Let $\text{QR}(n)$ denote the set of quadratic residues modulo n.

A seed s_0 is any element of $\text{QR}(n)$. For $i \geq 0$, define

$$s_{i+1} = s_i^2 \bmod n,$$

and then define
$$f(s_0) = (z_1, z_2, \ldots, z_\ell),$$

where
$$z_i = s_i \bmod 2,$$

$1 \leq i \leq \ell$. Then f is a (k, ℓ)-PRBG, called the **Blum-Blum-Shub Generator**, which we abbreviate to **BBS Generator**.

---

the **Quadratic Residues** problem requires us to distinguish quadratic residues modulo n from pseudo-squares modulo n. This can be no more difficult than factoring n. For if the factorization $n = pq$ can be computed, then it is a simple matter to compute $\left(\frac{x}{p}\right)$, say. Given that $\left(\frac{x}{n}\right) = 1$, it follows that x is a quadratic residue if and only if $\left(\frac{x}{p}\right) = 1$.

There does not appear to be any way to solve the **Quadratic Residues** problem efficiently if the factorization of n is not known. So this problem appears to be intractible if it is infeasible to factor n.

The **Blum-Blum-Shub Generator** is presented in Figure 12.6.   The generator works quite simply. Given a seed $s_0 \in \text{QR}(n)$, we compute the sequence $s_1, s_2, \ldots, s_\ell$ by successive squaring modulo n, and then reduce each s_i modulo

TABLE 12.3
Bits produced by BBS generator

i	s_i	z_i
0	20749	
1	143135	1
2	177671	1
3	97048	0
4	89992	0
5	174051	1
6	80649	1
7	45663	1
8	69442	0
9	186894	0
10	177046	0
11	137922	0
12	123175	1
13	8630	0
14	114386	0
15	14863	1
16	133015	1
17	106065	1
18	45870	0
19	137171	1
20	48060	0

2 to obtain z_i. It follows that

$$z_i = \left(s_0^{2^i} \bmod n \right) \bmod 2,$$

$1 \le i \le \ell$.

We now give an example of the **BBS Generator**.

Example 12.4
Suppose $n = 192649 = 383 \times 503$ and $s_0 = 101355^2 \bmod n = 20749$. The first 20 bits produced by the **BBS Generator** are computed as shown in Table 12.3. Hence the bit-string resulting from this seed is

$$11001110000100111010.$$

$\square$

Here is a feature of the **BBS Generator** that is useful when we look at its security. Since $n = pq$ where $p \equiv q \equiv 3 \bmod 4$, it follows that for any quadratic residue x, there is a unique square root of x that is also a quadratic residue. This square root is called the *principal* square root of x. It follows the mapping $x \mapsto x^2 \bmod n$ used to define the **BBS Generator** is a permutation on $\mathrm{QR}(n)$, the set of quadratic residues modulo n.

12.3.1 Security of the BBS Generator

In this section, we look at the security of the **BBS Generator** in detail. We begin by supposing that the pseudo-random bits produced by the **BBS Generator** are ϵ-distinguishable from ℓ random bits and then see where that leads us. Throughout this section, $n = pq$, where p and q are primes such that $p \equiv q \equiv 3 \bmod 4$, and the factorization $n = pq$ is unknown.

We have already discussed the idea of a next bit predictor. In this section we consider a similar concept that we call a *previous bit predictor*. A previous bit predictor for a (k, ℓ)-**BBS Generator** will take as input ℓ pseudo-random bits produced by the generator (as determined by an unknown random seed $s_0 \in \mathrm{QR}(n)$), and attempt to predict the value $z_0 = s_0 \bmod 2$. A previous bit predictor can be a probabilistic algorithm, and we say that a previous bit predictor $\mathbf{B}_0$ is an ϵ-previous bit predictor if its probability of correctly guessing z_0 is at least $1/2 + \epsilon$, where this probability is computed over all possible seeds s_0.

We state the following theorem, which is similar to Theorem 12.3, without proof.

THEOREM 12.4
Suppose $\mathbf{A}$, *is an ϵ-distinguisher of p_1 and p_0, where p_1 is the probability distribution induced on $(\mathbb{Z}_2)^\ell$ by the (k, ℓ)-***BBS Generator***, f, and p_0 is the uniform probability distribution on $(\mathbb{Z}_2)^\ell$. Then there exists an (ϵ/ℓ)-previous bit predictor* $\mathbf{B}_0$ *for f.*

We now show how to use an (ϵ/ℓ)-previous bit predictor, $\mathbf{B}_0$, to construct a probabilistic algorithm that distinguishes quadratic residues modulo n from pseudo-squares modulo n with probability $1/2 + \epsilon$. This algorithm $\mathbf{A}$, presented in Figure 12.7, uses $\mathbf{B}_0$ as a subroutine, or oracle.

THEOREM 12.5
Suppose $\mathbf{B}_0$ *is an ϵ-previous bit predictor for the (k, ℓ)-***BBS Generator*** f. Then the algorithm* $\mathbf{A}$, *as described in Figure 12.7, determines quadratic residuosity correctly with probability at least $1/2 + \epsilon$, where this probability is computed over all possible inputs $x \in \mathrm{QR}(n) \cup \tilde{\mathrm{QR}}(n)$.*

PROOF Since $n = pq$ and $p \equiv q \equiv 3 \bmod 4$, it follows that $\left(\frac{-1}{n}\right) = 1$, so $-1 \in \tilde{\mathrm{QR}}(n)$. Hence, if $\left(\frac{x}{n}\right) = 1$, then the principal square root of $s_0 = x^2$ is x if $x \in \mathrm{QR}(n)$; and $-x$ if $x \in \tilde{\mathrm{QR}}(n)$. But

$$(-x \bmod n) \bmod 2 \neq (x \bmod n) \bmod 2,$$

FIGURE 12.7
Constructing a quadratic residue distinguisher from a previous bit predictor

---

Input: $x \in \mathbb{Z}_n^*$ such that $\left(\frac{x}{n}\right) = 1$

1.   compute $s_0 = x^2 \bmod n$ and compute $z_0 = s_0 \bmod 2$

2.   use the **BBS Generator** to compute $z_1, \ldots, z_{\ell-1}$ from seed s_0

3.   compute $z = \mathbf{B}_0(z_0, \ldots, z_{\ell-1})$

4.   **if** $(x \bmod 2) = z$ **then**

           answer "$x \in QR(n)$"

   else

           answer "$x \in \tilde{QR}(n)$."

---

so it follows that algorithm **A** gives the correct answer if and only if $\mathbf{B}_0$ correctly predicts z. The result then follows immediately.   ∎

Theorem 12.5 shows how we can distinguish pseudo-squares from quadratic residues with probability at least $1/2 + \epsilon$. We now show that this leads to a Monte Carlo algorithm that gives the correct answer with probability at least $1/2 + \epsilon$. In other words, for any $x \in QR(n) \cup \tilde{QR}(n)$, the Monte Carlo algorithm gives the correct answer with probabilty at least $1/2 + \epsilon$. Note that this algorithm is an *unbiased* algorithm (it may give an incorrect answer for any input) in contrast to the Monte Carlo algorithms that we studied in Section 4.5 which were all biased algorithms.

The Monte Carlo algorithm $\mathbf{A}_1$ is presented in Figure 12.8. It calls the previous algorithm **A** as a subroutine.

THEOREM 12.6

Suppose that algorithm **A** *determines quadratic residuosity correctly with probability at least* $1/2 + \epsilon$. *Then the algorithm* $\mathbf{A}_1$, *as described in Figure 12.8, is a Monte Carlo algorithm for* **Quadratic Residues** *with error probability at most* $1/2 - \epsilon$.

PROOF   For any given input $x \in QR(n) \cup \tilde{QR}(n)$, the effect of step 2 in algorithm $\mathbf{A}_1$ is to produce an element x' that is a random element of $QR(n) \cup \tilde{QR}(n)$ whose status as a quadratic residue is known.   ∎

The last step is to show that any (unbiased) Monte Carlo algorithm that has

FIGURE 12.8
A Monte Carlo algorithm for Quadratic Residues

Input: $x \in \mathbb{Z}_n{}^*$ such that $\left(\frac{x}{n}\right) = 1$

1. choose $r \in \mathbb{Z}_n^*$ at random
2. with probability $1/2$, compute

$$x' = r^2 x \bmod n,$$

otherwise compute
$$x' = -r^2 x \bmod n.$$

3. call $\mathbf{A}(x')$, obtaining an answer "QR" or "$\tilde{\text{QR}}$"
4. **if**

$$\mathbf{A}(x') = \text{QR and } x' = r^2 x \bmod n$$

or

$$\mathbf{A}(x') = \tilde{\text{QR}} \text{ and } x' = -r^2 x \bmod n$$

then
    answer "$x \in \text{QR}$"
else
    answer "$x \in \tilde{\text{QR}}$."

---

error probability at most $1/2 - \epsilon$ can be used to construct an unbiased Monte Carlo algorithm with error probability at most δ, for any $\delta > 0$. In other words, we can make the probability of correctness arbitrarily close to 1. The idea is to run the given Monte Carlo algorithm $2m + 1$ times, for some integer m, and take the "majority vote" as the answer. By computing the error probability of this algorithm, we can also see how m depends on δ. This dependence is stated in the following theorem.

THEOREM 12.7
Suppose $\mathbf{A}_1$ is an unbiased Monte Carlo algorithm with error probability at most $1/2 - \epsilon$. Suppose we run $\mathbf{A}_1$ $n = 2m + 1$ times on a given instance I, and we take the most frequent answer. Then the error probability of the resulting algorithm is at most

$$\frac{(1 - 4\epsilon^2)^m}{2}.$$

PROOF   The probability of obtaining exactly i correct answers in the n trials is at most

$$\binom{n}{i}\left(\frac{1}{2}+\epsilon\right)^i\left(\frac{1}{2}-\epsilon\right)^{n-i}.$$

The probability that the most frequent answer is incorrect is equal to the probability that the number of correct answers in the n trials is at most m. Hence, we compute as follows

$$\begin{aligned}
p(\text{error}) &\leq \sum_{i=0}^{m}\binom{n}{i}\left(\frac{1}{2}+\epsilon\right)^i\left(\frac{1}{2}-\epsilon\right)^{2m+1-i}\\
&= \left(\frac{1}{2}+\epsilon\right)^m\left(\frac{1}{2}-\epsilon\right)^{m+1}\sum_{i=0}^{m}\binom{n}{i}\left(\frac{1/2-\epsilon}{1/2+\epsilon}\right)^{m-i}\\
&\leq \left(\frac{1}{2}+\epsilon\right)^m\left(\frac{1}{2}-\epsilon\right)^{m+1}\sum_{i=0}^{m}\binom{n}{i}\\
&= \left(\frac{1}{2}+\epsilon\right)^m\left(\frac{1}{2}-\epsilon\right)^{m+1}2^{2m}\\
&= \left(\frac{1}{4}-\epsilon^2\right)^m\left(\frac{1}{2}-\epsilon\right)2^{2m}\\
&= (1-4\epsilon^2)^m\left(\frac{1}{2}-\epsilon\right)\\
&\leq \frac{(1-4\epsilon^2)^m}{2},
\end{aligned}$$

as required.   ∎

Suppose we want to lower the probability of error to some value δ, where $0 < \delta < 1/2 - \epsilon$. We need to choose m so that

$$\frac{(1-4\epsilon^2)^m}{2} \leq \delta.$$

Hence, it suffices to take

$$m = \left\lceil \frac{1+\log_2\delta}{\log_2(1-4\epsilon^2)} \right\rceil.$$

Then, if algorithm A is run $2m+1$ times, the majority vote yields the correct answer with probability at least $1 - \delta$. It is not hard to show that this value of m is at most $c/(\delta\epsilon^2)$ for some constant c. Hence, the number of times that the algorithm must be run is polynomial in $1/\delta$ and $1/\epsilon$.

Example 12.5

Suppose we start with a Monte Carlo algorithm that returns the correct answer with probability at least .55, so $\epsilon = .05$. If we desire a Monte Carlo algorithm in which the probability of error is at most .05, then it suffices to take $m = 230$ and $n = 461$. ⬚

Let us combine all the reductions we have done. We have the following sequence of implications:

(k, ℓ)-**BBS Generator** can be ϵ-distinguished from ℓ random bits

$$\Downarrow$$

(ϵ/ℓ)-previous bit predictor for (k, ℓ)-**BBS Generator**

$$\Downarrow$$

distinguishing algorithm for **Quadratic Residues** that is correct with probability at least $1/2 + \epsilon/\ell$

$$\Downarrow$$

unbiased Monte Carlo algorithm for **Quadratic Residues** having error probability at most $1/2 - \epsilon/\ell$

$$\Downarrow$$

unbiased Monte Carlo algorithm for **Quadratic Residues** having error probability at most δ, for any $\delta > 0$

Since it is widely believed that there is no polynomial-time Monte Carlo algorithm for **Quadratic Residues** with small error probability, we have some evidence that the **BBS Generator** is secure.

We close this section by mentioning a way of improving the efficiency of the **BBS Generator**. The sequence of pseudo-random bits is constructed by taking the least significant bit of each s_i, where $s_i = s_0^{2^i} \bmod n$. Suppose instead that we extract the m least significant bits from each s_i. This will improve the efficiency of the PRBG by a factor of m, but we need to ask if the PRBG will remain secure. It has been shown that this approach will remain secure provided that $m \leq \log_2 \log_2 n$. So we can extract about $\log_2 \log_2 n$ pseudo-random bits per modular squaring. In a realistic implementation of the **BBS Generator**, $n \approx 10^{160}$, so we can extract nine bits per squaring.

12.4 Probabilistic Encryption

Probabilistic encryption is an idea of Goldwasser and Micali. One motivation is as follows. Suppose we have a public-key cryptosystem, and we wish to encrypt a single bit, i.e., $x = 0$ or 1. Since anyone can compute $e_K(0)$ and $e_K(1)$, it is a simple matter for an opponent to determine if a ciphertext y is an encryption of 0 or an encryption of 1. More generally, an opponent can always determine if the plaintext has a specified value by encrypting a hypothesized plaintext, hoping to match a given ciphertext.

The goal of probabilistic encryption is that "no information" about the plaintext should be computable from the ciphertext (in polynomial time). This objective can be realized by a public-key cryptosystem in which encryption is probabilistic rather than deterministic. Since there are "many" possible encryptions of each plaintext, it is not feasible to test whether a given ciphertext is an encryption of a particular plaintext.

Here is a formal mathematical definition of this concept.

DEFINITION 12.3 *A probabilistic public-key cryptosystem is defined to be a six-tuple $(\mathcal{P}, \mathcal{C}, \mathcal{K}, \mathcal{E}, \mathcal{D}, \mathcal{R})$, where $\mathcal{P}$ is the set of plaintexts, $\mathcal{C}$ is the set of ciphertexts, $\mathcal{K}$ is the keyspace, $\mathcal{R}$ is a set of randomizers, and for each key $K \in \mathcal{K}$, $e_K \in \mathcal{E}$ is a public encryption rule and $d_K \in \mathcal{D}$ is a secret decryption rule. The following properties should be satisfied:*

1. *Each $e_K : \mathcal{P} \times \mathcal{R} \to \mathcal{C}$ and $d_K : \mathcal{C} \to \mathcal{P}$ are functions such that*

$$d_K(e_K(b, r)) = b$$

   for every plaintext $b \in \mathcal{P}$ and every $r \in \mathcal{R}$. (In particular, this implies that $e_K(x, r) \neq e_K(x', r,)$ if $x \neq x'$.)

2. *Let ϵ be a specified security parameter. For any fixed $K \in \mathcal{K}$ and for any $x \in \mathcal{P}$, define a probability distribution $p_{K,x}$ on $\mathcal{C}$, where $p_{K,x}(y)$ denotes the probability that y is the ciphertext given that K is the key and x is the plaintext (this probability is computed over all $r \in \mathcal{R}$). Suppose $x, x' \in \mathcal{P}$, $x \neq x'$, and $K \in \mathcal{K}$. Then the probability distributions $p_{K,x}$ and $p_{K,x'}$ are not ϵ-distinguishable.*

Here is how the system works. To encrypt a plaintext x, choose a randomizer $r \in \mathcal{R}$ and compute $y = e_K(x, r)$. Any such value $y = e_K(x, r)$ can be decrypted to x. Property 2 is stating that the probability distribution of all encryptions of x cannot be distinguished from the probability distribution of all encryptions of x' if $x' \neq x$. Informally, an encryption of x "looks like" an encryption of x'. The security parameter ϵ should be small: in practice we would want to have $\epsilon = c/|\mathcal{R}|$ for some small $c > 0$.

We now present the **Goldwasser-Micali Probabilistic Public-key Cryptosystem** in Figure 12.9. This system encrypts one bit at a time. A 0 bit is encrypted to a random quadratic residue modulo n; a 1 bit is encrypted to a random pseudosquare modulo n. When Bob recieves an element $y \in QR(n) \cup \tilde{Q}R(n)$, he can use his knowledge of the factorization of n to determine whether $y \in QR(n)$ or whether $y \in \tilde{Q}R(n)$. He does this by computing

$$\left(\frac{y}{p}\right) = (y)^{(p-1)/2} \bmod p;$$

FIGURE 12.9
Goldwasser-Micali Probabilistic Public-key Cryptosystem

Let $n = pq$, where p and q are primes, and let $m \in \tilde{\mathrm{QR}}(n)$. The integers n and m are public; the factorization $n = pq$ is secret. Let $\mathcal{P} = \{0, 1\}$, $\mathcal{C} = \mathcal{R} = \mathbb{Z}_n{}^*$, and define

$$\mathcal{K} = \left\{ (n, p, q, m) : n = pq, p, q \text{ prime}, m \in \tilde{\mathrm{QR}}(n) \right\}.$$

For $K = (n, p, q, m)$, define

$$e_K(x, r) = m^x r^2 \bmod n$$

and

$$d_K(y) = \begin{cases} 0 & \text{if } y \in \mathrm{QR}(n) \\ 1 & \text{if } y \notin \mathrm{QR}(n), \end{cases}$$

where $x = 0$ or 1 and $r, y \in \mathbb{Z}_n{}^*$.

then

$$y \in \mathrm{QR}(n) \Leftrightarrow \left(\frac{y}{p}\right) = 1.$$

A more efficient probabilistic public-key cryptosystem was given by Blum and Goldwasser. The **Blum-Goldwasser Probabilistic Public-key Cryptosystem** is presented in Figure 12.10. The basic idea is as follows. A random seed s_0 generates a sequence of ℓ psuedorandom bits $z_1, \ldots, z_\ell$ using the **BBS Generator**. The z_i's are used as a keystream, i.e., they are exclusive-ored with the ℓ plaintext bits to form the ciphertext. As well, the $(\ell + 1)$st element $s_{\ell+1} = s_0{}^{2^{\ell+1}} \bmod n$ is transmitted as part of the ciphertext.

When Bob receives the ciphertext, he can compute s_0 from $s_{\ell+1}$, then reconstruct the keystream, and finally exclusive-or the keystream with the ℓ ciphertext bits to obtain the plaintext. We should explain how Bob derives s_0 from $s_{\ell+1}$. Recall that each s_{i-1} is the principal square root of s_i. Now, $n = pq$ with $p \equiv q \equiv 3 \bmod 4$, so the square roots of any quadratic residue x modulo p are $\pm x^{(p+1)/4}$. Using properties of Jacobi symbols, we have that

$$\left(\frac{x^{(p+1)/4}}{p}\right) = \left(\frac{x}{p}\right)^{(p+1)/4}$$
$$= 1.$$

It follows that $x^{(p+1)/4}$ is the principal square root of x modulo p. Similarly,

FIGURE 12.10
Blum-Goldwasser Probabilistic Public-key Cryptosystem

Let $n = pq$, where p and q are primes, $p \equiv q \equiv 3 \bmod 4$. The integer n is public; the factorization $n = pq$ is secret. Let $\mathcal{P} = (\mathbb{Z}_2)^\ell$, $\mathcal{C} = (\mathbb{Z}_2)^\ell \times \mathbb{Z}_n^*$ and $\mathcal{R} = \mathbb{Z}_n^*$. Define

$$\mathcal{K} = \{(n, p, q) : n = pq, p, q \text{ prime}\}.$$

For $K = (n, p, q)$ and $x \in (\mathbb{Z}_2)^\ell$ and $r \in \mathbb{Z}_n^*$, encrypt x as follows:

1. Compute $z_1, \ldots, z_\ell$ from seed $s_0 = r$ using the **BBS Generator**.
2. Compute $s_{\ell+1} = s_0^{2^{\ell+1}} \bmod n$.
3. Compute $y_i = (x_i + z_i) \bmod 2$ for $1 \le i \le \ell$.
4. Define $e_K(x, r) = (y_1, \ldots, y_\ell, s_{\ell+1})$.

To decrypt y, Bob performs the following steps:

1. Compute $a_1 = ((p+1)/4)^{\ell+1} \bmod (p-1)$.
2. Compute $a_2 = ((q+1)/4)^{\ell+1} \bmod (q-1)$.
3. Compute $b_1 = s_{\ell+1}^{a_1} \bmod p$.
4. Compute $b_2 = s_{\ell+1}^{a_2} \bmod q$.
5. Use the Chinese remainder theorem to find s_0 such that

$$s_0 \equiv b_1 \bmod p$$

   and

$$s_0 \equiv b_2 \bmod q.$$

6. Compute $z_1, \ldots, z_\ell$ from seed $s_0 = r$ using the **BBS Generator**.
7. Compute $x_i = (y_i + z_i) \bmod 2$ for $1 \le i \le \ell$.
8. The plaintext is $x = (x_1, \ldots, x_\ell)$.

$x^{(q+1)/4}$ is the principal square root of x modulo q. Then, using the the Chinese remainder theorem, we can find the principal square root of x modulo n.

More generally, $x^{((p+1)/4)^{\ell+1}}$ will be the principal $2^{\ell+1}$st root of x modulo p and $x^{((p+1)/4)^{\ell+1}}$ will be the principal $2^{\ell+1}$st root of x modulo q. Since $\mathbb{Z}_p^*$ has order $p - 1$, we can reduce the exponent $((p + 1)/4)^{\ell+1}$ modulo $p - 1$ in the computation $x^{((p+1)/4)^{\ell+1}} \bmod p$. In a similar fashion, we can reduce the exponent $((q + 1)/4)^{\ell+1}$ modulo $q - 1$. In Figure 12.10, having obtained the principal $2^{\ell+1}$st roots of $s_{\ell+1}$ modulo p and modulo q (steps 1-4 of the decryption process), the Chinese remainder theorem is used to compute the principal $2^{\ell+1}$st root of $s_{\ell+1}$ modulo n.

Here is an example to illustrate.

Example 12.6
Suppose $n = 192649$, as in Example 12.4. Suppose further that Alice chooses $r = 20749$ and wants to encrypt the 20-bit plaintext string

$$x = 11010011010011101101.$$

She will first compute the keystream

$$z = 11001110000100111010,$$

exactly as in Example 12.4, and then exclusive-or it with the plaintext, to obtain the ciphertext
$$y = 00011101010111010111$$

which she transmits to Bob. She also computes

$$s_{21} = s_{20}^2 \bmod n = 94739$$

and sends it to Bob.

Of course Bob knows the factorization $n = 383 \times 503$, so $(p + 1)/4 = 96$ and $(q + 1)/4 = 126$. He begins by computing

$$a_1 = ((p + 1)/4)^{\ell+1} \bmod (p - 1)$$
$$= 96^{21} \bmod 382$$
$$= 266$$

and

$$a_2 = ((q + 1)/4)^{\ell+1} \bmod (q - 1)$$
$$= 126^{21} \bmod 502$$
$$= 486.$$

Next, he calculates

$$b_1 = s_{21}{}^{a_1} \bmod p$$
$$= 94739^{266} \bmod 383$$
$$= 67$$

and

$$b_2 = s_{21}{}^{a_2} \bmod q$$
$$= 94739^{486} \bmod 503$$
$$= 126.$$

Now Bob proceeds to solve the system of congruences

$$r \equiv 67 \;(\bmod\; 383)$$
$$r \equiv 126 \;(\bmod\; 503)$$

to obtain Alice's seed $r = 20749$. Then he constructs Alice's keystream from r. Finally, he exclusive-ors the keystream with the ciphertext to get the plaintext. □

12.5 Notes and References

A lengthy treatment of PRBGs can be found in the book by Kranakis [KR86]. See also the survey paper by Lagarias [LA90].

The **Shrinking Generator** is due to Coppersmith, Krawczyk, and Mansour [CKM94]; another practical method of constructing PBRGs using LFSRs has been given by Gunther [GU88]. For methods of breaking the **Linear Congruential Generator**, see Boyar [BO89].

The basic theory of secure PRBGs is due to Yao [YA82], who proved the universality of the next bit test. Further basic results can be found in Blum and Micali [BM84]. The **BBS Generator** is described in [BBS86]. The security of the **Quadratic Residues** problem is studied by Goldwasser and Micali [GM84], on which we based much of Section 12.3.1. We have, however, used the approach of Brassard and Bratley [BB88A, Section 8.6] to reduce the error probability of an unbiased Monte Carlo algorithm.

Properties of the **RSA Generator** are studied in Alexi, Chor, Goldreich, and Schnorr [ACGS88]. PRBGs based on the **Discrete Logarithm** problem are treated in Blum and Micali [BM84], Long and Wigderson [LW88], and Håstad,

FIGURE 12.11
Discrete Logarithm Generator

Let p be a k-bit prime, and let α be a primitive element modulo p.

A seed x_0 is any element of $\mathbb{Z}_p^*$. For $i \geq 0$, define

$$x_{i+1} = \alpha^{x_i} \bmod p,$$

and then define

$$f(x_0) = (z_1, z_2, \ldots, z_\ell),$$

where

$$z_i = \begin{cases} 1 & \text{if } x_i > p/2 \\ 0 & \text{if } x_i < p/2. \end{cases}$$

Then f is called a (k, ℓ)-**Discrete Logarithm Generator**.

Schrift, and Shamir [HSS93]. A sufficient condition for the secure extraction of multiple bits per iteration of a PRBG was proved by Vazirani and Vazirani [VV84].

The idea of probabilistic encryption is due to Goldwasser and Micali [GM84]; the **Blum-Goldwasser Cryptosystem** is presented in [BG85].

Exercises

12.1 Consider the **Linear Congruential Generator** defined by $s_i = (as_{i-1} + b) \bmod M$. Suppose that $M = qa + 1$ where a is odd and q is even, and suppose that $b = 1$. Show that the next bit predictor $\mathbf{B}_i(z) = 1 - z$ for the i bit is an ϵ-next bit predictor, where

$$\frac{1}{2} + \epsilon = \frac{q(a+1)}{2M}.$$

12.2 Suppose we have an **RSA Generator** with $n = 36863$, $b = 229$ and seed $s_0 = 25$. Compute the first 100 bits produced by this generator.

12.3 A PRBG based on the **Discrete Logarithm** problem is given in Figure 12.11. Suppose $p = 21383$, the primitive element $\alpha = 5$ and the seed $s_0 = 15886$. Compute the first 100 bits produced by this generator.

12.4 Suppose that Bob has knowledge of the factorization $n = pq$ in the **BBS Generator**.
   (a) Show how Bob can use this knowledge to compute any s_i from s_0 with $2k$ multiplications modulo $\phi(n)$ and $2k$ multiplications modulo n, where n has k bits in its binary representation. (If i is large compared to k, then this approach represents a substantial improvement over the i multiplications required to sequentially compute $s_0, \ldots, s_i$.)
   (b) Use this method to compute s_{10000} if $n = 59701 = 227 \times 263$ and $s_0 = 17995$.

TABLE 12.4
Blum-Goldwasser Ciphertext

```
E1866663F17FDBD1DC8C8FD2EEBC36AD7F53795DBA3C9CE22D
C9A9C7E2A56455501399CA6B98AED22C346A529A09C1936C61
ECDE10B43D226EC683A669929F2FFB912BFA96A8302188C083
46119E4F61AD8D0829BD1CDE1E37DBA9BCE65F40C0BCE48A80
0B3D087D76ECD1805C65D9DB730B8D0943266D942CF04D7D4D
76BFA891FA21BE76F767F1D5DCC7E3F1D86E39A9348B3
```

12.5 We proved that, in order to reduce the error probability of an unbiased Monte Carlo algorithm from $1/2 - \epsilon$ to δ, where $\delta + \epsilon < 1/2$, it suffices to run the algorithm m times, where

$$m = \left\lceil \frac{1 + \log_2 \delta}{\log_2 (1 - 4\epsilon^2)} \right\rceil .$$

Prove that this value of m is $O(1/(\delta \epsilon^2))$.

12.6 Suppose Bob receives some ciphertext which was encrypted with the **Blum-Goldwasser Probabilistic Public-key Cryptosystem**. The original plaintext consisted of English text. Each alphabetic character was converted to a bitstring of length five in the obvious way: $A \leftrightarrow 00000, B \leftrightarrow 00001, \ldots, Z \leftrightarrow 11001$. The plaintext consisted of 236 alphabetic characters, so a bitstring of length 1180 resulted. This bitstring was then encrypted. The resulting ciphertext bitstring was then converted to a hexadecimal representation, to save space. The final string of 295 hexadecimal characters is presented in Table 12.4. Also, $s_{1181} = 20291$ is part of the ciphertext, and $n = 29893$ is Bob's public key. Bob's secret factorization of n is $n = pq$, where $p = 167$ and $q = 179$.

Your task is to decrypt the given ciphertext and restore the original English plaintext, which was taken from "Under the Hammer," by John Mortimer, Penguin Books, 1994.

13

Zero-knowledge Proofs

13.1 Interactive Proof Systems

Very informally, a zero-knowledge proof system allows one person to convince another person of some fact without revealing any information about the proof. We first discuss the idea of an interactive proof system. In an interactive proof system, there are two participants, Peggy and Vic. Peggy is the *prover* and Vic is the *verifier*. Peggy knows some fact, and she wishes to prove to Vic that she does.

It is necessary to describe the kinds of computations that Peggy and Vic will be allowed to perform, and also to describe the interaction that takes place. It is convenient to think of both Peggy and Vic as being probabilistic algorithms. Peggy and Vic will each perform private computations, and each of them has a private random number generator. They will communicate to each other through a communication channel. Initially, Peggy and Vic both possess an input x. The object of the interactive proof is for Peggy to convince Vic that x has some specified property. More precisely, x will be a yes-instance of a specified decision problem Π.

The interactive proof, which is a challenge-and-response protocol, consists of a specified number of rounds. During each round, Peggy and Vic alternately do the following:

1. receive a message from the other party

2. perform a private computation

3. send a message to the other party.

A typical *round* of the protocol will consist of a *challenge* by Vic, and a *response* by Peggy. At the end of the proof, Vic either *accepts* or *rejects*, depending on whether or not Peggy successfully replies to all of Vic's challenges. We define the protocol to be an *interactive proof system* for the decision problem Π if the following two properties are satisfied whenever Vic follows the protocol:

FIGURE 13.1
Graph Isomorphism

Problem Instance   Two graphs on n vertices, $G_1 = (V_1, E_1)$ and $G_2 = (V_2, E_2)$.

Question   Is there a bijection $\pi : V_1 \to V_2$ such that $\{u, v\} \in E_1$ if and only if $\{\pi(u), \pi(v)\} \in E_2$? (In other words, are G_1 and G_2 *isomorphic?*)

---

completeness

If x is a yes-instance of the decision problem Π, then Vic will always accept Peggy's proof.

soundness

If x is a no-instance of Π, then the probability that Vic accepts the proof is very small.

We will restrict our attention to interactive proof systems in which the computations performed by Vic can be done in polynomial time. On the other hand, we do not place any bound on the computation time required by Peggy (informally, Peggy is "all-powerful").

We begin by presenting an interactive proof system for the problem of **Graph Non-isomorphism**. The **Graph Isomorphism** problem is described in Figure 13.1. This is an interesting problem since no polynomial-time algorithm to solve it is known, but it is not known to be NP-complete.

We will present an interactive proof system which will allow Peggy to "prove" to Vic that two specified graphs are not isomorphic. For simplicity, let us suppose that G_1 and G_2 each have vertex set $\{1, \ldots, n\}$. The interactive proof system for **Graph Non-isomorphism** is presented in Figure 13.2.

We present a toy example.

Example 13.1
Suppose $G_1 = (V, E_1)$ and $G_2 = (V, E_2)$, where $V = \{1, 2, 3, 4\}$, $E_1 = \{12, 14, 23, 34\}$ and $E_2 = \{12, 13, 14, 34\}$.

Suppose in some round of the protocol that Vic gives Peggy the graph $H = (V, E_3)$, where $E_3 = \{13, 14, 23, 24\}$ (see Figure 13.3). The graph H is isomorphic to G_1 (one isomorphism from H to G_1 is the permutation $(1\ 3\ 4\ 2)$). So Peggy answers "1."   □

It is easy to see that this proof system satisfies the completeness and soundness

FIGURE 13.2
An interactive proof system for Graph Non-isomorphism

Input: two graphs G_1 and G_2, each having vertex set $\{1, \ldots, n\}$

1. Repeat the following steps n times:
2.   Vic chooses a random integer $i = 1$ or 2 and a random permutation π of $\{1, \ldots, n\}$. Vic computes H to be the image of G_i under the permutation π, and sends H to Peggy.
3.   Peggy determines the value j such that G_j is isomorphic to H, and sends j to Vic.
4.   Vic checks to see if $i = j$.
5. Vic accepts Peggy's proof if $i = j$ in each of the n rounds.

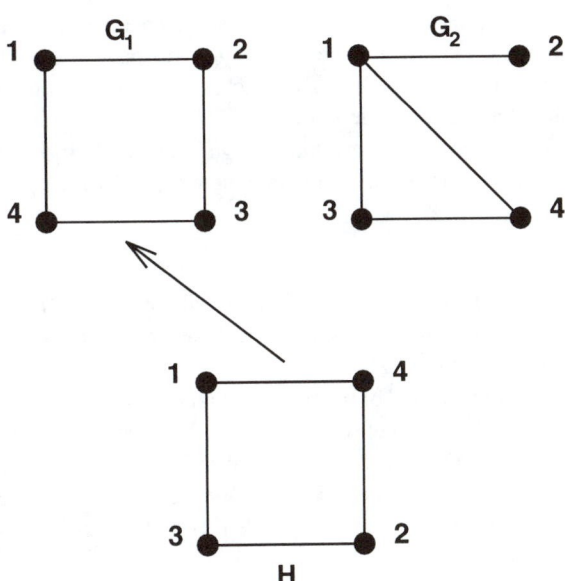

FIGURE 13.3
Peggy's non-isomorphic graphs and Vic's challenge

properties. If G_1 is not isomorphic to G_2, then j will equal i in every round, and Vic will accept with probability 1. Hence, the protocol is complete.

On the other hand, suppose that G_1 is isomorphic to G_2. Then any challenge graph H submitted by Vic is isomorphic to both G_1 and G_2. Peggy has no way of determining if Vic constructed H as an isomorphic copy of G_1 or of G_2, so she can do no better than make a guess $j = 1$ or 2 for her response. The only way that Vic will accept is if Peggy is able to guess all n choices of i made by Vic. Her probability of doing this is 2^{-n}. Hence, the protocol is sound.

Notice that Vic's computations are all polynomial-time. We cannot say anything about Peggy's computation time since the **Graph Isomorphism** problem is not known to be solvable in polynomial time. However, recall that we assumed that Peggy has infinite computing power, so this is allowed under the "rules of the game."

13.2   Perfect Zero-knowledge Proofs

Although interactive proof systems are of interest in their own right, the most interesting type of interactive proof is a zero-knowledge proof. This is one in which Peggy convinces Vic that x possesses some specified property, but at the end of the protocol, Vic still has no idea of how to prove (himself) that x has this property. This is a very tricky concept to define formally, and we present an example before attempting any definitions.

In Figure 13.4, we present a zero-knowledge interactive proof for **Graph Isomorphism**. A small example will illustrate the workings of the protocol.

Example 13.2
Suppose $G_1 = (V, E_1)$ and $G_2 = (V, E_2)$, where $V = \{1, 2, 3, 4\}$, $E_1 = \{12, 13, 14, 34\}$ and $E_2 = \{12, 13, 23, 24\}$. One isomorphism from G_2 to G_1 is the permutation $\sigma = (4\ 1\ 3\ 2)$.

Now suppose in some round of the protocol that Peggy chooses the permutation $\pi = (2\ 4\ 1\ 3)$. Then H has edge set $\{12, 13, 23, 24\}$ (see Figure 13.5).

If Vic's challenge is $i = 1$, then Peggy gives Vic the permutation π and Vic checks that the image of G_1 under π is H. If Vic's challenge is $i = 2$, then Peggy gives Vic the composition $\rho = \pi \circ \sigma = (3\ 2\ 1\ 4)$ and Vic checks that the image of G_2 under ρ is H. $\Box$

Completeness and soundness of the protocol are easy to verify. It is easy to see that the probablity that Vic accepts is 1 if G_1 is isomorphic to G_2. On the other hand, if G_1 is not isomorphic to G_2, then the only way for Peggy to deceive Vic is for her to correctly guess the value i that Vic will choose in each round, and write

FIGURE 13.4
A perfect zero-knowledge interactive proof system for Graph Isomorphism

---

Input: two graphs G_1 and G_2, each having vertex set $\{1, \ldots, n\}$

1. Repeat the following steps n times:

2.   Peggy chooses a random permutation π of $\{1, \ldots, n\}$. She computes H to be the image of G_1 under the permutation π, and sends H to Vic.

3.   Vic chooses a random integer $i = 1$ or 2 and sends it to Peggy.

4.   Peggy computes a permutation ρ of $\{1, \ldots, n\}$ such that H is the image of G_i under ρ. Peggy sends ρ to Vic. (If $i = 1$, then Peggy defines $\rho = \pi$; and if $i = 2$, then Peggy defines ρ to be the composition of σ and π, where σ is some fixed permutation such that the image of G_2 under σ is G_1.)

5.   Vic checks to see if H is the image of G_i under ρ.

6. Vic accepts Peggy's proof if H is the image of G_i in each of the n rounds.

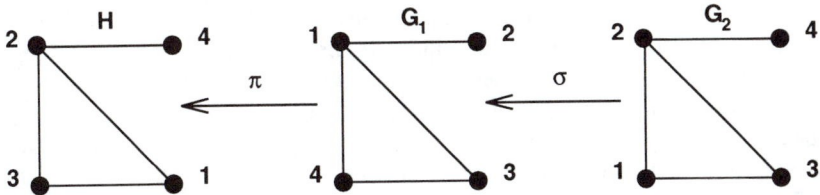

FIGURE 13.5
Peggy's isomorphic graphs

a (random) isomorphic copy of G_i on the communication tape. Her probability of correctly guessing Vic's n random challenges is 2^{-n}.

All of Vic's computations can be done in polynomial time (as a function of n, the number of vertices in G_1 and G_2). Although it is not necessary, notice that Peggy's computations can also be done in polynomial time provided that she knows the existence of one permutation σ such that the image of G_2 under σ is G_1.

Why would we refer to this proof system as a zero-knowledge proof? The reason is that, although Vic is convinced that G_1 is isomorphic to G_2, he does not gain any "knowledge" that would help him find a permutation σ that carries G_2 to G_1. All he sees in each round of the proof is a random isomorphic copy H of the graphs G_1 and G_2, together with a permutation that carries G_1 to H or G_2 to H (but not both!). But Vic can compute random isomorphic copies of these graphs by himself, without any help from Peggy. Since the graphs H are chosen independently and at random in each round of the proof, it seems unlikely that this will help Vic find an isomorphism from G_1 to G_2.

Let us look carefully at the information that Vic obtains by participating in the interactive proof system. We can represent Vic's view of the interactive proof by means of a *transcript* that contains the following information:

1. the graphs G_1 and G_2

2. all the messages that are transmitted by both Peggy and Vic

3. the random numbers used by Vic to generate his challenges.

Hence, a transcript T for the above interactive proof of **Graph Isomorphism** would have the following form:

$$T = ((G_1, G_2); (H_1, i_1, \rho_1); \ldots; (H_n, i_n, \rho_n)).$$

The essential point, which is the basis for the formal definition of zero-knowledge proof, is that Vic (or anyone else) can forge transcripts — without participating in the interactive proof — that "look like" real transcripts. This can be done provided that the input graphs G_1 and G_2 are isomorphic. Forging is accomplished by means of the algorithm presented in Figure 13.6. The forging algorithm is a polynomial-time probabilistic algorithm. In the vernacular of zero-knowledge proofs, a forging algorithm is often called a *simulator*.

The fact that a simulator can forge transcripts has a very important consequence. Anything that Vic (or anyone else) can compute from the transcript could also be computed from a forged transcript. Hence, participating in the proof system does not increase Vic's ability to perform any computation; and in particular, it does not enable Vic himself to "prove" that G_1 and G_2 are isomorphic. Moreover, Vic cannot subsequently convince someone else that G_1 and G_2 are isomorphic by showing them the transcript T, since there is no way to distinguish a legitimate transcript from one that has been forged.

We still have to make precise the idea that a forged transcript "looks like" a real one. We give a rigorous definition in terms of probability distributions.

FIGURE 13.6
Forging algorithm for transcripts for Graph Isomorphism

> Input: two isomorphic graphs G_1 and G_2, each having vertex set $\{1, \ldots, n\}$
> 1. $T = (G_1, G_2)$
> 2. **for** $j = 1$ **to** n **do**
> 3.   Choose $i_j = 1$ or 2 at random;
> 4.   Choose ρ_j to be a random permutation of $\{1, \ldots, n\}$;
> 5.   Compute H_j to be the image of G_{i_j} under ρ_j;
> 6.   concatenate (H_j, i_j, ρ_j) onto the end of T

DEFINITION 13.1 *Suppose that we have a polynomial-time interactive proof system for a decision problem Π, and a polynomial-time simulator S. Denote the set of all possible transcripts that could be produced as a result of Peggy and Vic carrying out the interactive proof with a yes-instance x by $\mathcal{T}(x)$, and denote the the set of all possible forged transcripts that could be produced by S by $\mathcal{F}(x)$. For any transcript $T \in \mathcal{T}(x)$, let $p_T(T)$ denote the probability that T is the transcript produced from the interactive proof. Similarly, for $T \in \mathcal{F}(x)$, let $p_{\mathcal{F}}(T)$ denote the probability that T is the (forged) transcript produced by S. Suppose that $\mathcal{T}(x) = \mathcal{F}(x)$, and for any $T \in \mathcal{T}(x)$, suppose that $p_T(T) = p_{\mathcal{F}}(T)$. (In other words, the set of real transcripts is identical to the set of forged transcripts, and the two probability distributions are identical.) Then we define the interactive proof system to be perfect zero-knowledge for Vic.*

Of course we can define zero-knowledge however we like. But it is important that the definition captures our intuitive concept of what "zero-knowledge" should mean. We are saying that an interactive proof system is zero-knowlege for Vic if there exists a simulator that produces transcripts with an identical probability distribution to those produced when Vic actually takes part in the protocol. (This is a related but stronger concept than that of indistinguishable probability distributions that we studied in Chapter 12.) We have observed that a transcript contains all the information gained by Vic by taking part in the protocol. So it should seem reasonable to say that whatever Vic might be able to do after taking part in the protocol he could equally well do by just using the simulator to generate a forged transcript. We are perhaps not defining "knowledge" by this approach; but whatever "knowledge" might be, Vic doesn't gain any!

We will now prove that the interactive proof system for **Graph Isomorphism** is perfect zero-knowledge for Vic.

THEOREM 13.1

The interactive proof system for **Graph Isomorphism** *is perfect zero-knowledge for Vic.*

PROOF   Suppose that G_1 and G_2 are isomorphic graphs on n vertices. A transcript T (real or forged) contains n triples of the form (H, i, ρ), where $i = 1$ or 2, ρ is a permutation of $\{1, \ldots, n\}$, and H is the image of G_i under the permutation ρ. Call such a triple a *valid* triple and denote by $\mathcal{R}$ the set of all valid triples. We begin by computing $|\mathcal{R}|$, the number of valid triples. Evidently $|\mathcal{R}| = 2 \times n!$ since each choice of i and ρ determines a unique graph H.

In any given round, say j, of the forging algorithm, it is clear that each valid triple (H, i, ρ) occurs with equal probability $1/(2 \times n!)$. What is the probability that the valid triple (H, i, ρ) is the jth triple on a real transcript? In the interactive proof system, Peggy first chooses a random permutation π and then computes H to be the image of G_1 under π. The permutation ρ is defined to be π if $i = 1$, and it is defined to be the composition of the two permutations π and σ if $i = 2$.

We are assuming that the value of i is chosen at random by Vic. If $i = 1$, then all $n!$ permutations ρ are equiprobable, since $\rho = \pi$ in this case and π was chosen to be a random permutation. On the other hand, if $i = 2$, then $\rho = \pi \circ \sigma$, where π is random and σ is fixed. In this case as well, every possible permutation ρ is equally probable. Now, since the two cases $i = 1$ and 2 are equally probable, and each permutation ρ is equally probable (independent of the value of i), and since i and ρ together determine H, it follows that all triples in $\mathcal{R}$ are equally likely.

Since a transcript consists of the concatenation of n independent random triples, it follows that

$$p_T(T) = p_\mathcal{F}(T) = \frac{1}{(2 \times n!)^n}$$

for every possible transcript T.   ∎

The proof of Theorem 13.1 assumes that Vic follows the protocol when he takes part in the interactive proof system. The situation is much more subtle if Vic does not follow the protocol. Is it true that an interactive proof remains zero-knowledge even if Vic deviates from the protocol?

In the case of **Graph Isomorphism**, the only way that Vic can deviate from the protocol is to choose his challenges i in a non-random way. Intuitively, it seems that this does not provide Vic with any "knowledge." However, transcripts produced by the simulator will not "look like" transcripts produced by Vic if he deviates from the protocol. For example, suppose Vic chooses $i = 1$ in every round of the proof. Then a transcript of the interactive proof will have $i_j = 1$ for $1 \leq j \leq n$; whereas a transcript produced by the simulator will have $i_j = 1$ for $1 \leq j \leq n$ only with probability 2^{-n}.

The way around this difficulty is to show that, no matter how a "cheating" Vic deviates from the protocol, there exists a polynomial-time simulator that will produce forged transcripts that "look like" the transcripts produced by Peggy and (the cheating) Vic during the interactive proof. As before, the phrase "looks like" is formalized by saying that two probability distributions are identical.

Here is a more formal definition.

DEFINITION 13.2 *Suppose that we have a polynomial-time interactive proof system for a given decision problem Π. Let V^* be any polynomial-time probabilistic algorithm that (a possibly cheating) verifier uses to generate his challenges. (That is, V^* represents either an honest or cheating verifier.) Denote the set of all possible transcripts that could be produced as a result of Peggy and V^* carrying out the interactive proof with a yes-instance x of Π by $\mathcal{T}(V^*, x)$. Suppose that, for every such V^*, there exists an expected polynomial-time probabilistic algorithm $S^* = S^*(V^*)$ (the simulator) which will produce a forged transcript. Denote the set of possible forged transcripts by $\mathcal{F}(V^*, x)$. For any transcript $T \in \mathcal{T}(V^*, x)$, let $p_T(T)$ denote the probability that T is the transcript produced by V^* taking part in the interactive proof. Similarly, for $T \in \mathcal{F}(x)$, let $p_{\mathcal{F}}(T)$ denote the probability that T is the (forged) transcript produced by S^*. Suppose that $\mathcal{T}(V^*, x) = \mathcal{F}(V^*, x)$, and for any $T \in \mathcal{T}(V^*, x)$, suppose that $p_{\mathcal{F},V^*}(T) = p_{T,V^*}(T)$. Then the interactive proof system is said to be* perfect zero-knowledge *(without qualification).*

In the special case where V^* is the same as Vic (i.e., when Vic is honest), the above definition is exactly the same as what we defined as "perfect zero-knowledge for Vic."

In order to prove that a proof system is perfect zero-knowledge, we need a generic transformation which will construct a simulator S^* from any V^*. We proceed to do this for the proof system for **Graph Isomorphism**. The simulator will play the part of Peggy, using V^* as a "restartable subroutine." Informally, S^* tries to guess the challenge i_j that V^* will make in each round j. That is, S^* generates a random valid triple of the form (H_j, i_j, ρ_j), and then executes the algorithm V^* to see what its challenge is for round j. If the guess i_j is the same as the challenge i'_j (as produced by V^*), then the triple (H_j, i_j, ρ_j) is appended to the forged transcript. If not, then this triple is discarded, S^* guesses a new challenge i_j, and the algorithm V^* is restarted after resetting its "state" to the way it was at the beginning of the current round. By the term "state" we mean the values of all variables used by the algorithm.

We now give a more detailed description of the simulation algorithm S^*. At any given time during the execution of the program V^*, the current state of V^* will be denoted by $\text{state}(V^*)$. A pseudo-code description of the simulation algorithm is given in Figure 13.7.

It is possible that the simulator will run forever, if it never happens that $i_j = i'_j$. However, we can show that the average running time of the simulator is polynomial, and that the two probability distributions $p_{\mathcal{F},V^*}(T)$ and $p_{T,V^*}(T)$ are identical.

FIGURE 13.7
Forging algorithm for V^* for transcripts for Graph Isomorphism

---

Input: two isomorphic graphs G_1 and G_2, each having vertex
set $\{1, \ldots, n\}$

1. $T = (G_1, G_2)$

2.   **for** $j = 1$ **to** n **do**

3.   define oldstate $=$ state(V^*)

4.   **repeat**

5.   Choose $i_j = 1$ or 2 at random

6.   Choose ρ_j to be a random permutation of $\{1, \ldots, n\}$

7.   Compute H_j to be the image of G_i under ρ

8.   call V^* with input H_j, obtaining a challenge i'_j

9.   **if** $i_j = i'_j$ **then**
                         concatenate (H_j, i_j, ρ_j) onto the end of T
                  else
                         reset V^* by defining state(V^*) $=$ oldstate

10.   **until** $i_j = i'_j$

THEOREM 13.2

The interactive proof system for **Graph Isomorphism** *is perfect zero-knowledge.*

PROOF   First, we observe that, regardless of how V^* generates its challenges, the probability that the guess i_j of S^* is the same as the challenge i'_j is $1/2$. Hence, on average, S^* will generate two triples for every triple that it concatenates to the forged transcript. Hence, the average running time is polynomial in n.

The more difficult task is to show that the two probability distributions $p_{\mathcal{F}, V^*}(T)$ and $p_{T, V^*}(T)$ are identical. In Theorem 13.1, where Vic was honest, we were able to compute the two probability distributions and see that they were identical. We also used the fact that triples (H, i, ρ) generated in different rounds of the proof are independent. However, in the current setting, we have no way of explicitly computing the two probability distributions. Further, triples generated in different rounds of the proof need not be independent. For example, the challenge that V^* presents in round j may depend in some very complicated way on challenges from previous rounds and on the way Peggy replied to those challenges.

The way to handle these difficulties is to look at the probability distributions on the possible partial transcripts during the course of the simulation or interactive proof, and proceed by induction on the number of rounds. For $0 \le j \le n$, we define probability distributions $p_{T,V^*,j}$ and $p_{F,V^*,j}$ on the set of partial transcripts T_j that could occur at the end of round j. Notice that $p_{T,V^*,n} = p_{T,V^*}$ and $p_{F,V^*,n} = p_{F,V^*}$. Hence, if we can show that the two distributions $p_{T,V^*,j}$ and $p_{F,V^*,j}$ are identical for all j, then we will be done.

The case $j = 0$ corresponds to the beginning of the algorithm; at this point the transcript contains only the two graphs G_1 and G_2. Hence, the probability distributions are identical when $j = 0$. We use this for the start of the induction.

We make an inductive hypothesis that the two probability distributions $p_{T,V^*,j-1}$ and $p_{F,V^*,j-1}$ on T_{j-1} are identical, for some $j \geq 1$. We now prove that the two probability distributions $p_{T,V^*,j}$ and $p_{F,V^*,j}$ on T_j are identical.

Consider what happens during round j of the interactive proof. The probability that V^*'s challenge $i'_j = 1$ is some real number p_1 and the probability that his challenge $i'_j = 2$ is $1 - p_1$, where p_1 depends on the state of the algorithm V^* at the beginning of round j. We noted earlier that in the interactive proof, all possible graphs H are chosen by Peggy with equal probability. As well, any permutation ρ occurs with equal probability, independent of the value of p_1, since all permutations are equally likely for either possible challenge i'_j. Hence, the probability that the jth triple on the transcript is (H, i, ρ) is $p_1/n!$ if $i = 1$, and $(1 - p_1)/n!$ if $i = 2$.

Next, let's do a similar analysis for the simulation. In any given iteration of the **repeat** loop, S^* will choose any graph H with probability $1/n!$. The probability that $i_j = 1$ and V^*'s challenge is 1 is $p_1/2$; and the probability that $i_j = 2$ and V^*'s challenge is 2 is $(1 - p_1)/2$. In each of these situations, (H, i_j, ρ) is written as the jth triple on the transcript. With probability $1/2$, nothing is written on the tape during any given iteration of the **repeat** loop.

Let us first consider the case $i_j = 1$. As mentioned above, the probability that V^*'s challenge is 1 is p_1. The probability that a triple $(H, 1, \rho)$ is written as the jth triple on the transcript during the ℓth iteration of the **repeat** loop is

$$\frac{p_1}{2^\ell \times n!}.$$

Hence, the probability that $(H, 1, \rho)$ is the jth triple on the transcript is

$$\frac{p_1}{2 \times n!} \left(1 + \frac{1}{2} + \frac{1}{4} + \dots \right) = \frac{p_1}{n!}.$$

The case $i_j = 2$ is analyzed in a similar fashion: the probability that $(H, 2, \rho)$ is written as the jth triple on the transcript is $(1 - p_1)/n!$

Hence, the two probability distributions on the partial transcripts at the end of round j are identical. By induction, the two probability distributions $p_{F,V^*}(T)$ and $p_{T,V^*}(T)$ are identical, and the proof is complete. ∎

It is interesting also to look at the interactive proof system for **Graph Non-isomorphism**. It is not too difficult to prove that this proof is perfect zero-knowledge if Vic follows the protocol (i.e., if Vic chooses each challenge graph

FIGURE 13.8
A perfect zero-knowledge interactive proof system for Quadratic Residues

---

Input: an integer n with unknown factorization $n = pq$, where p and q are prime, and $x \in \mathrm{QR}(n)$

1.   Repeat the following steps $\log_2 n$ times:

2.   Peggy chooses a random $v \in \mathbb{Z}_n{}^*$, and computes

$$y = v^2 \bmod n.$$

     Peggy sends y to Vic.

3.   Vic chooses a random integer $i = 0$ or 1 and sends it to Peggy.

4.   Peggy computes

$$z = u^i v \bmod n,$$

     where u is a square root of x, and sends z to Vic.

5.   Vic checks to see if

$$z^2 \equiv x^i y \pmod{n}.$$

6.   Vic accepts Peggy's proof if the computation of step 5 is verified in each of the $\log_2 n$ rounds.

---

to be a random isomorphic copy of G_i where $i = 1$ or 2 is chosen at random). Further, provided that Vic constructs each challenge graph by taking an isomorphic copy of either G_1 or G_2, the protocol remains zero-knowledge even if Vic chooses his challenges in a non-random fashion. However, suppose that our ubiquitous troublemaker, Oscar, gives a graph H to Vic which is isomorphic to one of G_1 or G_2, but Vic does not know which G_i is isomorphic to H. If Vic uses this H as one of his challenge graphs in the interactive proof system, then Peggy will give Vic an isomorphism he didn't previously know, and (possibly) couldn't figure out for himself. In this situation, the proof system is (intuitively) not zero-knowledge, and it does not seem likely that a transcript could be forged by a simulator.

It is possible to alter the proof of **Graph Non-isomorphism** so it is perfect zero-knowledge, but we will not go into the details.

We now present some other examples of perfect zero-knowledge proofs. A perfect zero-knowledge proof for **Quadratic Residues** (modulo $n = pq$, where p and q are prime) is given in Figure 13.8. Peggy is proving that x is a quadratic

FIGURE 13.9
Subgroup Membership

Problem Instance Two positive integers n and ℓ, and two distinct elements $\alpha, \beta \in \mathbb{Z}_n{}^*$, where α has order ℓ in $\mathbb{Z}_n{}^*$.

Question Is $\beta = \alpha^k$ for some integer k such that $0 \le k \le \ell - 1$? (In other words, is β a member of the subgroup of $\mathbb{Z}_n{}^*$ generated by α?)

---

residue. In each round, she generates a random quadratic residue y and sends it to Vic. Then, depending on Vic's challenge, Peggy either gives Vic a square root of y or a square root of xy.

It is clear that the protocol is complete. To prove soundness, observe that if x is not a quadratic residue, then Peggy can answer only one of the two possible challenges since, in this case, y is a quadratic residue if and only if xy is not a quadratic residue. So Peggy will be caught in any given round of the protocol with probability $1/2$, and her probability of deceiving Vic in all $\log_2 n$ rounds is only $2^{-\log_2 n} = 1/n$. (The reason for having $\log_2 n$ rounds is that the size of the problem instance is proportional to the number of bits in the binary representation of n, which is $\log_2 n$. Hence, the deception probability for Peggy is exponentially small as a function of the size of the problem instance, as in the zero-knowledge proof for **Graph Isomorphism**.)

Perfect zero-knowledge for Vic can be shown in a similar manner as was done for **Graph Isomorphism**. Vic can generate a triple (y, i, z) by first choosing i and z, and then defining

$$y = z^2 (x^i)^{-1} \bmod n.$$

Triples generated in this fashion have exactly the same probability distribution as those generated during the protocol, assuming Vic chooses his challenges at random. Perfect zero-knowledge (for an arbitrary V^*) is proved by following the same strategy as for **Graph Isomorphism**. It requires building a simulator S^* that guesses V^*'s challenges and keeps only the triples where the guesses are correct.

We now present one more example of a perfect zero-knowledge proof, this one for a decision problem related to the **Discrete Logarithm** problem. The problem, which we call **Subgroup Membership**, is defined in Figure 13.9. Of course, the integer k (if it exists) is just the discrete logarithm of β.

We present a perfect zero-knowledge proof for **Subgroup Membership** in Figure 13.10. The analysis of this protocol is similar to the others that we have looked at; the details are left to the reader.

FIGURE 13.10
A perfect zero-knowledge interactive proof system for Subgroup Membership

---

Input: A positive integer n, and two distinct elements $\alpha, \beta \in \mathbb{Z}_n^*$, where the order of α is denoted by ℓ and is publicly known

1. Repeat the following steps $\log_2 n$ times:

2.   Peggy chooses a random j such that $0 \leq j \leq \ell - 1$, and computes

$$\gamma = \alpha^j \bmod n.$$

   Peggy sends γ to Vic.

3.   Vic chooses a random integer $i = 0$ or 1 and sends it to Peggy.

4.   Peggy computes

$$h = j + ik \bmod \ell,$$

   where

$$k = \log_\alpha \beta,$$

   and sends h to Vic.

5.   Vic checks to see if

$$\alpha^h \equiv \beta^i \gamma \pmod{n}.$$

6. Vic accepts Peggy's proof if the computation of step 5 is verified in each of the $\log_2 n$ rounds.

13.3   Bit Commitments

The zero-knowledge proof system for **Graph Isomorphism** is interesting, but it would be more useful to have zero-knowledge proof systems for problems that are known to be NP-complete. There is theoretical evidence that perfect zero-knowledge proofs do not exist for NP-complete problems. However, we can describe proof systems that attain a slightly weaker form of zero-knowledge called *computational* zero-knowledge. The actual proof systems are described in the next section; in this section we describe the technique of bit commitment that is an essential tool used in the proof system.

Suppose Peggy writes a message on a piece of paper, and then places the message in a safe for which she knows the combination. Peggy then gives the safe to Vic. Even though Vic doesn't know what the message is until the safe is opened, we would agree that Peggy is *committed* to her message because she cannot change it. Further, Vic cannot learn what the message is (assuming he doesn't know the combination of the safe) unless Peggy opens the safe for him. (Recall that we used a similar analogy in Chapter 4 to describe the idea of a public-key cryptosystem, but in that case, it was the recipient of the message, Vic, who could open the safe.)

Suppose the message is a bit $b = 0$ or 1, and Peggy encrypts b in some way. The encrypted form of b is sometimes called a *blob* and the encryption method is called a *bit commitment scheme*. In general, a bit commitment scheme will be a function $f : \{0,1\} \times X \to Y$, where X and Y are finite sets. An encryption of b is any value $f(b, x)$, $x \in X$. We can informally define two properties that a bit commitment scheme should satisfy:

concealing

> For a bit $b = 0$ or 1, Vic cannot determine the value of b from the blob $f(b, x)$.

binding

> Peggy can later "open" the blob, by revealing the value of x used to encrypt b, to convince Vic that b was the value encrypted. Peggy should not be able to open a blob as both a 0 and a 1.

If Peggy wants to commit any bitstring, she simply commits every bit independently.

One way to perform bit commitment is to use the **Goldwasser-Micali Probabilistic Cryptosystem** described in Section 12.4. Recall that in this system, $n = pq$, where p and q are primes, and $m \in \tilde{Q}R(n)$. The integers n and m are public; the factorization $n = pq$ is known only to Peggy. In our bit commitment scheme, we have $X = Y = \mathbb{Z}_n^*$ and

$$f(b, x) = m^b x^2 \bmod n.$$

Peggy encrypts a value b by choosing a random x and computing $y = f(b, x)$; the value y comprises the blob.

Later, when Peggy wants to open y, she reveals the values b and x. Then Vic can verify that

$$y \equiv m^b x^2 \pmod{n}.$$

Let us think about the concealing and binding properties. A blob is an encryption of 0 or of 1, and reveals no information about the plaintext value x provided that the **Quadratic Residues** problem is infeasible (we discussed this at length in Chapter 12). Hence, the scheme is concealing.

Is the scheme binding? Let us suppose not; then

$$m x_1^2 \equiv x_2^2 \pmod{n}$$

for some $x_1, x_2 \in \mathbb{Z}_n{}^*$. But then

$$m \equiv (x_2 x_1{}^{-1})^2 \pmod{n},$$

which is a contradiction since $m \in \tilde{\mathrm{QR}}(n)$.

We will be using bit commitment schemes to construct zero-knowledge proofs. However, they have another nice application, to the problem of *coin-flipping by telephone*. Suppose Alice and Bob want to make some decision based on a random coin flip, but they are not in the same place. This means that it is impossible for one of them to flip a real coin and have the other verify it. A bit commitment scheme provides a way out of this dilemma. One of them, say Alice, chooses a random bit b, and computes a blob, y. She gives y to Bob. Now Bob guesses the value of b, and then Alice opens the blob to reveal b. The concealing property means that it is infeasible for Bob to compute b given y, and the binding property means that Alice can't "change her mind" after Bob reveals his guess.

We now give another example of a bit commitment scheme, this time based on the **Discrete Logarithm** problem. Recall from Section 5.1.2 that if $p \equiv 3$ (mod 4) is a prime such that the **Discrete Logarithm** problem in $\mathbb{Z}_p{}^*$ is infeasible, then the second least significant bit of a discrete logarithm is secure. Actually, it has been proved for primes $p \equiv 3 \pmod{4}$ that any Monte Carlo algorithm for the **Second Bit** problem having error probability $1/2 - \epsilon$ with $\epsilon > 0$ can be used to solve the **Discrete Log** problem in $\mathbb{Z}_p{}^*$. This much stronger result is the basis for the bit commitment scheme.

This bit commitment scheme will have $X = \{1, \ldots, p-1\}$ and $Y = \mathbb{Z}_p{}^*$. The second least significant bit of an integer x, denoted by $\mathrm{SLB}(x)$, is defined as follows:

$$\mathrm{SLB}(x) = \begin{cases} 0 & \text{if } x \equiv 0, 1 \pmod{4} \\ 1 & \text{if } x \equiv 2, 3 \pmod{4}. \end{cases}$$

The bit commitment scheme f is defined by

$$f(b, x) = \begin{cases} \alpha^x \bmod p & \text{if } \mathrm{SLB}(x) = b \\ \alpha^{p-x} \bmod p & \text{if } \mathrm{SLB}(x) \neq b. \end{cases}$$

In other words, a bit b is encrypted by choosing a random element having second last bit b, and raising α to that power modulo p. (Note that $\mathrm{SLB}(p-x) \neq \mathrm{SLB}(x)$ since $p \equiv 3 \pmod{4}$.)

The scheme is binding, and by the remarks made above, it is concealing provided that the **Discrete Logarithm** problem in $\mathbb{Z}_p{}^*$ is infeasible.

13.4   Computational Zero-knowledge Proofs

In this section, we give a zero-knowledge proof system for the NP-complete decision problem **Graph 3-Colorability**, which is defined in Figure 13.11. The

FIGURE 13.11
Graph 3-Colorability

Problem Instance   A graph $G = (V, E)$ on n vertices.

Question   Is there a *proper 3-coloring* of G? (In mathematical terms, is there a function $\phi : V(G) \rightarrow \{1, 2, 3\}$ such that $\{u, v\} \in E$ implies $\phi(u) \neq \phi(v)$?)

---

proof system uses a bit commitment scheme; to be specific, we will employ the bit commitment scheme presented in Section 13.3 that is based on probabilistic encryption. We assume that Peggy knows a 3-coloring ϕ of a graph G, and she wants to convince Vic that G is 3-colorable in a zero-knowledge fashion. Without loss of generality, we assume that G has vertex set $V = \{1, \ldots, n\}$. Denote $m = |E|$. The proof system will be described in terms of a commitment scheme $f : \{0, 1\} \times X \rightarrow Y$ which is made public. Since we want to encrypt a color rather than a bit, we will replace the color 1 by the two bits 01, the color 2 by 10 and the color 3 by 11. Then we encrypt each of the two bits representing the color by using f.

The interactive proof system is presented in Figure 13.12. Informally, what happens is the following. In each round, Peggy commits a coloring that is a permutation of the fixed coloring ϕ. Vic requests that Peggy open the blobs corresponding to the endpoints of some randomly chosen edge. Peggy does so, and then Vic checks that the commitments are as claimed and that the two colors are different. Notice that all Vic's computations are polynomial-time, and so are Peggy's, provided that she knows the existence of one 3-coloring ϕ.

Here is a very small example to illustrate.

Example 13.3
Suppose G is the graph (V, E), where

$$V = \{1, 2, 3, 4, 5\}$$

and

$$E = \{12, 14, 15, 23, 34, 45\}.$$

Suppose that Peggy knows the 3-coloring ϕ where $\phi(1) = 1$, $\phi(2) = \phi(4) = 2$ and $\phi(3) = \phi(5) = 3$. Suppose also that the bit commitment scheme is defined as $f(b, x) = 156897^b x^2 \bmod 321389$, where $b = 0, 1$ and $x \in \mathbb{Z}_{321389}{}^*$.

Suppose that Peggy chooses the permutation $\pi = (1\ 3\ 2)$ in some round of the proof. Then she computes:

$$c_1 = 1$$

FIGURE 13.12
A computational zero-knowledge interactive proof system for Graph 3-colorability

---

Input: a graph $G = (E, V)$ on vertex set $\{1, \ldots, n\}$

1. Repeat the following steps m^2 times:

2.   Let ϕ be a 3-coloring of G. Peggy chooses a random permutation π of $\{1, 2, 3\}$. For $1 \leq i \leq n$, she defines

$$c_i = \pi(\phi(i)),$$

and writes c_i as a bitstring of length two:

$$c_i = c_{i,1} c_{i,2}.$$

Then, for $1 \leq i \leq n$, she chooses two random elements $r_{i,1}, r_{i,2} \in X$, and computes

$$R_{i,j} = f(c_{i,j}, r_{i,j}),$$

$j = 1, 2$. She sends the list

$$(R_{1,1}, R_{1,2}, \ldots, R_{n,1}, R_{n,2})$$

to Vic.

3.   Vic chooses a random edge $\{u, v\} \in E$ and sends it to Peggy.

4.   Peggy sends $(c_{u,1}, c_{u,2}, r_{u,1}, r_{u,2})$ and $(c_{v,1}, c_{v,2}, r_{v,1}, r_{v,2})$ to Vic.

5.   Vic checks that

$$(c_{u,1}, c_{u,2}) \neq (c_{v,1}, c_{v,2}),$$
$$(c_{u,1}, c_{u,2}) \neq (0, 0),$$
$$(c_{v,1}, c_{v,2}) \neq (0, 0),$$
$$R_{u,j} = f(c_{u,j}, r_{u,j}), j = 1, 2, \text{ and}$$
$$R_{v,j} = f(c_{v,j}, r_{v,j}), j = 1, 2.$$

6. Vic accepts Peggy's proof if the computation of step 5 is verified in each of the m^2 rounds.

$$c_2 = 3$$
$$c_3 = 2$$
$$c_4 = 3$$
$$c_5 = 2.$$

She will encode this coloring in binary as the 10-tuple

$$0111101110$$

and then compute commitments of these ten bits. Suppose that she does this as follows:

b	x	$f(b, x)$
0	147658	176593
1	318856	205585
1	14497	189102
1	285764	294039
1	128589	230968
0	228569	77477
1	53369	305090
1	194634	276484
1	202445	292707
0	177561	290599

Then Peggy gives Vic the ten values $f(b, x)$ computed above.

Next, suppose that Vic chooses the edge 34 as his challenge. Then Peggy opens four blobs: the two that correspond to vertex 3 and the two that correspond to vertex 4. So Peggy gives Vic the ordered pairs

$$(b, x) = (1, 128589), (0, 228569), (1, 53369), (1, 194634).$$

Vic will first check that the two colors are distinct: 10 encodes color 2 and 11 encodes color 3, so this is all right. Next, Vic verifies that the four commitments are valid and hence this round of the proof is completed successfully. $\square$

As in previous proof systems we have studied, Vic will accept a valid proof with probability 1, so we have completeness. What is the probability that Vic will accept if G is not 3-colorable? In this case, for any coloring, there must be at least one edge ij such that i and j have the same color. Vic's chances of choosing such an edge are at least $1/m$. Peggy's probability of fooling Vic in all m^2 rounds is at most

$$\left(1 - \frac{1}{m}\right)^{m^2}.$$

Since $(1 - 1/m)^m \to e^{-1}$ as $m \to \infty$, there exists an integer m_0 such that $(1 - 1/m)^m \le 2/e$ for $m \ge m_0$. Hence $(1 - 1/m)^{m^2} \le (2/e)^m$ for $m \ge m_0$.

Since $(2/e)^m$ approaches zero exponentially quickly as a function of $m = |E|$, we have soundness as well.

Let's now turn to the zero-knowledge aspect of the proof system. All that Vic sees in any given round of the protocol is an encrypted 3-colouring of G, together with the two distinct colours of the endpoints of one particular edge, as previously committed by Peggy. Since the colors are permuted in each round, it seems that Vic cannot combine information from different rounds to reconstruct the 3-coloring.

The proof system is not perfect zero-knowledge, but it does provide a weaker form of zero-knowledge called *computational zero-knowledge*. Computational zero-knowledge is defined exactly as perfect zero-knowledge, except that the relevant probability distributions of transcripts are required only to be polynomially indistinguishable (in the sense of Chapter 12) rather than identical.

We begin by showing how transcripts can be forged. We give an explicit algorithm that will forge transcripts that cannot be distinguished from those produced by an honest Vic. If Vic deviates from the protocol, then it is possible to construct a simulator which uses the algorithm V^* as a restartable subroutine to construct forged transcripts. Both forging algorithms follow the pattern of the related algorithms for the **Graph Isomorphism** proof system.

Here, we consider only the case where Vic follows the protocol. A transcript T for the interactive proof of **Graph 3-colorability** would have the form

$$(G; A_1; \ldots; A_{m^2}),$$

where A_j consists of $2n$ blobs computed by Peggy, the edge uv chosen by Vic, the colors assigned by Peggy in round j to u and v, and the four random numbers used by Peggy to encrypt the colors of these two vertices. A transcript is forged by means of the forging algorithm presented in Figure 13.13.

Proving (computational) zero-knowledge for Vic requires showing that the two probability distributions on transcripts (as produced by the Vic taking part in the protocol, and as produced by the simulator) are indistinguishable. We will not do this here, but we will make a couple of comments. Notice that the two probability distributions are not identical. This is because virtually all the R_{ij}'s in a forged transcript are blobs encrypting 1; whereas the R_{ij}'s on a real transcript will (usually) be encryptions of more equal numbers of 0's and 1's. However, it is possible to show that the two probability distributions cannot be distinguished in polynomial time, provided that the underlying bit commitment scheme is secure. More precisely, this means that the probability distribution on blobs encrypting color c are indistinguishable from the probability distribution on blobs encrypting color d if $c \neq d$.

Readers familiar with NP-completeness theory will realize that, having given a zero-knowledge proof for one particular NP-complete problem, we can obtain a zero-knowledge proof for any other problem in NP. This can be done by applying a polynomial transformation from a given problem in NP to the **Graph 3-coloring** problem.

FIGURE 13.13
Forging algorithm for transcripts for Graph 3-colorability

Input: a graph $G = (V, E)$ having vertex set $V = \{1, \ldots, n\}$
1. $T = (G)$
2. **for** $j = 1$ to m^2 **do**
3.   Choose an edge $\{u, v\} \in E$ at random
4.   Choose $d = d_1 d_2$ and $e = e_1 e_2$ to be random, distinct colors, where $d_1, d_2, e_1, e_2 \in \{0, 1\}$
5.   Choose $r_{i,j}$ to be a random element of X, for $1 \leq i \leq n$, $j = 1, 2$
6.   For $1 \leq i \leq n$ and $j = 1, 2$, define
$$R_{i,j} = \begin{cases} f(1, r_{i,j}) & \text{if } i \neq u, v \\ f(d_j, r_{i,j}) & \text{if } i = u \\ f(e_j, r_{i,j}) & \text{if } i = v. \end{cases}$$
7.   concatenate
$$(R_{1,1}, \ldots, R_{n,2}, u, v, d_1, d_2, r_{d,1}, r_{d,2}, e_1, e_2, r_{e,1}, r_{e,2})$$
onto the end of T.

13.5 Zero-knowledge Arguments

Let us recap the basic properties of the computational zero-knowledge proof for **Graph 3-colorability** presented in the last section. No assumptions are needed to prove completeness and soundness of the protocol. A computational assumption is needed to prove zero-knowledge, namely that the underlying bit commitment scheme is secure. Observe that if Peggy and Vic take part in the protocol, then Vic may later try to break the bit commitment scheme that was used in the protocol (for example, if the scheme based on quadratic residuosity were used, then Vic would try to factor the modulus). If at any future time Vic can break the bit commitment scheme, then he can decrypt the blobs used by Peggy in the protocol and extract the 3-coloring.

This analysis depends on the properties of the blobs that were used in the protocol. Although the binding property of the blobs is unconditional, the concealing property relies on a computational assumption.

An interesting variation is to use blobs in which the concealing property is unconditional but the binding property requires a computational assumption. This leads to a protocol that is known as a zero-knowledge *argument* rather than a zero-knowledge proof. The reader will recall that we have assumed up until now that Peggy is all-powerful; in a zero-knowledge argument we will assume that Peggy's computations are required to be polynomial-time. (In fact, this assumption creates no difficulties, for we have already observed that Peggy's computations are polynomial-time provided she knows one 3-coloring of G.)

Let us begin by describing a couple of bit commitment schemes of this type and then examine the ramifications of using them in the protocol for **Graph 3-coloring**.

The first scheme is (again) based on the **Quadratic Residues** problem. Suppose $n = pq$, where p and q are prime, and let $m \in \text{QR}(n)$ (note that in the previous scheme m was a pseudo-square). In this scheme neither the factorization of n nor the square root of m should be known to Peggy. So either Vic should construct these values or they should be obtained from a (trusted) third party.

Let $X = \mathbb{Z}_n^*$ and $Y = \text{QR}(n)$, and define

$$f(b, x) = m^b x^2 \bmod n.$$

As before, Peggy encrypts a value b by choosing a random x and computing the blob $y = f(b, x)$. In this scheme all the blobs are quadratic residues. Further, any $y \in \text{QR}(n)$ is both an encryption of 0 and an encryption of 1. For suppose $y = x^2 \bmod n$ and $m = k^2 \bmod n$. Then

$$y = f(0, x) = f(1, xk^{-1} \bmod n).$$

This means that the concealing property is achieved unconditionally. On the other hand, what happens to the binding property? Peggy can open any given blob both as a 0 and as a 1 if and only if she can compute k, a square root of m. So, in order for the scheme to be (computationally) binding, we need to make the assumption that it is infeasible for Peggy to compute a square root of m. (If Peggy were all-powerful, then she could, of course, do this. This is one reason why we are now assuming that Peggy is computationally bounded.)

As a second bit commitment scheme of this type, we give an example of a scheme based on the **Discrete Logarithm** problem. Let p be a prime such that the discrete log problem in $\mathbb{Z}_p^*$ is infeasible, let α be a primitive element of $\mathbb{Z}_p^*$ and let $\beta \in \mathbb{Z}_p^*$. The value of β should be chosen by Vic, or by a trusted third party, rather than by Peggy. This scheme will have $X = \{0, \dots, p-1\}$, $Y = \mathbb{Z}_p^*$, and f is defined by

$$f(b, x) = \beta^b \alpha^x \bmod p.$$

It is not hard to see that this scheme is unconditionally concealing, and it is binding if and only if it is infeasible for Peggy to compute the discrete logarithm $\log_\alpha \beta$.

Now, suppose we use one of these two bit commitment schemes in the protocol for **Graph 3-colorability**. It is easy to see that the protocol remains complete. But

TABLE 13.1
Comparison of Properties of Proofs and Arguments

property	zero-knowledge proof	zero-knowledge argument
completeness	unconditional	unconditional
soundness	unconditional	computational
zero-knowledge	computational	perfect
binding blobs	unconditional	computational
concealing blobs	computational	unconditional

now the soundness condition depends on a computational assumption: the protocol is sound if and only if the bit commitment scheme is binding. What happens to the zero-knowledge aspect of the protocol? Because the bit commitment scheme is unconditionally concealing, the protocol is now perfect zero-knowledge rather than just computational zero-knowledge. Thus we have a perfect zero-knowledge argument.

Whether one prefers an argument to a proof depends on the application, and whether one wants to make a computational assumption regarding Peggy or Vic. A comparison of the properties of proofs and arguments is summarized in Table 13.1. In the column "zero-knowledge proof," the computational assumptions pertain to Peggy's computing power; in the column "zero-knowledge argument," the computational assumptions refer to Vic's computing power.

13.6 Notes and References

Most of the material in this chapter is based on Brassard, Chaum, and Crépeau [BCC88] and on Goldreich, Micali, and Wigderson [GMW91]. The bit commitment schemes we present, and a thorough discussion of the differences between proofs and arguments, can be found in [BCC88] (however, note that the term "argument" was first used in [BC90]). Zero-knowledge proofs for **Graph Isomorphism**, **Graph Non-isomorphism** and **Graph 3-colorability** can be found in [GMW91]. Another relevant paper is Goldwasser, Micali, and Rackoff [GMR89], in which interactive proof systems are first defined formally. The zero-knowledge proof for **Quadratic Residues** is from this paper.

The idea of coin-flipping by telephone is due to Blum [BL82].

A very informal and entertaining illustration of the concept of zero-knowledge is presented by Quisquater and Guillou [QG90]. Also, see Johnson [Jo88] for a more mathematical survey of interactive proof systems.

FIGURE 13.14
An interactive proof system for Quadratic Non-residues

Input: an integer n with unknown factorization $n = pq$, where p and q are prime, and $x \in \widetilde{QR}(n)$

1.   Repeat the following steps $\log_2 n$ times:

2.   Vic chooses a random $v \in \mathbb{Z}_n{}^*$, and computes

$$y = v^2 \bmod n.$$

Vic chooses $i = 0$ or 1 at random, and he sends

$$z = x^i y \bmod n$$

to Peggy.

3.   If $z \in QR(n)$, then Peggy defines $j = 0$, otherwise she defines $j = 1$. Then she sends j to Vic.

4.   Vic checks to see if $i = j$.

5.   Vic accepts Peggy's proof if the computation of step 4 is verified in each of the $\log_2 n$ rounds.

Exercises

13.1 Consider the interactive proof system for the problem **Quadratic Non-residues** presented in Figure 13.14. Prove that the system is sound and complete, and explain why the protocol is not zero-knowledge.

13.2 Devise an interactive proof system for the problem **Subgroup Non-membership**. Prove that your protocol is sound and complete.

13.3 Consider the zero-knowledge proof for **Quadratic Residues** that was presented in Figure 13.8.

(a) Define a **valid triple** to be one having the form (y, i, z), where $y \in QR(n)$, $i = 0$ or 1, $z \in \mathbb{Z}_n{}^*$ and $z^2 \equiv x^i y \pmod{n}$. Show that the number of valid triples is $2(p - 1)(q - 1)$, and each such triple is generated with equal probability if Peggy and Vic follow the protocol.

(b) Show that Vic can generate triples having the same probability distribution without knowing the factorization $n = pq$.

(c) Prove that the protocol is perfect zero-knowledge for Vic.

13.4 Consider the zero-knowledge proof for **Subgroup Membership** that was presented in Figure 13.10.

(a) Prove that the protocol is sound and complete.

(b) Define a **valid triple** to be one having the form (γ, i, h), where $\gamma \in \mathbb{Z}_n{}^*$, $i = 0$ or 1, $0 \le h \le \ell - 1$ and $\alpha^h \equiv \beta^i \gamma \pmod{n}$. Show that the number of valid triples is 2ℓ, and each such triple is generated with equal probability if Peggy and Vic follow the protocol.

(c) Show that Vic can generate triples having the same probability distribution without knowing the discrete logarithm $\log_\alpha \beta$.

    (d) Prove that the protocol is perfect zero-knowledge for Vic.

13.5 Prove that the **Discrete Logarithm** bit commitment scheme presented in Section 13.5 is unconditionally concealing, and prove that it is binding if and only if Peggy cannot compute $\log_\alpha \beta$.

13.6 Suppose we use the **Quadratic Residues** bit commitment scheme presented in Section 13.5 to obtain a zero-knowledge argument for **Graph 3-coloring**. Using the forging algorithm presented in Figure 13.13, prove that this protocol is perfect zero-knowledge for Vic.

Further Reading

Other recommended textbooks and monographs on cryptography include the following:

Beker and Piper [BP82]	Beutelspacher [BE94]
Brassard [BR88]	Biham and Shamir [BS93]
Denning [DE82]	Kahn [KA67]
Kaufman, Perlman and Speciner [KPS95]	Koblitz [KO94]
Konheim [KO81]	Kranakis [KR86]
Menezes [ME93]	Meyer and Matyas [MM82]
Patterson [PA87]	Pomerance [PO90A]
Rhee [RH94]	Rueppel [RU86]
Salomaa [SA90]	Schneier [SC95]
Seberry and Pieprzyk [SP89]	Simmons [SI92B]
Stallings [ST95]	van Tilborg [VT88]
Wayner [WA96]	Welsh [WE88]

For a thorough and highly recommended reference on all aspects of practical cryptogrpahy, see Menezes, Van Oorschot and Vanstone [MVV96].

The main research journals in cryptography are the *Journal of Cryptology, Designs, Codes and Cryptography* and *Cryptologia*. The *Journal of Cryptology* is the journal of the International Association for Cryptologic Research (or IACR) which also sponsors the two main annual cryptology conferences, CRYPTO and EUROCRYPT.

CRYPTO has been held since 1981 in Santa Barabara. The proceedings of CRYPTO have been published annually since 1982:

CRYPTO '82 [CRS83]	CRYPTO '83 [CH84]
CRYPTO '84 [BC85]	CRYPTO '85 [WI86]
CRYPTO '86 [OD87]	CRYPTO '87 [PO88]
CRYPTO '88 [GO90]	CRYPTO '89 [BR90]
CRYPTO '90 [MV91]	CRYPTO '91 [FE92]
CRYPTO '92 [BR93]	CRYPTO '93 [ST94]
CRYPTO '94 [DE94]	CRYPTO '95 [CO95]
CRYPTO '96 [KO96]	

EUROCRYPT has been held annually since 1982, and except for 1983 and 1986, its proceedings have been published, as follows:

EUROCRYPT '82 [BE83]   EUROCRYPT '84 [BCI85]
EUROCRYPT '85 [PI86]   EUROCRYPT '87 [CP88]
EUROCRYPT '88 [GU88A]   EUROCRYPT '89 [QV90]
EUROCRYPT '90 [DA91]   EUROCRYPT '91 [DA91A]
EUROCRYPT '92 [RU93]   EUROCRYPT '93 [HE94]
EUROCRYPT '94 [DE95]   EUROCRYPT '95 [GQ95]
EUROCRYPT '96 [MA96]

A third conference series, AUSCRYPT/ASIACRYPT, has been held "in association with" the IACR. Its conference proceedings have also been published:

AUSCRYPT '90 [SP90]   ASIACRYPT '91 [IRM93]
AUSCRYPT '92 [SZ92]   ASIACRYPT '94 [PS95]

Bibliography

[ACGS88] W. ALEXI, B. CHOR, O. GOLDREICH AND C. P. SCHNORR. RSA and Rabin functions: certain parts are as hard as the whole. *SIAM Jounal on Computing*, **17** (1988), 194–209.

[AN91] H. ANTON. *Elementary Linear Algebra* (Sixth Edition). John Wiley and Sons, 1991.

[BHS93] D. BAYER, S. HABER AND W. S. STORNETTA. Improving the efficiency and reliability of digital time-stamping. In *Sequences II, Methods in Communication, Security, and Computer Science*, pages 329–334. Springer-Verlag, 1993.

[BB88] P. BEAUCHEMIN AND G. BRASSARD. A generalization of Hellman's extension to Shannon's approach to cryptography. *Journal of Cryptology*, **1** (1988), 129–131.

[BBCGP88] P. BEAUCHEMIN, G. BRASSARD, C. CRÉPEAU, C. GOUTIER AND C. POMERANCE. The generation of random numbers that are probably prime. *Journal of Cryptology*, **1** (1988), 53–64.

[BC94] A. BEIMEL AND B. CHOR. Interaction in key distribution schemes. *Lecture Notes in Computer Science*, **773** (1994), 444–455. (Advances in Cryptology – CRYPTO '93.)

[BP82] H. BEKER AND F. PIPER. *Cipher Systems, The Protection of Communications*. John Wiley and Sons, 1982.

[BL90] J. BENALOH AND J. LEICHTER. Generalized secret sharing and monotone functions. *Lecture Notes in Computer Science*, **403** (1990), 27–35. (Advances in Cryptology – CRYPTO '88.)

[BE83] T. BETH (ED.) *Cryptography Proceedings, 1982. Lecture Notes in Computer Science*, vol. 149, Springer-Verlag, 1983.

[BCI85] T. BETH, N. COT AND I. INGEMARSSON (EDS.) *Advances in Cryptology: Proceedings of EUROCRYPT '84. Lecture Notes in Computer Science*, vol. 209, Springer-Verlag, 1985.

[BJL85] T. BETH, D. JUNGNICKEL, AND H. LENZ. *Design Theory*.

Bibliographisches Institut, Zurich, 1985.

[BE94] A. BEUTELSPACHER. *Cryptology.* Mathematical Association of America, 1994.

[BS91] E. BIHAM AND A. SHAMIR. Differential cryptanalysis of DES-like cryptosystems. *Journal of Cryptology*, **4** (1991), 3–72.

[BS93] E. BIHAM AND A. SHAMIR. *Differential Cryptanalysis of the Data Encryption Standard.* Springer-Verlag, 1993.

[BS93A] E. BIHAM AND A. SHAMIR. Differential cryptanalysis of the full 16-round DES. *Lecture Notes in Computer Science*, **740** (1993), 494–502. (Advances in Cryptology – CRYPTO '92.)

[BL79] G. R. BLAKLEY. Safeguarding cryptographic keys. *AFIPS Conference Proceedings*, **48** (1979), 313–317.

[BC85] G. R. BLAKLEY AND D. CHAUM (EDS.) *Advances in Cryptology: Proceedings of CRYPTO '84. Lecture Notes in Computer Science*, vol. 196, Springer-Verlag, 1985.

[BL85] R. BLOM An optimal class of symmetric key generation schemes. *Lecture Notes in Computer Science*, **209** (1985), 335–338. (Advances in Cryptology – EUROCRYPT '84.)

[BBS86] L. BLUM, M. BLUM AND M. SHUB. A simple unpredictable random number generator. *SIAM Jounal on Computing*, **15** (1986), 364–383.

[BL82] M. BLUM. Coin flipping by telephone: a protocol for solving impossible problems In *24th IEEE Spring Computer Conference*, pages 133–137. IEEE Press, 1982.

[BG85] M. BLUM AND S. GOLDWASSER. An efficient probabilistic public-key cryptosystem that hides all partial information. *Lecture Notes in Computer Science*, **196** (1985), 289–302. (Advances in Cryptology – CRYPTO '84.)

[BM84] M. BLUM AND S. MICALI. How to generate cryptographically strong sequences of pseudo-random bits. *SIAM Jounal on Computing*, **13** (1984), 850–864.

[Bo89] J. BOYAR. Inferring sequences produced by pseudo-random number generators. *Journal of Association for Computing Machinery*, **36** (1989), 129–141.

[BDSV93] C. BLUNDO, A. DE SANTIS, D. R. STINSON, AND U. VACCARO. Graph decompositions and secret sharing schemes. *Lecture Notes in Computer Science*, **658** (1993), 1–24. (Advances in Cryptology – EUROCRYPT '92.)

[BDSHKVY93] C. BLUNDO, A. DE SANTIS, A. HERZBERG, S. KUTTEN,

U. VACCARO AND M. YUNG. Perfectly-secure key distribution for dynamic conferences. *Lecture Notes in Computer Science*, **740** (1993), 471–486. (Advances in Cryptology – CRYPTO '92.)

[BC93]   J. N. E. BOS AND D. CHAUM. Provably unforgeable signatures. *Lecture Notes in Computer Science*, **740** (1993), 1–14. (Advances in Cryptology – CRYPTO '92.)

[BR88]   G. BRASSARD. *Modern Cryptology – A Tutorial. Lecture Notes in Computer Science*, vol. 325, Springer-Verlag, 1988.

[BR90]   G. BRASSARD (ED.) *Advances in Cryptology – CRYPTO '89 Proceedings. Lecture Notes in Computer Science*, vol. 435, Springer-Verlag, 1990.

[BB88A]   G. BRASSARD AND P. BRATLEY. *Algorithmics, Theory and Practice*. Prentice Hall, 1988.

[BCC88]   G. BRASSARD, D. CHAUM AND C. CRÉPEAU. Minimum disclosure proofs of knowledge. *Journal of Computer and Systems Science*, **37** (1988), 156–189.

[BC90]   G. BRASSARD AND C. CRÉPEAU. Sorting out zero-knowledge. *Lecture Notes in Computer Science*, **434** (1990), 181–191. (Advances in Cryptology – EUROCRYPT '89.)

[BR89]   D. M. BRESSOUD. *Factorization and Primality Testing*. Springer-Verlag, 1989.

[BR85]   E. F. BRICKELL. Breaking iterated knapsacks. *Lecture Notes in Computer Science*, **218** (1986), 342–358. (Advances in Cryptology – CRYPTO '85.)

[BR89A]   E. F. BRICKELL. Some ideal secret sharing schemes. *Journal of Combinatorial Mathematics and Combinatorial Computing*, **9** (1989), 105–113.

[BR93]   E. F. BRICKELL (ED.) *Advances in Cryptology – CRYPTO '92 Proceedings. Lecture Notes in Computer Science*, vol. 740, Springer-Verlag, 1993.

[BD91]   E. F. BRICKELL AND D. M. DAVENPORT. On the classification of ideal secret sharing schemes. *Journal of Cryptology*, **4** (1991), 123–134.

[BM92]   E. F. BRICKELL AND K. S. MCCURLEY. An interactive identification scheme based on discrete logarithms and factoring. *Journal of Cryptology*, **5** (1992), 29–39.

[BMP87]   E. F. BRICKELL, J. H. MOORE AND M. R. PURTILL. Structure in the *S*-boxes of DES. *Lecture Notes in Computer Science*, **263** (1987), 3–8. (Advances in Cryptology – CRYPTO

'86.)

[BO92]   E. F. BRICKELL AND A. M. ODLYZKO. Cryptanalysis, a survey of recent results. In *Contemporary Cryptology, The Science of Information Integrity*, pages 501–540. IEEE Press, 1992.

[BS92]   E. F. BRICKELL AND D. R. STINSON. Some improved bounds on the information rate of perfect secret sharing schemes. *Journal of Cryptology*, **5** (1992), 153–166.

[BKPS90]   L. BROWN, M. KWAN, J. PIEPRZYK AND J. SEBERRY. LOKI – A cryptographic primitive for authentication and secrecy applications. *Lecture Notes in Computer Science*, **453** (1990), 229–236. (Advances in Cryptology – AUSCRYPT '90.)

[BDB92]   M. BURMESTER, Y. DESMEDT AND T. BETH. Efficient zero-knowledge identification schemes for smart cards. *The Computer Journal*, **35** (1992), 21–29.

[CDGV93]   R. M. CAPOCELLI, A. DE SANTIS, L. GARGANO, AND U. VACCARO. On the size of shares for secret sharing schemes. *Journal of Cryptology*, **6** (1993), 157–167.

[CH95]   F. CHABAUD. On the security of some cryptosystems based on error-correcting codes. *Lecture Notes in Computer Science*, to appear. (Advances in Cryptology – EUROCRYPT '94.)

[CH84]   D. CHAUM (ED.) *Advances in Cryptology: Proceedings of CRYPTO '83*. Plenum Press, 1984.

[CP88]   D. CHAUM AND W. L. PRICE (EDS.) *Advances in Cryptology – EUROCRYPT '87 Proceedings. Lecture Notes in Computer Science*, vol. 304, Springer-Verlag, 1988.

[CRS83]   D. CHAUM, R. L. RIVEST AND A. T. SHERMAN (EDS.) *Advances in Cryptology: Proceedings of CRYPTO '82*. Plenum Press, 1983.

[CvA90]   D. CHAUM AND H. VAN ANTWERPEN. Undeniable signatures. *Lecture Notes in Computer Science*, **435** (1990), 212–216. (Advances in Cryptology – CRYPTO '89.)

[CvHP92]   D. CHAUM, E. VAN HEIJST AND B. PFITZMANN. Cryptographically strong undeniable signatures, unconditionally secure for the signer. *Lecture Notes in Computer Science*, **576** (1992), 470–484. (Advances in Cryptology – CRYPTO '91.)

[CR88]   B. CHOR AND R. L. RIVEST. A knapsack-type public key cryptosystem based on arithmetic in finite fields. *IEEE Transactions on Information Theory*, **45** (1988), 901–909.

[Co95]   D. COPPERSMITH (ED.) *Advances in Cryptology – CRYPTO '95 Proceedings. Lecture Notes in Computer Science*, vol. 963, Springer-Verlag, 1995.

[CKM94]   D. COPPERSMITH, H. KRAWCZYZ AND Y. MANSOUR. The shrinking generator. *Lecture Notes in Computer Science*, **773** (1994), 22–39. (Advances in Cryptology – CRYPTO '93.)

[CSV94]   D. COPPERSMITH, J. STERN AND S. VAUDENAY. Attacks on the birational permutation signature schemes. *Lecture Notes in Computer Science*, **773** (1994), 435–443. (Advances in Cryptology – CRYPTO '93.)

[CW91]   T. W. CUSICK AND M. C. WOOD. The REDOC-II cryptosystem. *Lecture Notes in Computer Science*, **537** (1991), 545–563. (Advances in Cryptology – CRYPTO '90.)

[DA90]   I. B. DAMGÅRD. A design principle for hash functions. *Lecture Notes in Computer Science*, **435** (1990), 416–427. (Advances in Cryptology – CRYPTO '89.)

[DA91]   I. B. DAMGÅRD (ED.) *Advances in Cryptology – EUROCRYPT '90 Proceedings. Lecture Notes in Computer Science*, vol. 473, Springer-Verlag, 1991.

[DLP93]   I. DAMGÅRD, P. LANDROCK AND C. POMERANCE. Average case error estimates for the strong probable prime test. *Mathematics of Computation*, **61** (1993), 177–194.

[DA91A]   D. W. DAVIES (ED.) *Advances in Cryptology – EUROCRYPT '91 Proceedings. Lecture Notes in Computer Science*, vol. 547, Springer-Verlag, 1991.

[DE84]   J. M. DELAURENTIS. A further weakness in the common modulus protocol for the RSA cryptosystem. *Cryptologia*, **8** (1984), 253–259.

[DBB92]   B. DEN BOER AND A. BOSSALAERS. An attack on the last two rounds of MD4. *Lecture Notes in Computer Science*, **576** (1992), 194–203. (Advances in Cryptology – CRYPTO '91.)

[DE82]   D. E. R. DENNING. *Cryptography and Data Security*. Addison-Wesley, 1982.

[DE95]   A. DE SANTIS (ED.) *Advances in Cryptology – EUROCRYPT '94 Proceedings. Lecture Notes in Computer Science*, vol. 950, Springer-Verlag, 1995.

[DE94]   Y. G. DESMEDT (ED.) *Advances in Cryptology – CRYPTO '94 Proceedings. Lecture Notes in Computer Science*, vol. 839, Springer-Verlag, 1994.

[DWQ93]   D. DE WALEFFE AND J.-J. QUISQUATER. Better login protocols for computer networks. *Lecture Notes in Computer Science*, **741** (1993), 50–70. (Computer Security and Industrial Cryptography, State of the Art and Evolution, ESAT Course, May 1991.)

[DI92] W. DIFFIE. The first ten years of public-key cryptography. In *Contemporary Cryptology, The Science of Information Integrity*, pages 135–175. IEEE Press, 1992.

[DH76] W. DIFFIE AND M. E. HELLMAN. Multiuser cryptographic techniques. *AFIPS Conference Proceedings*, **45** (1976), 109–112.

[DH76A] W. DIFFIE AND M. E. HELLMAN. New directions in cryptography. *IEEE Transactions on Information Theory*, **22** (1976), 644–654.

[DVW92] W. DIFFIE, P. C. VAN OORSCHOT AND M. J. WIENER. Authentication and authenticated key exchanges. *Designs, Codes and Cryptography*, **2** (1992), 107–125.

[EB93] H. EBERLE. A high-speed DES implementation for network applications. *Lecture Notes in Computer Science*, **740** (1993), 527–545. (Advances in Cryptology – CRYPTO '92.)

[EL85] T. ELGAMAL. A public key cryptosystem and a signature scheme based on discrete logarithms. *IEEE Transactions on Information Theory*, **31** (1985), 469–472.

[EAKMM86] D. ESTES, L. M. ADLEMAN, K. KOMPELLA, K. S. MC-CURLEY AND G. L. MILLER. Breaking the Ong-Schnorr-Shamir signature schemes for quadratic number fields. *Lecture Notes in Computer Science*, **218** (1986), 3–13. (Advances in Cryptology – CRYPTO '85.)

[FFS88] U. FEIGE, A. FIAT AND A. SHAMIR. Zero-knowledge proofs of identity. *Journal of Cryptology*, **1** (1988), 77–94.

[FE92] J. FEIGENBAUM (ED.) *Advances in Cryptology – CRYPTO '91 Proceedings. Lecture Notes in Computer Science*, vol. 576, Springer-Verlag, 1992.

[FE73] H. FEISTEL. Cryptography and computer privacy. *Scientific American*, **228**(5) (1973), 15–23.

[FN91] A. FIAT AND M. NAOR. Rigorous time/space trade-offs for inverting functions. In *Proceedings of the 23rd Symposium on the Theory of Computing*, pages 534–541. ACM Press, 1991.

[FS87] A. FIAT AND A. SHAMIR. How to prove yourself: practical solutions to identification and signature problems. *Lecture Notes in Computer Science*, **263** (1987), 186–194. (Advances in Cryptology – CRYPTO '86.)

[FOM91]   A. FUJIOKA, T. OKAMOTO AND S. MIYAGUCHI. ESIGN: an efficient digital signature implementation for smart cards. *Lecture Notes in Computer Science*, **547** (1991), 446–457. (Advances in Cryptology – EUROCRYPT '91.)

[GIB91]   J. K. GIBSON. Discrete logarithm hash function that is collision free and one way. *IEE Proceedings-E*, **138** (1991), 407–410.

[GMS74]   E. N. GILBERT, F. J. MACWILLIAMS AND N. J. A. SLOANE. Codes which detect deception. *Bell Systems Technical Journal*, **53** (1974), 405–424.

[GIR91]   M. GIRAULT. Self-certified public keys. *Lecture Notes in Computer Science*, **547** (1991), 490–497. (Advances in Cryptology – EUROCRYPT '91.)

[GP91]   C. M. GOLDIE AND R. G. E. PINCH. *Communication Theory*. Cambridge University Press, 1991.

[GMW91]   O. GOLDREICH, S. MICALI AND A. WIGDERSON. Proofs that yield nothing but their validity or all languages in NP have zero-knowledge proof systems. *Journal of the ACM*, **38** (1991), 691–729.

[Go90]   S. GOLDWASSER (ED.) *Advances in Cryptology – CRYPTO '88 Proceedings. Lecture Notes in Computer Science*, vol. 403, Springer-Verlag, 1990.

[GM84]   S. GOLDWASSER AND S. MICALI. Probabilistic encryption. *Journal of Computer and Systems Science*, **28** (1984), 270–299.

[GMR89]   S. GOLDWASSER, S. MICALI AND C. RACKOFF. The knowledge complexity of interactive proof systems. *SIAM Journal on Computing*, **18** (1989), 186–208.

[GMT82]   S. GOLDWASSER, S. MICALI AND P. TONG. Why and how to establish a common code on a public network. In *23rd Annual Symposium on the Foundations of Computer Science*, pages 134–144. IEEE Press, 1982.

[GM93]   D. M. GORDON AND K. S. MCCURLEY. Massively parallel computation of discrete logarithms. *Lecture Notes in Computer Science*, **740** (1993), 312–323. (Advances in Cryptology – CRYPTO '92.)

[GQ88]   L. C. GUILLOU AND J.-J. QUISQUATER. A practical zero-knowledge protocol fitted to security microprocessor minimizing both transmission and memory. *Lecture Notes in Computer Science*, **330** (1988), 123–128. (Advances in Cryptology – EUROCRYPT '88.)

[GQ95]   L. C. GUILLOU AND J.-J. QUISQUATER (EDS.) *Advances in Cryptology – EUROCRYPT '95 Proceedings. Lecture Notes in Computer Science*, vol. 921, Springer-Verlag, 1995.

[GU88]   C. G. GUNTHER Alternating step generators controlled by de Bruijn sequences. *Lecture Notes in Computer Science*, **304** (1988), 88–92. (Advances in Cryptology – EUROCRYPT '87.)

[GU88A]   C. G. GUNTHER (ED.) *Advances in Cryptology – EUROCRYPT '88 Proceedings. Lecture Notes in Computer Science*, vol. 330, Springer-Verlag, 1988.

[HS91]   S. HABER AND W. S. STORNETTA. How to timestamp a digital document. *Journal of Cryptology*, **3** (1991), 99–111.

[HSS93]   J. HÅSTAD, A. W. SCHRIFT AND A. SHAMIR. The discrete logarithm modulo a composite hides $O(n)$ bits. *Journal of Computer and Systems Science*, **47** (1993), 376–404.

[HE80]   M. E. HELLMAN. A cryptanalytic time-memory trade-off. *IEEE Transactions on Information Theory*, **26** (1980), 401–406.

[HI29]   L. S. HILL. Cryptogaphy in an algebraic alphabet. *American Mathematical Monthly*, **36** (1929), 306–312.

[HE94]   T. HELLESETH (ED.) *Advances in Cryptology – EUROCRYPT '93 Proceedings. Lecture Notes in Computer Science*, vol. 765, Springer-Verlag, 1994.

[HLLPRW91] D. G. HOFFMAN, D. A. LEONARD, C. C. LINDNER, K. T. PHELPS, C. A. RODGER AND J. R. WALL. *Coding Theory, The Essentials.* Marcel Dekker, 1991.

[IRM93]   H. IMAI, R. L. RIVEST AND T. MATSUMOTO (EDS.) *Advances in Cryptology – ASIACRYPT '91 Proceedings. Lecture Notes in Computer Science*, vol. 739, Springer-Verlag, 1993.

[ISN87]   M. ITO, A. SAITO, AND T. NISHIZEKI. Secret sharing scheme realizing general access structure. *Proceedings IEEE Globecom '87*, pages 99–102, 1987.

[JO88]   D. S. JOHNSON. The NP-completeness column: an ongoing guide. *Journal of Algorithms*, **9** (1988), 426–444.

[KA67]   D. KAHN. *The Codebreakers. The Story of Secret Writing.* Macmillan, 1967.

[KPS95]   C. KAUFMAN, R. PERLMAN AND M. SPECINER. *Network Security. Private Communication in a Public World.* Prentice Hall, 1995.

[KO87]   N. KOBLITZ. Elliptic curve cryptosystems. *Mathematics of*

Computation, **48** (1987), 203–209.

[KO94]   N. KOBLITZ. *A Course in Number Theory and Cryptography (Second Edition)*. Springer-Verlag, 1994.

[KO96]   N. KOBLITZ (ED.) *Advances in Cryptology – CRYPTO '96 Proceedings. Lecture Notes in Computer Science*, vol. 1109, Springer-Verlag, 1996.

[KN93]   J. KOHL AND C. NEUMAN. *The Kerboros Network Authentication Service*. Network Working Group Request for Comments: 1510, September 1993.

[KO81]   A. G. KONHEIM. *Cryptography, A Primer*. John Wiley and Sons, 1981.

[KR86]   E. KRANAKIS. *Primality and Cryptography*. John Wiley and Sons, 1986.

[LA90]   J. C. LAGARIAS Pseudo-random number generators in cryptography and number theory. In *Cryptology and Computational Number Theory*, pages 115–143. American Mathematical Society, 1990.

[LO91]   B. A. LAMACCHIA AND A. M. ODLYZKO. Computation of discrete logarithms in prime fields. *Designs, Codes and Cryptography*, **1** (1991), 47–62.

[LL93]   A. K. LENSTRA AND H. W. LENSTRA, JR. (EDS.) *The Development of the Number Field Sieve. Lecture Notes in Mathematics*, vol. 1554. Springer-Verlag, 1993.

[LL90]   A. K. LENSTRA AND H. W. LENSTRA, JR. Algorithms in number theory. In *Handbook of Theoretical Computer Science, Volume A: Algorithms and Complexity*, pages 673–715. Elsevier Science Publishers, 1990.

[LN83]   R. LIDL AND H. NIEDERREITER. *Finite Fields*. Addison-Wesley, 1983.

[LW88]   D. L. LONG AND A. WIGDERSON. The discrete log hides $O(\log n)$ bits. *SIAM Jounal on Computing*, **17** (1988), 363–372.

[MS77]   F. J. MACWILLIAMS AND N. J. A. SLOANE. *The Theory of Error-Correcting Codes*. North-Holland, 1977.

[MA86]   J. L. MASSEY. Cryptography – a selective survey. In *Digital Communications*, pages 3–21. North-Holland, 1986.

[MA94]   M. MATSUI. Linear cryptanalysis method for DES cipher. *Lecture Notes in Computer Science*, **765** (1994), 386–397. (Advances in Cryptology – EUROCRYPT '93.)

[MA94A]   M. MATSUI. The first experimental cryptanalysis of the data encryption standard. *Lecture Notes in Computer Science*, **839** (1994), 1–11. (Advances in Cryptology – CRYPTO '94.)

[MTI86]   T. MATSUMOTO, Y. TAKASHIMA AND H. IMAI. On seeking smart public-key distribution systems. *Transactions of the IECE (Japan)*, **69** (1986), 99–106.

[MA96]   U. MAURER (ED.) *Advances in Cryptology – EUROCRYPT '96 Proceedings. Lecture Notes in Computer Science*, vol. 1070, Springer-Verlag, 1996.

[MC90]   K. MCCURLEY The discrete logarithm problem. In *Cryptology and Computational Number Theory*, pages 49–74. American Mathematical Society, 1990.

[MC78]   R. MCELIECE. A public-key cryptosystem based on algebraic coding theory. *DSN Progress Report*, **42–44** (1978), 114–116.

[MC87]   R. MCELIECE. *Finite Fields for Computer Scientists and Engineers*. Kluwer Academic Publishers, 1987.

[ME93]   A. J. MENEZES. *Elliptic Curve Public Key Cryptosystems*. Kluwer Academic Publishers, 1993.

[MBGMVY93]   A. J. MENEZES, I. F. BLAKE, X. GAO, R. C. MULLIN, S. A. VANSTONE AND T. YAGHOOBIAN. *Applications of Finite Fields*. Kluwer Academic Publishers, 1993.

[MOV94]   A. J. MENEZES, T. OKAMOTO AND S. A. VANSTONE. Reducing elliptic curve logarithms to logarithms in a finite field. *IEEE Transactions on Information Theory*, **39** (1993), 1639–1646.

[MV91]   A. J. MENEZES AND S. A. VANSTONE (EDS.) *Advances in Cryptology – CRYPTO '90 Proceedings. Lecture Notes in Computer Science*, vol. 537, Springer-Verlag, 1991.

[MV93]   A. J. MENEZES AND S. A. VANSTONE. Elliptic curve cryptosystems and their implementation. *Journal of Cryptology*, **6** (1993), 209–224.

[MVV96]   A. J. MENEZES, P. C. VAN OORSCHOT AND S. A. VANSTONE. *Handbook of Applied Cryptography*. CRC Press, 1996.

[ME78]   R. C. MERKLE. Secure communications over insecure channels. *Communications of the ACM*, **21** (1978), 294–299.

[ME90]   R. C. MERKLE. One way hash functions and DES. *Lecture Notes in Computer Science*, **435** (1990), 428–446. (Advances in Cryptology – CRYPTO '89.)

[ME90A]   R. C. MERKLE. A fast software one-way hash function. *Jour-*

nal of Cryptology, **3** (1990), 43–58.

[MH78] R. C. MERKLE AND M. E. HELLMAN. Hiding information and signatures in trapdoor knapsacks. *IEEE Transactions on Information Theory*, **24** (1978), 525–530.

[MM82] C. MEYER AND S. MATYAS. *Cryptography: A New Dimension in Computer Security*. John Wiley and Sons, 1982.

[MI76] G. L. MILLER. Riemann's hypothesis and tests for primality. *Journal of Computer and Systems Science*, **13** (1976), 300–317.

[MI86] V. MILLER. Uses of elliptic curves in cryptography. *Lecture Notes in Computer Science*, **218** (1986), 417–426. (Advances in Cryptology – CRYPTO '85.)

[MPW92] C. J. MITCHELL, F. PIPER AND P. WILD. Digital signatures. In *Contemporary Cryptology, The Science of Information Integrity*, pages 325–378. IEEE Press, 1992.

[MI91] S. MIYAGUCHI. The FEAL cipher family. *Lecture Notes in Computer Science*, **537** (1991), 627–638. (Advances in Cryptology – CRYPTO '90.)

[MOI90] S. MIYAGUCHI, K. OHTA AND M. IWATA. 128-bit hash function (*N*-hash). *Proceedings of SECURICOM 1990*, 127–137.

[MO92] J. H. MOORE. Protocol failures in cryptosystems. In *Contemporary Cryptology, The Science of Information Integrity*, pages 541–558. IEEE Press, 1992.

[NBS77] *Data Encryption Standard (DES)*. National Bureau of Standards FIPS Publication 46, 1977.

[NBS80] *DES modes of operation*. National Bureau of Standards FIPS Publication 81, 1980.

[NBS81] *Guidelines for implementing and using the NBS data encryption standard*. National Bureau of Standards FIPS Publication 74, 1981.

[NBS85] *Computer data authentication*. National Bureau of Standards FIPS Publication 113, 1985.

[NBS93] *Secure hash standard*. National Bureau of Standards FIPS Publication 180, 1993.

[NBS94] *Digital signature standard*. National Bureau of Standards FIPS Publication 186, 1994.

[OD87] A. M. ODLYZKO (ED.) *Advances in Cryptology – CRYPTO '86 Proceedings. Lecture Notes in Computer Science*, vol. 263, Springer-Verlag, 1987.

[OK93]   T. OKAMOTO. Provably secure and practical identification
            schemes and corresponding signature schemes. *Lecture Notes
            in Computer Science*, **740** (1993), 31–53. (Advances in Cryp-
            tology – CRYPTO '92.)

[OSS85]   H. ONG, C. P. SCHNORR AND A. SHAMIR. Efficient signa-
            ture schemes based on polynomial equations. *Lecture Notes
            in Computer Science*, **196** (1985), 37–46. (Advances in Cryp-
            tology – CRYPTO '84.)

[PA87]   W. PATTERSON. *Mathematical Cryptology for Computer Sci-
            entists and Mathematicians*. Rowman and Littlefield, 1987.

[PE86]   R. PERALTA. Simultaneous security of bits in the discrete log.
            Lecture Notes in Computer Science, **219** (1986), 62–72. (Ad-
            vances in Cryptology – EUROCRYPT '85.)

[PI86]   F. PICHLER (ED.) *Advances in Cryptology – EUROCRYPT
            '85 Proceedings. Lecture Notes in Computer Science*, vol.
            219, Springer-Verlag, 1986.

[PS95]   J. PIEPRYZK AND R. SAFAVI-NAINI (EDS.) *Advances in
            Cryptology – ASIACRYPT '94 Proceedings. Lecture Notes in
            Computer Science*, vol. 917, Springer-Verlag, 1995.

[PB45]   R. L. PLACKETT AND J. P. BURMAN. The design of op-
            timum multi-factorial experiments. *Biometrika*, **33** (1945),
            305–325.

[PH78]   S. C. POHLIG AND M. E. HELLMAN. An improved algo-
            rithm for computing logarithms over $GF(p)$ and its crypto-
            graphic significance. *IEEE Transactions on Information The-
            ory*, **24** (1978), 106–110.

[PO88]   C. POMERANCE (ED.) *Advances in Cryptology – CRYPTO
            '87 Proceedings. Lecture Notes in Computer Science*, vol.
            293, Springer-Verlag, 1988.

[PO90]   C. POMERANCE. Factoring. In *Cryptology and Computa-
            tional Number Theory*, pages 27–47. American Mathematical
            Society, 1990.

[PO90A]   C. POMERANCE (ED.) *Cryptology and Computational Num-
            ber Theory*, American Mathematical Society, 1990.

[PGV93]   B. PRENEEL, R. GOVAERTS AND J. VANDEWALLE. Infor-
            mation authentication: hash functions and digital signatures.
            Lecture Notes in Computer Science, **741** (1993), 87–131.
            (Computer Security and Industrial Cryptography, State of the
            Art and Evolution, ESAT Course, May 1991.)

[PGV94]   B. PRENEEL, R. GOVAERTS AND J. VANDEWALLE. Hash
            functions based on block ciphers: a synthetic approach. *Lec-*

ture Notes in Computer Science, **773** (1994), 368–378. (Advances in Cryptology – CRYPTO '93.)

[QG90]   J.-J. QUISQUATER AND L. GUILLOU. How to explain zero-knowledge protocols to your children. *Lecture Notes in Computer Science*, **435** (1990), 628–631. (Advances in Cryptology – CRYPTO '89.)

[QV90]   J.-J. QUISQUATER AND J. VANDEWALLE (EDS.) *Advances in Cryptology – EUROCRYPT '89 Proceedings. Lecture Notes in Computer Science*, vol. 434, Springer-Verlag, 1990.

[RA79]   M. O. RABIN. Digitized signatures and public-key functions as intractible as factorization. *MIT Laboratory for Computer Science Technical Report*, LCS/TR-212, 1979.

[RA80]   M. O. RABIN. Probabilistic algorithms for testing primality. *Journal of Number Theory*, **12** (1980), 128–138.

[RH94]   M. Y. RHEE. *Cryptography and Secure Communications*. McGraw-Hill, 1994.

[RI91]   R. L. RIVEST. The MD4 message digest algorithm. *Lecture Notes in Computer Science*, **537** (1991), 303–311. (Advances in Cryptology – CRYPTO '90.)

[RSA78]   R. L. RIVEST, A. SHAMIR, AND L. ADLEMAN. A method for obtaining digital signatures and public key cryptosystems. *Commununications of the ACM*, **21** (1978), 120–126.

[RO93]   K. H. ROSEN. *Elementary Number Theory and its Applications* (Third Edition). Addison Wesley, 1993.

[RU86]   R. A. RUEPPEL. *Analysis and Design of Stream Ciphers*. Springer-Verlag, 1986.

[RU93]   R. A. RUEPPEL (ED.) *Advances in Cryptology – EUROCRYPT '92 Proceedings. Lecture Notes in Computer Science*, vol. 658, Springer-Verlag, 1993.

[RV94]   R. A. RUEPPEL AND P. C. VAN OORSCHOT Modern key agreement techniques. To appear in *Computer Communications*, 1994.

[SA90]   A. SALOMAA. *Public-Key Cryptography*. Springer-Verlag, 1990.

[SC94]   J. I. SCHILLER. Secure distributed computing. *Scientific American*, **271**(5) (1994), 72–76.

[SC95]   B. SCHNEIER. *Applied Cryptography, Protocols, Algorithms and Source Code in C (Second Edition)*. John Wiley and Sons, 1995.

[SC91]   C. P. SCHNORR. Efficient signature generation by smart cards. *Journal of Cryptology*, **4** (1991), 161–174.

[SP89] J. SEBERRY AND J. PIEPRZYK *Cryptography: An Introduction to Computer Security.* Prentice-Hall, 1989.

[SP90] J. SEBERRY AND J. PIEPRZYK (EDS.) *Advances in Cryptology – AUSCRYPT '90 Proceedings. Lecture Notes in Computer Science*, vol. 453, Springer-Verlag, 1990.

[SZ92] J. SEBERRY AND Y. ZHENG (EDS.) *Advances in Cryptology – AUSCRYPT '92 Proceedings. Lecture Notes in Computer Science*, vol. 718, Springer-Verlag, 1993.

[SH79] A. SHAMIR. How to share a secret. *Communications of the ACM*, **22** (1979), 612–613.

[SH84] A. SHAMIR. A polynomial-time algorithm for breaking the basic Merkle-Hellman cryptosystem. *IEEE Transactions on Information Theory*, **30** (1984), 699–704.

[SH90] A. SHAMIR. An efficient identification scheme based on permuted kernels. *Lecture Notes in Computer Science*, **435** (1990), 606–609. (Advances in Cryptology – CRYPTO '89.)

[SH94] A. SHAMIR. Efficient signature schemes based on birational permutations. *Lecture Notes in Computer Science*, **773** (1994), 1–12. (Advances in Cryptology – CRYPTO '93.)

[SH48] C. E. SHANNON. A mathematical theory of communication. *Bell Systems Technical Journal*, **27** (1948), 379–423, 623–656.

[SH49] C. E. SHANNON. Communication theory of secrecy systems. *Bell Systems Technical Journal*, **28** (1949), 656–715.

[ST92] J. H. SILVERMAN AND J. TATE. *Rational Points on Elliptic Curves.* Springer-Verlag, 1992.

[SI85] G. J. SIMMONS. Authentication theory / coding theory. *Lecture Notes in Computer Science*, **196** (1985), 411–432. (Advances in Cryptology – CRYPTO '84.)

[SI88] G. J. SIMMONS. A natural taxonomy for digital information authentication schemes. *Lecture Notes in Computer Science*, **293** (1988), 269–288. (Advances in Cryptology – CRYPTO '87.)

[SI92] G. J. SIMMONS. A survey of information authentication. In *Contemporary Cryptology, The Science of Information Integrity*, pages 379–419. IEEE Press, 1992.

[SI92A] G. J. SIMMONS. An introduction to shared secret and/or shared control schemes and their application. In *Contemporary Cryptology, The Science of Information Integrity*, pages

441–497. IEEE Press, 1992.

[SI92B] G. J. SIMMONS (ED.) *Contemporary Cryptology, The Science of Information Integrity.* IEEE Press, 1992.

[SB92] M. E. SMID AND D. K. BRANSTAD. The data encryption standard: past and future. In *Contemporary Cryptology, The Science of Information Integrity*, pages 43–64. IEEE Press, 1992.

[SB93] M. E. SMID AND D. K. BRANSTAD. Response to comments on the NIST proposed digital signature standard. *Lecture Notes in Computer Science*, **740** (1993), 76–88. (Advances in Cryptology – CRYPTO '92.)

[SS77] R. SOLOVAY AND V. STRASSEN. A fast Monte Carlo test for primality. *SIAM Journal on Computing*, **6** (1977), 84–85.

[ST95] W. STALLINGS. *Network and Internetwork Security. Principles and Practice.* Prentice Hall, 1995.

[ST88] D. R. STINSON. Some constructions and bounds for authentication codes. *Journal of Cryptology*, **1** (1988), 37–51.

[ST90] D. R. STINSON. The combinatorics of authentication and secrecy codes. *Journal of Cryptology*, **2** (1990), 23–49.

[ST92] D. R. STINSON. Combinatorial characterizations of authentication codes. *Designs, Codes and Cryptography*, **2** (1992), 175–187.

[ST92A] D. R. STINSON. An explication of secret sharing schemes. *Designs, Codes and Cryptography*, **2** (1992), 357–390.

[ST94] D. R. STINSON (ED.) *Advances in Cryptology – CRYPTO '93 Proceedings. Lecture Notes in Computer Science*, vol. 773, Springer-Verlag, 1994.

[vHP93] E. VAN HEYST AND T. P. PEDERSEN. How to make efficient fail-stop signatures. *Lecture Notes in Computer Science*, **658** (1993), 366–377. (Advances in Cryptology – EUROCRYPT '92.)

[VV89] S. A. VANSTONE AND P. C. VAN OORSCHOT. *An Introduction to Error Correcting Codes with Applications.* Kluwer Academic Publishers, 1989.

[vT88] H. C. A. VAN TILBORG. *An Introduction to Cryptology.* Kluwer Academic Publishers, 1988.

[vT93] J. VAN TILBURG. Secret-key exchange with authentication. *Lecture Notes in Computer Science*, **741** (1993), 71–86. (Computer Security and Industrial Cryptography, State of the Art and Evolution, ESAT Course, May 1991.)

[VV84] U. VAZIRANI AND V. VAZIRANI. Efficient and secure pseudorandom number generation. In *Proceedings of the 25th Annual Symposium on the Foundations of Computer Science*, pages 458–463. IEEE Press, 1984.

[WA90] M. WALKER. Information-theoretic bounds for authentication systems. *Journal of Cryptology*, **2** (1990), 131–143.

[WA96] P. WAYNER. *Disappearing Cryptography*. Academic Press, 1996.

[WE88] D. WELSH. *Codes and Cryptography*. Oxford Science Publications, 1988.

[WI94] M. J. WIENER. Efficient DES key search. Technical report TR-244, School of Computer Science, Carleton University, Ottawa, Canada, May 1994 (also presented at CRYPTO '93 Rump Session).

[WI80] H. C. WILLIAMS. A modification of the RSA public-key encryption procedure. *IEEE Transactions on Information Theory*, **26** (1980), 726–729.

[WI86] H. C. WILLIAMS (ED.) *Advances in Cryptology – CRYPTO '85 Proceedings. Lecture Notes in Computer Science*, vol. 218, Springer-Verlag, 1986.

[YA82] A. YAO. Theory and applications of trapdoor functions. In *Proceedings of the 23rd Annual Symposium on the Foundations of Computer Science*, pages 80–91. IEEE Press, 1982.

Index